DICTIONARY OF ENVIRONMENT AND DEVELOPMENT

People, Places, Ideas and Organizations

Andy Crump

Earthscan Publications Ltd, London

First published 1991 by
Earthscan Publications Ltd
3 Endsleigh Street, London WC1H 0DD

Copyright © 1991 Andy Crump

All rights reserved

British Library Cataloguing in Publication Data
available

ISBN 1–85383–078–X

Production by Bob Towell
Design by Mick Keates

Typeset by Florencetype Ltd, Kewstoke, Avon
Printed in Great Britain by
The Guernsey Press Co. Ltd, Guernsey, Channel Islands.

Earthscan Publications Ltd is an editorially
independent subsidiary of the International
Institute for Environment and Development
(Charity No. 800066).

ABOUT THE AUTHOR

Andy Crump has worked in various parts of the world as an ecologist, biological pest-control specialist, teacher and writer.

His career has ranged from pure scientific research to the critical analysis of the social impact and value of such research and the presentation of this information to all levels of audience, from governments to the general public. As a freelance writer, researcher and audio-visual consultant, he has dealt with all aspects of environment and development using the full range of communications media.

His published work includes computer software programmes, videos, exhibitions, books and numerous articles in professional and academic journals covering environmental and development topics.

Formerly with the Centre for World Development Education and the PANOS Institute in London, he is currently in the TDR Communications Unit of the World Health Organization in Geneva.

ACKNOWLEDGEMENTS

There have been many influences on the production and content of this book, too numerous and diverse to mention. Of the various colleagues on whose enthusiasm and knowledge I have drawn, I should, in particular, like to extend my thanks to Stephen Setford of the Catholic Fund for Overseas Development and Derek Walker of the Centre for World Development Education for their contributions and suggestions during the formative stages.

I would also like to express my special and undying gratitude to my wife, Silvia, for her unwavering support and without whom . . .

Cross-references are indicated by ⌁ (see above) and ⌁ (see below).

FOREWORD

Why a *Dictionary of Environment and Development*?
Surely it is a bit like compiling a *Dictionary of Choral Singing and Construction Techniques*, or any two other improbably paired concerns.

There was indeed a time when Environment and Development were not on speaking terms, and certainly would not have appeared in the same dictionary. Those who loved the environment regarded those who championed progress and 'development' as the enemy; and those who wanted to develop the poorest peoples and nations felt that much of 'the environment' would have to be sacrificed in this worthwhile endeavour.

But these attitudes – except among the most intractable of both camps – have changed, and changed most rapidly over only the past few years. Andy Crump's dictionary is right on time.

Most Greens at last realize that poor people and poor nations are going to destroy their environment to survive. Development – not old-fashioned, mindless economic growth but a redistribution of the ability to produce – is required. For their part, most developers have realized that progress cannot be measured in rising GNP figures produced by development policies which destroy the basis of all progress: topsoil, clean water, breathable air, genetic resources, predictable climate, and all other renewable environmental resources.

The dictionary is more than just a big book of accurate, concise definitions. First, it remains unbiased; it lets the reader decide. Second, it offers many examples of misuse of resources from the wealthy, industrialized, 'developed' nations. The poor are destroying their own patches. The wealthy are destroying the planet, both through their over-consumption of resources and through pollution which warms the atmosphere and destroys the ozone layer.

To leave a liveable world for our children, we will have to make hard choices in changing the ways in which we use energy, get from place to place, dispose of our garbage and generally pursue happiness.

This *Dictionary of Environment and Development* offers us the words with which to discuss with one another what those changes are and how they can be made. It can help us educate our children for sustainable life-styles. It can help us tell our politicians what to do.

The 1990s are the 'Decade of Decision'. We are all the decision-makers. This dictionary gives us a common vocabulary for those decisions.

Lloyd Timberlake   London, 1991

A

ABORIGINES Although the word 'aborigine' is now most commonly used to describe the native inhabitants of Australia, the direct descendants of the original inhabitants of any country or region can be referred to as aborigines or indigenous people.

Over the last two centuries millions of aboriginal people have died as a result of the rapid developments seen in transport and weaponry and following the invasion of native lands by a succession of explorers, colonists and traders, mostly of European origin. Less than two hundred years after the Spanish conquest of Latin America, the number of indigenous people in the region had fallen from 70 million to fewer than 4 million. Before European explorers invaded North America, there were an estimated 12 million indigenous Indians on the continent. Now the number is closer to 3 million. In Australia, the aboriginal population was reduced by 80 per cent during the first hundred years of colonization, and only 116,000 aborigines survive today. Many of the indigenous people who have perished died in the struggle to repel the invaders or in slavery. Others succumbed to imported diseases – such as measles⋄, smallpox⋄ and typhus – to which they had never been previously exposed and had thus never developed any form of resistance.

Today, around 250 million indigenous, aboriginal people (estimates range from 200 to 500 million) live in some 20,000 or more distinct groups around the world. The exact total of known aboriginals is uncertain; furthermore, tribal groups live in some of the more remote and least explored areas of the world, and there is a possibility, albeit slight, that small isolated groups may yet remain to be discovered.

Even where significant numbers of indigenous people have survived, their traditional culture and way of life have been suppressed, extinguished or irrevocably altered by foreign influence.

Whereas most of the world's states have been artificially created since the Second World War, indigenous or tribal people, usually characterized by distinct language, culture, territory and political organization, have existed for centuries. Furthermore, most of the world's states have been created by Europeans or their descendants to suit their own economic or political interests, with little regard to aboriginal peoples. As a result of decisions taken in Europe over sixty years ago, the Kurds, some 20 million people, were divided between five countries: Iran, Iraq, the Soviet Union, Syria and Turkey.

During the late 1980s major government-backed programmes designed to assert control over aboriginal populations or their resources continued in various parts of the world including Ethiopia, Guatemala, India, Indonesia and Iraq. In 1988 at least 200,000 indigenous people from around the world were killed and two million forced to flee their homes. In both Borneo and Brazil the leaders of indigenous groups who were fighting to stop logging companies and other commercial operations from destroying their forest homes and traditional lifestyles were arrested for their activities.

ABSORBTIVE CAPACITY According to the experts on the Independent Commission on International Development Issues (ICIDI)⋄, developing countries⋄ are considered to have an ability to deal effectively with only a finite amount of financial aid⋄ and investment. They are thought to become unable to manage any extra funds or resources properly when this limit is exceeded. This limit, or 'absorbtive capacity', of a nation was much referred to by the Independent Commission on International Development Issues (ICIDI)⋄ (also known as the Brandt Commission⋄), which considered it to be determined and regulated by the degree of technical expertise and managerial capabilities within the country in question and, in particular, dependent on the pool of trained and qualified personnel available to help plan and execute development projects and programmes. Improved transfer of technology and support for educational initiatives and institution-building within the developing countries themselves, provided as part of the aid programme from donor countries, will help to increase the absorbtive capacity of any recipient nation.

ABU DHABI FUND FOR ECONOMIC DEVELOPMENT (ADFAED) The ADFAED was established in 1971 with the objective of playing a constructive role in the provision of loans and economic assistance to Arab, Asian and African countries. The Fund, which became operational in 1974, provides finance and resources to support the economies of the countries in question and improve their prospects of external trade and internal development so that they may achieve prosperity, security and a satisfactory standard of living.

ACACIA A genus of evergreen trees and shrubs with 800 different species. Found throughout the tropics and subtropics, they are particularly abundant in savannahs and arid regions in Africa, the Americas, Australia and India. They are leguminous and consequently help to improve or maintain soil fertility. Most species are fast-growing, robust and grow well in unfavourable conditions, especially in semi-arid areas, and are consequently now widely used in land reclamation and agroforestry⬦ projects. The trees and shrubs are multipurpose. They are grown as shade trees and soil-improvers as well as for helping to control soil erosion⬦, and in several regions they are indispensable as they form a prime source of forage for livestock⬦. The leaves and the nutritive pods can be browsed, or the trees can be lopped to provide fodder. Acacia trees also yield a variety of useful products, including gums – especially the commercially important gum arabic – tannins, dyes, and wood⬦ which is suitable for both construction and furniture-making. Gum arabic comes from *Acacia senegal*, a tree native to North Africa where it is grown commercially. The gum is used to make jellies and sweets; Over 90 per cent of world supplies come from Sudan.

ACCEPTABLE DAILY INTAKE (ADI) This represents a measurement of the amount of any chemical that can safely be consumed each day by an average human being. It is usually defined in terms of milligrams per kilogram of body weight and is occasionally defined as the provisional tolerable weekly intake (PTWI). There are three international agencies which discuss and set these safety standards. They are the Joint Expert Committee on Food Additives, formed by scientists from the United Nations Food and Agriculture Organization (FAO)⬦ and the World Health Organization (WHO)⬦, the European Community's⬦ Scientific Committee for Food, and the United States Food and Drug Administration (FDA). The levels of consumption regarded as safe which are set independently by each of these agencies do not necessarily coincide.

Calculation of the ADI usually involves finding out the maximum level of a substance that can be fed to animals without producing any detrimental physiological effect. This amount is then divided by a 'safety factor' – usually 100 – to compensate for the differences between animals and humans, and for variations in human diets. There is no substantiated evidence to determine the 'safety factor'; it is set arbitrarily by each committee and may be frequently changed.

ACID RAIN A term used to describe the atmospheric fallout of industrial pollutants. This occurs sometimes in the form of dry deposits but mostly as acidified natural precipitation. The pH scale is a measure of acidity or alkalinity. Distilled water has a pH of 7.0, and anything below this is acidic. Natural rainfall is slightly acidic owing to atmospheric carbon dioxide⬦ dissolving in water⬦ to form weak carbonic acid. In general, this gives rainwater a pH of approximately 5.6, and anything below this is defined as acid rain. The pH scale is logarithmic, so rain of pH 5.0, which is common in north-eastern parts of the United States, is ten times more acidic than that of pH 6. In areas in, or close to, heavily industrialized countries, or in proximity to nations which rely mainly on the burning of coal⬦ for power production, precipitation can contain both dilute sulphuric acid and nitric acids, formed as a result of the absorption of sulphur dioxide⬦ and nitrogen oxides⬦ in the atmosphere. These gases, together with most of the other pollutants contributing to the acidification process, are produced through the burning of fossil fuels⬦ without adequate treatment of the resultant products. Vehicle exhausts also contribute a significant amount of pollutants which exacerbate the degree of acidification.

Acid rain damages trees, plants and crops, acidifies lakes, rivers and groundwater⬦, and corrodes buildings. In many industrialized parts of Europe acidic rain has been falling for well over a century, decimating forests⬦, causing buildings to crumble and destroying aquatic life in lakes, especially those in Scandinavia. It has been estimated that the accumulating effects of acid rain in Europe will cause the loss of timber worth over $20 billion every year for most of the next century. Acidification also alters the chemistry of soil, releasing harmful chemicals such as aluminium⬦ and cadmium.⬦ The cadmium content of wheat⬦ in Sweden has doubled during this century.

In countries such as the United Kingdom and the United States, where the problem of pollution from

power stations was readily identified, tall smokes-
tacks were introduced in an effort to reduce pollution levels. These measures merely succeeded in
dispersing the pollutants by emitting them higher
into the atmosphere. As a result, the problem of acid
rain became one of the first widely recognized
examples of transboundary pollution⋄ damage
Countries suffering from acid rain were not necessarily those responsible for creating it. Attempts to
control the presence of the atmospheric oxides
which are the basis for acid rain, especially those
produced through industrial and vehicle emissions,
have been severely hampered by the lack of international agreement on both cause and effect, and on
measures that could be taken to enforce any internationally agreed clean-up programmes or directives.

ACP COUNTRIES The name applied to those
countries of Africa, the Caribbean and the Pacific –
hence ACP – which are signatories and contracting
parties to the Lomé Convention⋄, which attempts
to promote economic co-operation between these
countries and the European Community⋄. Mem-
bers are: Antigua and Barbuda, Bahamas, Barbados,
Belize, Benin, Botswana, Burkina Faso, Burundi,
Cameroon, Cape Verde, Central African Republic,
Chad, Comoros Islands, Congo, Djibouti,
Dominica, Equatorial Guinea, Ethiopia, Fiji,
Gabon, The Gambia, Ghana, Grenada, Guinea,
Guinea-Bissau, Guyana, Ivory Coast, Jamaica,
Kenya, Kiribati, Lesotho, Liberia, Madagascar,
Malawi, Mali, Mauritania, Mauritius, Mozambique,
Niger, Nigeria, Papua New Guinea, Rwanda, St
Kitts and Nevis, St Lucia, St Vincent and the
Grenadines, São Tomé e Principe, Senegal,
Seychelles, Sierra Leone, Solomon Islands, Somalia,
Sudan, Suriname, Swaziland, Tanzania, Togo,
Tonga, Trinidad and Tobago, Tuvalu, Uganda,
Vanuatu, Western Samoa, Zaire, Zambia, Zim-
babwe. Following independence, Namibia will also
join the group.

**ACQUIRED IMMUNE DEFICIENCY
SYNDROME (AIDS)** An – as yet – incurable
disease first identified in 1980. It is thought to be
caused by a blood-borne virus, the Human
Immunodeficiency Virus (HIV). The disease is generally marked by symptoms such as weight loss,
diarrhoea, and swollen glands, which follow a
lengthy incubation period. It leads to the destruction of the body's normal cell-mediated immune
system, due to a decreased number of T-lympho

cytes in the blood. These cells form a major part of
the body's natural defence mechanism attacking invading viruses and bacteria, and the loss leaves the
body vulnerable to infection from any micro-
organism. Many patients developing the disease suffer from severe pneumonia caused by the normally
harmless protozoan *Pneumocystis carinii*, or a virulent form of the skin cancer *Kaposi's sarcoma*, a normally mild cancer common in Africa.

By 1991, 300,000 cases of AIDS had been
reported to the World Health Organization
(WHO)⋄ from 150 countries. This total represented
an increase of more than 100,000 over a two-year
period. Although epidemiological data are scarce,
computer models indicate that retrospectively, before the disease was first recognized, only 103 cases
probably existed. The models also predict that in
reality there are probably about 600,000 cases
worldwide and that by the year 2000, 6 million
people will have developed AIDS. According to figures released in 1990, since the start of the epidemic
at least 8–10 million adults had been infected with
HIV, although many had not yet developed AIDS. By
the end of the century there will be an estimated
cumulative total of 15–20 million HIV-infected
adults and 10 million infected infants and children
in the world.

A massive and unprecedented global effort by
academic and commercial scientists in research institutions around the world has produced some
encouraging results in the development of therapeutic drugs and vaccines. Nevertheless, by the beginning of the 1990s only a single drug, Zidovudine
(formerly known as AZT), had been officially
licensed for treatment of patients. Zidovudine
appears to delay the onset of the disease in those
infected with HIV. In late 1989 the discovery of a
vaccine which appeared to immunize monkeys successfully against infection with Simian Immuno-
deficiency Virus (SIV), a virus which is both similar
to HIV and causes AIDS-like illness in monkeys,
raised hopes that a vaccine to combat HIV will
eventually be produced. Despite this, other advances
in the search for a cure for the disease, and a growing understanding of the disease and the role played
by HIV, it seems likely that the only effective weapon
against AIDS available for the next decade or so will
be preventive measures to avoid infection.

ADOBE In resource-poor developing countries⋄,
many dwellings, especially those in low-income rural
areas, are built from adobe, a material made by
mixing clay-based soil with water⋄, usually with

the addition of a binding material such as straw or animal dung. The mixture is fashioned into bricks which are dried hard in the sun, providing a minimal-cost building material. Adobe has several beneficial characteristics. It is relatively easy to produce, fire-resistant, a good absorber of humidity and acts as an efficient heat insulator. The inclusion of a lattice of bamboo⇔ or other similar materials in the walls of houses, schools and hospitals considerably strengthens the buildings and helps to protect them against damage from adverse weather or minor earth tremors.

ADRIATIC SEA   The Adriatic is the northern arm of the Mediterranean Sea extending for 750 km between Italy and Yugoslavia. Italy's longest inland waterway, the River Po, the valley of which contains the nation's most fertile and economically important region, flows into the Adriatic. Each year the river discharges a cocktail of pollutants – including 200 tonnes of arsenic, over 15,000 tonnes of phosphates⇔ and 130,000 tonnes of nitrates – into the sea. Between 1988 and 1991 large sectors of the waters of the Adriatic became starved of oxygen, eutrophication⇔ set in, and dense blankets of algae became common. A gelatinous soup of weeds and scum, up to 10 m thick in places and weighing millions of tonnes, effectively destroyed the region's economy and environment. Fishing nets broke under the weight of the algae, tourists who ventured to the area could not swim and were driven away by the stench. Overall, the annual losses to the fishing⇔ and tourist industry alone were estimated to be in excess of $1 million.

AEDES   A widely distributed genus of mosquito⇔ which occurs throughout the tropics and subtropics and transmit several diseases of great consequence to public health. Most species are easily recognized: black with white or silvery-yellow markings on the legs and thorax. *Aëdes aegypti* is the most important, as it is the principal vector responsible for the transmission of dengue⇔ and yellow fever⇔. The causative agents for both diseases are transmitted to humans through the bite of female mosquitoes. *A. aegypti* is a coastal and riverine species that rarely breeds far from human habitations. The female lays eggs in any area of stagnant water, ranging from holes in trees, puddles and old tins to discarded coconut shells and tyres. Other species of *Aedes* act as vectors of filariasis⇔ and other viruses causing various forms of encephalitis.

AFLATOXINS   Poisonous substances produced in the spores of a fungal mould, *Aspergillus flavus*, aflatoxins can damage the immune system and cause liver cancer⇔. They are also mutagenic, inducing chromosomal damage in a wide range of animals. Aflatoxins were first identified in the 1960s, after 100,000 turkeys mysteriously died on a farm in the United Kingdom. The birds were later found to have been fed meal contaminated with Aspergillus mould.

In general, moulds grow best in warm, moist conditions. Consequently, aflatoxins are a particular problem in the humid tropics and subtropics, where stored cereals and nuts are frequently contaminated. Peanuts (groundnuts⇔) are especially at risk of contamination with *A. flavus*. Modern agricultural techniques – notably the use of artificial fertilizers⇔, which tend to increase the water content of crops, thereby encouraging the growth of moulds – are suspected of being a major factor in promoting aflatoxin contamination, which is on the increase worldwide. Acute aflatoxin poisoning is fatal; the majority of cases are reported from India and eastern Africa where surveys in the Sudan have reported aflatoxins in 80 per cent of raw foods. Malnourished children are most at risk, since they become unable to excrete any aflatoxins ingested with their food.

AFRICAN DEVELOPMENT BANK   One of the four Multilateral Development Banks (MDB). A regional bank, established in 1963 with its headquarters in the Ivory Coast. In operation since 1966, the Bank aims to contribute to the economic development and social progress of its regional members by encouraging partnerships between African and non-African countries. Originally it restricted membership to regional countries but removed this restriction in 1982. It now has fifty African members and twenty-five non-regional members. Although the authorized capital stock of around $6 billion is subscribed by both types of members, loans are available only to African nations. In 1973, the African Development Fund was set up as the concessional loan implement of the Bank. Most of the major Western donor nations are members of the Fund.

AFRICAN NATIONAL CONGRESS (ANC) An organization formed in 1912 in South Africa to fight against racial discrimination and extend the franchise to native black Africans. Following the

Sharpeville incident in 1960, when African demonstrators were shot at by the police and sixty demonstrators were killed, the ANC was officially outlawed by the South African government. Despite the imprisonment of several leading members, the ANC carried on its activities as a guerrilla organization. Gradual reforms in South Africa during the 1980s, together with increasing international pressure, eventually saw the release of the ANC's leader, Nelson Mandela◇, in 1990. Within a few months of his release the ANC was holding official talks with the South African government, and will probably play an important role over the next decade towards achieving majority rule in South Africa. Oliver Tambo, another ANC leader, was allowed to return from exile at the end of 1990 to attend an official meeting of the Congress – the first allowed by the government for thirty years – at which there were suggestions that the sanctions levelled against South Africa by the international community as a result of the government's apartheid◇ policy should be reviewed.

AFRICA REAL-TIME ENVIRONMENTAL MONITORING USING IMAGING SATELLITES (ARTEMIS)
The ARTEMIS project is an attempt to keep track of changing environmental conditions and provide forecasts of impending threats to agricultural production in Africa using the most up-to-date remote-sensing technology. Coordinated by the Food and Agriculture Organization (FAO)◇, ARTEMIS provides early warnings of drought◇, locust◇ plagues and crop failures. The European Space Agency's◇ satellite, METEOSAT, provides the project with hourly cloud-temperature readings, and the United States' National Oceanic and Atmospheric Administration (NOAA) satellite provides data on the changing state of vegetation throughout Africa over ten-day periods. Analysis of the satellite data and comparison with agroclimatical statistics in FAO data banks and from ground stations allows rainfall and vegetation conditions across the continent to be closely monitored so that areas of abnormal rainfall or potential locust breeding sites can be rapidly identified.

AFRICAN TRANSPORT AND COMMUNICATIONS DECADE
During the 1970s it became obvious that one of the major problems facing the African continent was the lack of a basic or reliable transport and communications infrastructure. A ministerial meeting of the Economic Commission for Africa (ECA) launched the African Transport and Communications Decade in 1978, with the full backing of the United Nations◇. The intention of the Decade was to 'harmonize, coordinate, modernize and develop communications of all kinds, including road, rail, water transport, telecommunications, radio, television and postal services'. The intended programme of activities called for expenditure well in excess of $10 billion. This level of funding was nowhere near realized, and the Decade met with very limited success.

AGING
Advances in medical practices have brought about a decline in the annual birth rate in many countries and, at the same time, prolonged life expectancy◇. Improved health services and living standards, with the simultaneous decline in fertility and mortality, increase the proportion of the elderly in the population◇, and this is starting to have a significant impact in both developed and developing countries◇. There are no official definitions of what constitutes an 'aged' population, although a commonly employed yardstick is 10 per cent of a nation's population being over sixty. There are well in excess of 380 million people over sixty in the world. By 2000 the total is likely to be 600 million, with the situation in Asia being most acute. If China were to realize its intended goal of zero population growth by the year 2000, by 2035 a quarter of its population would be over 60 years old. Japan's advances in health care have already raised the number of 'elderly' people beyond the 10 per cent criterion.

A skewed, predominantly elderly population causes social and economic problems – more so in the industrialized world, where there is a marked dwindling in the labour pool and a massively increased burden on social welfare systems. In the Third World◇, where social welfare systems are generally absent, the World Health Organization (WHO)◇ reports that more than half of people over sixty-five remain gainfully employed, whereas in the industrialized world, where age is often a barrier to work, the figure is less than 25 per cent.

In the developed nations vast amounts of research and resources have been deployed to enable people to reach 'old age', but little thought has been devoted to what happens when they get there. Many of the problems they encounter, such as increasing dementia, loss of mobility and incontinence, are of a traditionally sensitive nature and have not been studied in any depth. In Japan, one of the world's most aged populations, mental health and other age-related problems have received very little attention, despite the fact that there are more than 1,000 centenarians in the country.

AGRARIAN REFORM   One of the major problems hindering the development process throughout the Third World⊘ is that of land-ownership. Of the world's total amount of land⊘ that can be owned, nearly three-quarters is controlled by just 2·5 per cent of all landowners. In some countries – most notably those in South America – a few wealthy landlords or large-scale farmers own half or more of the land. In Brazil, the wealthiest 2 per cent of the country's landowners control 60 per cent of the arable land. Poor people, especially in rural communities, have either relatively small patches of land or no land at all legally to call their own.

Over half the rural population in the Third World are effectively landless. They may occupy land as tenants or sharecroppers and are often exploited by the landowner because of their status as landless labourers⊘. Under these circumstances, and due to governmental influence, agricultural development programmes often fail to help those most in need, the landless poor. In Brazil, for example, 70 per cent of agricultural workers do not own land but between 1960 and 1970 the military regime running the country granted millions of hectares to transnational corporations to raise beef for export.

Agrarian reform, through which land will be distributed with greater equality, is recognized to be a crucial factor in promoting development in the Third World; this is evident in the concluding declaration of the 1979 World Conference on Agrarian Reform and Rural Development. Several successful examples of large-scale land reform have already been seen, including those in China, South Korea and North Vietnam, where at least 45 per cent of arable land was redistributed between 1954 and 1957 to the benefit of three-quarters of the nation's rural households. Yields of rice⊘ and other crops rose by 20–50 per cent as a result.

It has continually been shown that small-scale farmers are significantly more productive than large-scale agricultural operations. Furthermore, they tend to employ more sustainable forms of agricultural production.

In addition to being given access, and perhaps title, to land, poor farmers also need assistance with finance, resources and technology, and here international aid⊘ agencies have a major role to play.

AGRIBUSINESS   In many parts of the world agricultural production, the bulk of which used to be in the hands of peasants or small-scale farmers, has become industrialized and is now under the control of a relatively small number of transnational corporations. Agribusiness is the term used to describe these companies and their agricultural activities. The influence of these transnational businesses extends to every aspect of agricultural production, including ownership of land⊘ and plantations, agrochemical factories, shipping companies, marketing organizations, research facilities, banking institutions and even the outlets through which agricultural produce is sold to the consumer.

Agribusinesses such as Unilever, Tate & Lyle, Hoechst, and British American Tobacco (BAT) are profit-orientated corporate entities; their production methods are capital-intensive, with an emphasis on the use of expensive, complex technology and farming methods to increase and maximize yields in the pursuit of profits. Consequently, the traditional, more ecologically sound farming practices which recycled nutrients back into the soil, used crop rotations to avoid the build up of pests and helped replace nitrogen in the soil by growing leguminous crops, have been replaced in many areas where agribusiness operates. Instead, intensive farming systems have been introduced based on the use of expensive, energy-demanding inputs of fertilizers⊘, pesticides⊘, herbicides⊘ and irrigation.⊘

Agribusiness also has the power to make Third World⊘ farmers assume all the risks in agricultural production. Nestlé, a Swiss-based transnational organization, provides a good example. The company derives half its turnover from raw materials such as milk, coffee⊘ and cocoa⊘. But it does not own a cocoa plantation or a cow, simply contracting with local farmers throughout the Third World who must provide a fixed amount of produce by a certain date at a price set by the company.

AGRICULTURE   Plants and animals have been exploited as sources of food ever since the human race appeared on the earth. The systematic cultivation of land⊘ by humans probably originated in Egypt at least 17,000 years ago, although, as with many other natural environmental processes, the full impact of human activity has really been felt only over the past two centuries or so.

The industrialization of agriculture, movement away from natural farming systems⊘, and the application of scientific techniques began in earnest in eighteenth-century Britain, with the mechanization of farming developing rapidly in the United States during the 1800s. After the Second World War there was a swift and expansive growth in the synthesis and use of agricultural chemicals, followed in the 1960s by advances in plant breeding and the development of high-yield varieties (HYVS)⟡ of specific crops (mostly those preferred or widely traded by the industrialized countries) for growth under relatively specialized conditions. The so-called 'Green Revolution'⟡ in developing countries⟡, using these new HYVS, primarily on irrigated land, occurred at the same time as cattle- and poultry-rearing on 'production lines' was developing apace in the industrialized world.

In the 1970s, widespread realization of the environmental toll exacted by modern agriculture and its unsustainability encouraged those farmers who were able to afford the choice to revert back towards a more sustainable 'organic' system of farming, avoiding the use of agricultural chemicals and monoculture. This trend, which had started during the 1960s and has continued to grow, began to be supported by scientific evidence which showed that organic farming was considerably more sustainable and offered greater protection to both the environment and the natural resource base than industrialized, monocrop farming operations. In the 1980s, pest control also moved towards an integrated approach which called for a reduction in the use of hazardous man-made pesticides⟡ that persisted in the environment. Genetic engineering also began to be applied to crops, opening up the possibility of improved agricultural production without depleting the earth's natural resource base.

AGRICULTURAL EXTENSION AND RURAL DEVELOPMENT CENTRE

(AERDC)   The Centre is part of the University of Reading (UK). It provides information, advice and training focusing on extension services and all aspects of farming technology, systems and processes with a view to promoting rural development in agricultural communities around the world.

AGROFORESTRY   Agroforestry describes any system where trees are deliberately left, planted, or encouraged on land⟡ where crops are grown or animals grazed. It includes practices as diverse as slash-and-burn agriculture⟡ (shifting cultivation), the growth of shade trees in cash-crop⟡ systems, and the use of living fences either to contain or to exclude animals. Agroforestry helps to recycle nutrients, prevents or minimizes soil erosion⟡, improves tilth and raises crop yields. Deep-rooted trees tap nutrient sources that are out of reach of most crops;

these nutrients become readily available when the leaves fall. Leguminous trees, such as leucaena and acacia⟡, improve soil fertility directly through nitrogen fixation. Tree roots help to bind the soil and increase aeration. Mixtures of trees and crops provide more complete ground cover which helps to prevent soil erosion and weed invasion while making full and productive use of available solar radiation. Leaf litter from the trees adds organic matter to the soil and acts as a mulch, retaining soil moisture and so further enhancing the prevention of erosion. Tree cover also helps to regulate temperatures, reducing extremes and maintaining temperatures nearer optimum levels during both day and night. Farming communities also benefit from a regular supply of wood⟡ and other tree products. Multipurpose trees can provide fodder for livestock⟡, edible fruits and nuts, fuel, timber, supports for climbing vegetables and medicinal products.

Traditional, integrated agroforestry programmes have been followed for centuries around the world. Amongst the most successful are those practised in West Java and parts of West Africa. These generally incorporate cultivation of a wide range of crops of all types in a system with an overall structure that closely resembles the natural forest found in high-humidity tropical zones. Annuals such as rice⟡, maize⟡ and sweet potato are grown among perennials such as fruits, nuts and fibres, while forest trees are retained to provide fuel, timber, tannins, gums and medicinal materials. The leaves of the trees and the grasses that grow underneath them provide a valuable source of livestock fodder.

Agroforestry systems are now spreading and are being used more and more in Africa, Asia and Latin America. Over the last two decades the value of agroforestry has begun to be fully realized and over $1 billion has been spent on promoting forestry projects. The Kenyan-based International Council for Research In Agroforestry (ICRAF)⟡ reports that agroforestry projects are now under way in over a hundred developing countries⟡.

AHIMSA   An ethical practice of fundamental importance in Hinduism, Buddhism and Jainism. Because of their belief in reincarnation, followers of these religions respect all forms of life as being parts of the cycle of rebirth; thus their religion prohibits causing harm to all living things. Vegetarianism is common in areas where ahimsa is followed. The word itself is a Sanskrit word meaning non-injury.

AID A term used to denote the net flow of official development assistance (ODA). This comprises the transfer of capital – usually in the form of loans or grants – from governments, international agencies and public institutions of the industrialized world to governments of the less-developed nations, theoretically to assist these countries to improve their standard of living. The value and costs of the provision of other forms of assistance, consultancy or technology transfer and the provision of materials may also be included in the aid totals. There are a number of different categories of aid:

Net aid – the amount of assistance allocated under government overseas aid schemes. Net flows of official development assistance count towards the ODA-target figure of 0·7 per cent of Gross National Product (GNP)⋄, set as a minimum target by the United Nations⋄. Major donor governments have agreed to try to achieve this.

Gross aid – the amount in the net aid package before deductions have been made for amortization⋄ and interest payments on past aid loans.

Bilateral aid – aid provided on a country-to-country basis. It also encompasses financial assistance provided by a donor government to a national institution in a specific country concerned with development projects. Globally, of the total bilateral aid, over half is provided to finance specific projects, 10 per cent is given in the form of food aid and 2 per cent as disaster relief.

Multilateral aid – any contributions made to international institutions for use in, or on behalf of, a developing country, including national contributions to the regular budgets of certain multilateral organizations such as the World Bank⋄. The term is also used to describe the aid provided by these multilateral organizations.

Aid and Trade Provision – the proportion of a bilateral aid programme that is set aside to give higher priority to the commercial and industrial importance of a restricted number of development projects.

Budgetary aid – all financial assistance that is given to the government of a developing country to help its annual recurrent budget.

Financial aid – held to include all financial flows, both bilateral and multilateral, except those which finance technical co-operation. It includes both project and non-project aid.

Non-project aid – all financial aid except project aid, including programme aid, budgetary aid, debt⋄ relief, food aid and disaster relief.

Programme aid – the main form of non-project aid: non-specific support given to finance essential imports, usually provided to assist countries which have particularly acute balance-of-payments problems.

Project aid – grants or loans given to finance the establishment of new production and infrastructure facilities or the expansion and improvement of existing ones. This aid is generally tied to a specific investment or rehabilitation project – in contrast to programme aid.

Sector aid – assistance intended to benefit a designated sector of the economy, or key sectoral institutions, involving a flexible mix of maintenance goods, miscellaneous capital goods, project aid and technical co-operation.

Tied and untied aid – aid is said to be 'untied' when it is not subject to any geographical limitations on procurement. 'Tied' aid is much more common; in this case procurement is specifically restricted to the goods and services of the donor country. This form of aid, conditional on recipient nations buying goods and services from the donor country, is little more than an export subsidy.

Technical assistance – aid can be given in the form of technical co-operation or the provision of expert knowledge and guidance. This is commonly practised by the various specialized agencies of the UN.

The aid system, whereby wealthy nations help those less well off, is a complex, controversial and inequitable system with several major failings.

The global community has agreed a number of 'Aid Targets' over the years. Several goals have been set by the United Nations with regard to how much aid the wealthier nations should provide to the developing world, but few have ever been achieved. In 1968 the UN Conference on Trade and Development (UNCTAD)⋄ recommended that the annual net flow of resources from each economically advanced country to the developing world should be at least 1 per cent of their GNP. This target was held to include net official development assistance, other official flows and private flows. The UN's second Development Decade⋄ Strategy (1971–80) called upon donor countries to devote a minimum of 0.7 per cent of GNP to net official development assistance by 1985. The Paris Conference in 1981 recommended that donors direct at least 0.15 per cent of their net official development assistance specifically to those nations included in the list of least developed countries (LLDCS)⋄.

Between 1970 and 1980 the total of annual official aid rose by 50 per cent but it has declined steadily ever since. In 1980 Western nations gave $47.6 billion in aid to developing countries; this represented only a slender increase in real terms over previous contributions. The eighteen industrialized country members of the Organization for Economic Co-operation and Development (OECD)◇ increased their total contribution to development aid to a meagre average of 0.35 per cent of their GNP – half the target set and agreed upon by members of the United Nations. Only four industrialized countries – Denmark, the Netherlands, Norway and Sweden – are currently meeting the 0.7 per cent of GNP target. Aid from wealthy members of the Organization of Petroleum Exporting Countries (OPEC)◇ also continued to fall following a continued depression in world oil◇ prices.

By 1990 Japan had overtaken the United States to become the world's largest aid donor, providing $10.4 billion in overseas assistance. The Japanese government announced that by 1992 its aid should be running at $50 billion a year. This figure may be compared with that of the Japanese trade surplus with the rest of the world which, at the time of the announcement, amounted to over $300 billion annually.

Development aid is given for a variety of reasons, including political, idealistic, humanitarian, and commercial considerations. It is often directed by donor nations to countries that are of some importance to them for strategic, cultural or economic purposes. Japanese aid, for example, goes primarily to twenty-five countries, mostly in Asia, whereas European nations such as Britain and France heavily favour their ex-colonies.

As it currently operates, the overseas aid scheme in general has many critics who argue that the whole system is flawed, with aid being used simply as a political and economic weapon solely in the interests of the donor countries. Indeed, a large proportion of aid is of the 'tied' variety, and much of the official aid given never leaves the shores of the donor nation; it merely subsidizes specialized exports or expert services. Critics also argue that the bulk of aid given to the Third World◇ serves to increase exports of unprocessed food crops and raw materials which ultimately benefit the industrialized world by fuelling their processing industries.

In 1974, E.F. Schumacher◇ described overseas aid as 'a process where you collect money from the poor people in the rich countries and give it to the rich people in the poor countries'. It is estimated that 20–30 per cent of aid never actually reaches and benefits the poor in the developing world but is either siphoned off into private bank accounts through widespread corruption or lost in administrative costs. It has also, in the past, been directed toward large-scale development projects that provide very little real benefit for the poor and hungry for whom the aid was theoretically intended, serving instead to improve the living standards of the elite and relatively well-off urban dwellers in Third World countries. Many of these large-scale projects were initiated because they served the interests of the donor nations. The folly of this has now been recognized and smaller-scale projects, the positive impact of which can be felt more directly by poverty-stricken rural communities, are becoming more commonplace.

Perhaps the strongest criticism levelled against the aid system is that financial assistance proves counterproductive, encouraging indebtedness in the developing world without providing any real financial advantage. Despite massive amounts of financial assistance given by donor governments, in reality since 1982 the Third World has actually received less money in aid from the industrial world as a whole than it has paid back in interest and repayments on loans already received. Between 1983 and 1990, a net total of over $160 billion was transferred from the developing countries◇ of the South◇ to the wealthy, donor countries in the North◇. In 1989 the net transfer of financial resources from developing countries peaked at $52 billion, representing the difference between debt service and new investments or loans. Various plans to reverse the flow of funds and alleviate the debt problem faced by the Third World have been proposed, but none has been implemented with any major degree of success, although in 1990 debt relief measures and increased aid led to an overall net flow back to the South of around $9 billion.

AIDS *See* Acquired Immune Deficiency Syndrome

ALAR A trade name for daminozide, a chemical sprayed on to vegetables and fruits, notably apples, to boost yields while stunting the growth of the plants. During the late 1980s evidence emerged to indicate that this chemical causes cancer◇. A ban on the use of Alar was introduced in the United States in 1989 after studies revealed a direct correlation between its use and life-threatening cancerous tumours in animals. The principal manufacturer, Uniroyal, withdrew Alar worldwide after its own studies had detected that the chemicals may be capable of inducing cancerous tumours in mice; therefore it could no longer be considered safe. In the United Kingdom, the goverment – despite the links with cancer and the fact that Alar breaks down into a much more dangerous chemical known as UDMH – concluded that Alar was not an unacceptable hazard to health and cleared it for use. The cancer risk was deemed by government scientists to be relatively low for adults and children, although other British scientists, given the same data, concluded that there was an unequivocal risk that would result in several thousand people developing cancers if widespread use of the chemical continued. Owing to increasing concerns over food safety, and despite the government's safety findings, from 1989 onwards farmers in the UK began voluntarily to abandon the use of Alar.

ALGIERS ACTION PROGRAMME A conference of the Non-Aligned⟳ Countries meeting in the Algerian capital, Algiers, in 1973 adopted an Action Programme in which the idea of a New International Economic Order (NIEO)⟳ was first introduced. Following on from the conference, the United Nations⟳ General Assembly agreed in 1974 to give its support to the ideas and recommendations contained in the Programme of Action, and later formed and officially adopted the Charter of Economic Rights and Duties of States⟳, which drew heavily on the contents of the Algiers Action Programme.

ALLENDE GOSSENS, SALVADOR (1908–73) Chilean politician, the first Marxist head of state to be elected by a democratic vote in the Western world. He became president in 1970 as the candidate of the Popular Front Alliance, although during his term of office he failed to keep the alliance together. He advocated a 'quiet revolution' to socialism, but his failure to solve the nation's economic problems or to deal effectively with subversive political elements eventually led the armed forces to attempt to overthrow the government in 1973. Allende and many of his supporters died in the coup, which was widely believed to have been staged with the involvement of the United States's Central Intelligence Agency (CIA).

ALLEYCROPPING A system of agroforestry⟳ originating in Asia and designed to compensate for reductions in fallow periods. Modern, sophisticated alleycropping systems were developed in 1976 at the International Institute of Tropical Agriculture (IITA) in Nigeria in an attempt to solve soil fertility and soil erosion⟳ problems. Lines of trees or shrubs are grown with food crops being raised between the rows. The trees are trimmed back like a hedge while the food crops are growing, but are allowed to spread and cover the soil once they have been harvested. Wood⟳ can be used as fuel or as stakes, and small twigs and leaves that are pruned can be dug in as fertilizer⟳, applied as a surface mulch or used as livestock⟳ fodder. Cropping and fallowing thus occur more or less together, allowing almost continuous productive use of the same piece of land⟳ with no need for additional inputs, especially when soil-enriching species of trees or shrubs are used. The International Fund for Agricultural Development (IFAD)⟳ is actively promoting the widespread use of alleycropping, supporting experimental projects and growing systems under differing conditions in eighteen sub-Saharan countries.

ALLIANCE FOR PROGRESS Founded by US President John F. Kennedy at Punta del Este, Uruguay, in 1961 to help accelerate economic progress, improve co-operation and improve living standards throughout Latin America. Its Inter-American Committee examines and reports annually on the progress made by each country in the region, and on the need for external assistance. Most economic integration attempts in the Latin American region have met with, at best, only partial success, and there is a growing feeling that the Alliance was dominated by US self interest.

ALUMINIUM A light silvery-white metal, the most abundant metal in the earth's crust, the main source being bauxite⟳. It is remarkably light and easy to work with, forming alloys with several other metals and plastics. It has a wide range of uses, from kitchenware and foil wrapping to aircraft, electrical conductors and sheathing nuclear fuel rods.

The levels of aluminium in many water⟳ supplies are increasing from both natural causes and through man-made contamination. Acid rain⟳ has been shown to release free aluminium in the soil. The presence of aluminium in water supplies is thought to cause kidney damage and lead to increased incidence of Alzheimer's disease, a form of senile dementia. Several governments have taken steps to limit exposure to aluminium, and in 1990 the United States National Aeronautical and Space Administration (NASA) announced plans to try and re-use the external fuel tanks from the US space shuttles. This followed evidence that the shuttle's two tanks, both made from aluminium, burnt up on every launch, adding 60 tonnes of aluminium to the environment with every mission flown.

Worldwide, over 17 million tonnes of aluminium are produced each year the United States (3.9 million tonnes), the Soviet Union (2.4 million tonnes) and Canada (1.5 million tonnes) are the major producers.

AMARANTH   A widely used, red coal-tar dye known in the United States as Red 2 and in the European Community⋄ as E123. It was first synthesized in 1878, and large-scale production commenced in 1914. It is used in food products, drug preparations, cosmetics and inks. In the food industry it is one of the commonest colouring agents, employed to produce the colour of cherries, strawberries and raspberries. In 1970, research in the Soviet Union suggested that the chemical was carcinogenic. As a result of this evidence and further tests, the use of amaranth has since been banned in Austria, Finland, Greece, Norway, the Soviet Union, the USA and Yugoslavia. Its use is still permitted by over sixty other nations around the world.

AMAZONIA   The Amazon River in South America is the world's largest by volume and, at 6,518 km, the second longest. The Amazon Basin, covering some 5–7 million sq km, contains a network of waterways which drains half of the continent and a vast array of differing habitats, including grasslands, wetlands⋄, shrublands and lakes, Most notable, however, are the region's tropical forests⋄, ranging from the swampy mangroves⋄ in the east to the high mountain forests in the Andes.

Most of Amazonia is covered in luxuriant vegetation with a wide mix of trees including acacia⋄, rosewood, brazil nut, palm, mahogany, cedar and rubber⋄. The massive extent of rainforest⋄, covering 40 per cent of Brazil, has a major impact on the planet, producing an estimated 25 per cent of the world's oxygen. In addition, the area exhibits the greatest biodiversity⋄ of any in the world. There are more species of vascular plants, insects, terrestrial vertebrates, freshwater fish⋄, amphibians and primates in Amazonia than anywhere else in the world. Yet although the rainforest soil is poor, the forests are being cleared for timber and agriculture⋄ at an alarming rate, and an incalculable number of wildlife species are disappearing as a result.

Economic development and exploitation of Amazonia began in earnest in the 1960s, fuelled by tax incentives and the construction of the Trans-Amazonia Highway, the Belém–Brasilia Highway, and two railway lines. The area is known to contain large deposits of natural gas⋄, oil⋄, iron⋄, bauxite⋄, gold⋄, nickel, copper⋄ and tin⋄, as well as substantial supplies of timber. The accuracy of estimates of clearance of the Amazonian forest is notoriously difficult to establish, but best estimates conclude that 5–7 per cent of the total forested area has been lost, even though by 1980 some exploitation policies had begun to be reviewed. Between 1960 and 1970, at least 10 million hectares (ha) of land in Brazil, together with 1.5 million ha in Colombia and 500,000 ha in Peru had been cleared for pasture. Clearance has continued virtually unabated since, although a mixture of government policies and the clamour of the international community has managed to halt the rate of deforestation⋄ increase. In 1987, 8 million ha were cleared. This fell to 4.8 million ha in 1988 and was reduced by a further 40 per cent in 1989.

Through the destruction of the natural Amazonian ecosystem⋄, the world is losing millions of species of plants and animals. In addition, the lives and cultures of indigenous peoples in the area are under threat. Furthermore, secondary problems arising as a result of forest clearance, such as the spread of disease, are continuing to surface. Malaria⋄ in particular is becoming of major concern; the 51,000 cases reported in the region during 1970 rocketed to a million in 1990.

AMIN DADA, IDI (1925–)   Ugandan politician. After military training in Britain he rose rapidly in the Ugandan army and led the 1971 coup which deposed President Milton Obote. He served as the nation's president from 1971 until he was overthrown in 1979. He was also elected President of the Organization of African Unity (OAU)⋄ in 1975. During his term of office he exercised a reign of terror over his own people, and in 1972 expelled the 80,000 non-Ugandan Asian community from the country. Throughout his reign the Ugandan environment was overexploited and the degradation was so intense and widespread that the country has yet to fully recover. He fled when 'rebel' Ugandan and Tanzanian troops invaded the country in 1979, and went into exile in Saudi Arabia.

AMMONIA   Until fairly recently, attention was focused on sulphur dioxide⋄ and nitrogen oxides⋄ as the major precursors of acid rain⋄ and the main transboundary air pollutants. However, evidence is now emerging that ammonia is also of significant importance, causing the acidification of soils and aquatic ecosystems⟡ and encouraging the formation of acid rain.

Ammonia is a colourless, pungent and toxic gas (NH_3), thought to be so-named because a compound used in soldering, ammonium chloride, was once made from camel dung near the temple of the Egyptian god Ammon. Ammonia is used in the manufacture of fertilizers⟡, nitric acid, explosives and synthetic fibres, but it is the connection with livestock⟡ which remains of greatest importance.

The principal source of atmospheric ammonia is the decomposition of livestock wastes, and levels of ammonia in the atmosphere have been increasing in proportion to the increasing number of livestock being kept by farmers around the world. In the industrialized nations, emissions of ammonia produced during the manufacture and use of nitrogenous fertilizers have also been increasing steadily. Little is known about the levels and dynamics of natural emissions of the gas, but in the developed world man-made emissions have grown steadily and now stand at 40 million tonnes annually. In the European region, emissions have increased 70 per cent over the past century, with the contribution from fertilizer production rising sevenfold.

AMNESTY INTERNATIONAL An international organization, founded in Britain in 1961, which aims to defend both freedom of speech and the right of individuals to follow religious or political beliefs in every part of the world. Politically unaligned, it campaigns all over the world for the release of 'prisoners of conscience', for the provision of all human rights⟡ and against all forms of torture, and also strives to improve the welfare of refugees⟡. The organization, which is funded by voluntary contributions, now has more than 100,000 members in over seventy-five countries. Amnesty International was awarded the Nobel Peace Prize in 1977.

AMOCO CADIZ The *Amoco Cadiz*, a tanker belonging to the US-owned AMOCO Oil Corporation, ran aground off the French coast in 1978, releasing 223,000 tonnes of crude oil⟡ into the sea. More than 130 beaches were subsequently covered in oil up to a depth of 30 cm. Over 30,000 seabirds died, together with 230,000 tonnes of crabs, lobsters, sole and other fish⟡. The area's prized oyster and seaweed beds, covering 800 hectares, which provided a source of income for many of the local inhabitants,

were totally destroyed. In 1988, following ten years of litigation, a Federal Court judge in the United States awarded $85.2 million in damages to the ninety Breton communities, representing over 400,000 people, who had fought the case. The claimants had been seeking damages of $750 million and the amount awarded did not even meet their legal costs, calculated to be over $165 million. Many people, including France's President Mitterrand, strongly criticized the size of the settlement, drawing attention to the fact that during the first nine months of 1987 alone, AMOCO announced a net income of $983 million on a turnover of $16.5 billion. An appeal was immediately made against the judge's ruling, but it will probably be at least 1994 before a final decision is reached.

AMORTIZATION The repayment of the capital, or principal, of a loan rather than mere repayment of interest.

ANAEMIA A disease brought about by a reduction in the quantity of the oxygen-carrying pigment, haemoglobin, in the blood. The main symptoms are persistent fatigue, breathlessness following even the slightest exertion, pallor, and poor resistance to infection. Widespread in many countries, anaemia afflicts 5–15 per cent of all adult men and even higher proportions of women and children. It is particularly dangerous in pregnant women, as it significantly multiplies the risk of dying during childbirth.

There are many causes of anaemia. The disease arises as a result of the loss of blood caused by an accident, operation, or through chronic internal bleeding from an ulcer or invasion by intestinal worms. It can also be caused by a lack of iron⟡, an essential component for the production of haemoglobin. An estimated 800 to 900 million people in the world suffer from anaemia due to iron deficiency. The condition can also be brought on by destruction of red blood cells, as a result of either exposure to toxic chemicals or the action of parasites, most notably those that cause malaria⟡. Certain hereditary conditions which cause deformities in red blood cells can also bring on anaemia, as can diseases such as leukaemia which suppress the production of red blood cells in the bone marrow. Successful treatment regimes for most forms of anaemia exist, but vary according to the cause.

ANC *See* African National Congress

ANDEAN GROUP A subgroup of the Latin America Free Trade Association (LAFTA)⬦. Formed initially by Bolivia, Chile, Colombia, Ecuador, Peru and Venezuela, it was founded following the Cartegna⬦ Agreement in 1969 with the objective of promoting economic co-operation and tariff integration between member countries, with the intention of producing far better results than were being seen under LAFTA as a whole. Chile withdrew from the Group in 1977.

ANOPHELES A widely distributed genus of mosquito⬦ with some 350 species, occurring in both tropical and temperate regions. These insects have a massive impact on human health, for the malarial parasite, *Plasmodium*, is transmitted to humans solely through the bite of female *Anopheles* mosquitoes. Only the females seek blood meals; males feed on nectar and other plant juices. Some species of *Anopheles* also transmit the parasites which cause Bancroftian filariasis⬦, which can lead to blindness.

ANTARCTICA The southernmost continent, 14.2 million sq km in area, surrounding the South Pole and consisting mainly of a vast ice-covered plateau. The prevailing climate is the severest and most inhospitable in the world, and although the continent contains almost 90 per cent of the world's fresh water⬦, it is mostly in the form of ice. Because of the ice, the continent supports only a small range of cold-adapted land plants and animals, although the surrounding sea has a rich flora and fauna. Despite the presence of so much ice, Antarctica is one of the world's greatest deserts, the average precipitation being about 50 mm over the polar plateau. The specialized plants must survive cold, lack of water and lengthy winter periods of almost total darkness. Of the 800 species found on the continent, 350 are slow-growing lichens. The indigenous land animals are wholly invertebrate. Many, such as lice and fleas, are parasites of the millions of seabirds which visit the continent.

Antarctica forms part of what are known as the global commons, land and resources which are not legally owned. Some nations have made political claims to territory in Antarctica, several claims overlap and are therefore under dispute. Semi-permanent scientific research stations began to be established on the continent during International Geophysical Year (1957–8). The nations involved agreed to respect the terms laid out in the Antarctic Treaty signed the following year, which put a halt to all territorial claims and provided freedom for future scientific experimentation and observation.

The continent could prove to be of enormous economic value as significant quantities of minerals⬦, such as coal⬦, iron⬦ ore and oil⬦, have already been discovered there. Coal deposits under the Transantarctic Mountains are reported to be amongst the most extensive in the world. However, the problems of recovering these under the severe environmental conditions which exist for most of the year may prove insurmountable. Global concern over the probable environmental impact of any such mining or oil exploration work has also prevented any resources from being exploited. Indeed, in the late 1980s several nations, led by Australia and France, opted to press for Antarctica to be turned into a World Park free from any form of mining or exploitation of its mineral wealth.

Ownership or control of Antarctica and its resources is becoming of increasing importance, and no international agreement has ever been reached on how the resources in this 'common heritage' should be apportioned. Nor has any system been devised which would decide who should pay for any environmental damage that could arise if the continent's resources were ever to be exploited. Despite this, nations such as Japan and the United States are continuing to explore for oil, in violation of a 1977 moratorium agreed under the Convention on the Conservation of Antarctic Marine Living Resources (CCAMLR). The Ross and Weddell Seas are thought to contain 15 billion barrels of oil, only slightly fewer than the 20 billion barrels estimated to be in the British sector of the North Sea. Conservationists' concern over possible pollution⬦ problems in Antarctica are well founded for oil spills in Antarctica could prove catastrophic. Oil takes far longer to decompose in low temperatures, and an oil spill on ice would increase its capacity to absorb heat and thus cause it to melt.

The enormous marine potential in the waters of the southern oceans is now being exploited more fully as a result of a combination of the depletion of catches seen in northern waters due to overfishing and the moratorium on whaling⬦. Fleets from the Federal Republic of Germany, Japan, Poland and the Soviet Union are fishing extensively in Antarctic waters, mainly for krill.

ANTARCTIC TREATY An agreement signed in 1959 by Argentina, Australia, Belgium, Chile, France, Japan, New Zealand, Norway, South Africa, the Soviet Union, the United Kingdom and the

United States to maintain the Antarctic as a demili-tarized zone for thirty years. The Treaty now has thirty-nine signatories. The twelve original nations have actual territorial claims on the continent, though these claims are not internationally recog-nized. A further ten nations, including Brazil, China, the Federal Republic of Germany, India, Poland and Uruguay are consultative members be-cause of their scientific involvement. Seventeen other nations have simply acceded to the Treaty and attend meetings with observer status.

The Treaty permits Antarctica⇔ to be used for scientific or peaceful purposes. It prohibits all forms of military manoeuvres, weapons testing, mining or the disposal of radioactive wastes⇔. Under a comprehensive system of verification, observers appointed by each of the twelve original contracting parties have the right of aerial surveillance and com-plete access at all times to any area or installation.

ANZUS TREATY   A security treaty concluded in 1951 by Australia, New Zealand and the United States which required them to provide mutual aid in the event of aggression by foreign powers. Temporarily replaced by the South East Asia Treaty Organization (SEATO)⇔ between 1954 and 1977, the Treaty was effectively rendered inoperative in 1985 when New Zealand adopted a non-nuclear policy and refused to allow US warships into its ports because they might be carrying nuclear weapons.

APARTHEID   A policy adopted by the South African government which calls for the separate development of the white and black populations in South Africa. The word 'apartheid' was coined by the South African Bureau for Racial Affairs in the late 1930s and the policy was introduced by the ruling Afrikaner National Party in 1948. Since then, the country has evolved under a system of segre-gation and separate development for its peoples. Under the system, inhabitants are classified into four racial categories. African, Coloured, Indian and White. Special education, amenities and places to live were designated for each category. The original apartheid system had seven major components:

Group Areas Act – People are restricted to living in areas allocated to their particular racial category.
Separate education – Children may attend only schools allotted to their own race.
Homelands – All black South Africans were deemed to have a tribal homeland which became their official home, even though they might never have been there.

Voting rights – Black South Africans are not eligible to vote in parliamentary elections. They are allowed to vote for candidates of their own racial grouping in elections for local authorities, institutions in which precious little power is invested.

Separate amenities – Places of entertainment, shop-ping facilities and public transport were all segre-gated according to racial groupings, although local authorities have since been given the power to ease the regulations in many instances.

Mixed Marriage – An Act of Parliament made it illegal for people of different racial groups to marry.

Immorality Act – Sexual relations between people from different racial groups were prohibited under this legislation.

The obvious inequalities inherent in the apartheid system, and the international outcry that followed its introduction and continued implementation, eventually forced South Africa to withdraw from the Commonwealth⇔ in 1961.

Following years of national and international campaigning to persuade the South African govern-ment to abandon the apartheid policy, limited but ever-increasing reforms began to be seen through-out the 1980s. In 1985 both the the Mixed Marriage and Immorality Acts were repealed. Since 1985 non-Whites have won limited constitutional reforms, and increased internal unrest, coupled with sporting and financial pressures from outside the country, led to further concessions and a commit-ment among leading politicians to remove the inequalities.

In certain instances, however, the more conserva-tive members of the white community refused to comply with government reforms. In 1990 relax-ation of the Separate Amenities legislation, which opened up such facilities as swimming pools and cinemas to mixed race use, was nullified by conserva-tive local community leaders who either banned non-Whites from using local amenities or imposed a *de facto* ban by simply introducing exorbitantly high entrance fees which the majority of non-whites could not afford.

Despite the government's avowed intention to carry out further reforms, by the beginning of the 1990s racial groups in South Africa were still being forced to undergo independent and unequal pro-cesses of development.

APPROPRIATE TECHNOLOGY Low-cost technology designed for small-scale use, an idea originating in Mahatma Gandhi's ideas for improving living standards in rural regions of India. The concept was taken up and developed by E.F. Schumacher⋄, who founded the Intermediate Technology Development Group⋄, which helps to put some appropriate technology ideas into practice.

'Appropriate technology' describes the use of tools, machinery and systems that can be applied readily by the people who will benefit directly from their application. Moreover, these can be sustained and maintained by local inhabitants without recourse to imported supplies that may prove either difficult to obtain or prohibitively expensive. The term is usually applied to small-scale, decentralized development approaches in which locally available resources are identified and fully utilized with a view to long-term viability.

AQUACULTURE Aquaculture is the controlled cultivation and harvest of freshwater or marine aquatic species of plants and animals. Rearing of freshwater fish⋄ has been practised in Asia for over 4,000 years, usually on small farms of less than 1 hectare each. These farms generally produce over 4.5 tonnes of fish per hectare a protein production five times better than any agricultural crop. Fish have a food-to-flesh conversion rate of 1.5:1 – much better than that for beef, pigs or chickens. As the economic value of fish farming⋄ is becoming apparent, intensive farming of fish and other aquatic organisms is being actively encouraged.

Marine fisheries traditionally provide the bulk of the world's fish – around 90 million tonnes – but by the year 2000 demand will have outstripped marine supplies. At present, aquaculture accounts for only 10 per cent of global fish production, although over half the fish eaten in Israel and a quarter of the fish consumed in China and India now comes from fish farms. The Chinese – mostly using species of *Tilapia*, a carp-like fish – have developed a system of fish farming in which bottom-feeding fish and surface feeders, together with fish which feed at five other levels, are reared in the same ponds. In Kuwait, where fresh water is a scarce commodity, *Tilapia* are being reared successfully in brackish water.

Elsewhere in the world, various other forms of aquaculture are being developed and encouraged. Seaweed is being grown extensively as a source of food for humans and livestock⋄, and as a source of valuable industrial materials. Shellfish such as oysters and mussels have been exploited as a source of food or jewellery for centuries. The giant clam, native to coral⋄ reefs in the Pacific Ocean, has long been a source of meat delicacies for the Chinese and Pacific islanders. Large specimens can yield up to 25 kg of meat, and they are now being farmed in coastal waters off several islands in the region. The clam manufactures its own food, microscopic algae growing inside the molluscs producing carbohydrates through photosynthesis using the waste carbon dioxide⋄ produced by their hosts. Large-scale farming is feasible, giving yields of up 60 tonnes of meat per hectare of open sea, but complications arise, as the clam is regarded as an endangered species⋄ and trade is restricted under the Convention on International Trade in Endangered Species of Wild Fauna and Flora (CITES)⋄.

ARAB AID AGENCIES A grouping of national and regional institutions which, using the massive revenues generated by sales of oil⋄ produced by the Arabic countries in the Middle East, provide financial assistance on concessionary terms to developing countries⋄. Aid⋄ is usually given in the form of project aid or balance-of-payments support. The agencies' administrative resources are limited, and they work closely with other multilateral bodies such as the World Bank⋄. The various agencies have a diverse membership and exhibit a marked geographic, religious or political bias. The Arab Fund for Economic and Social Development (AFESD)⋄ limits its activities to Arab nations, and the Islamic Development Bank (ISDB)⋄ restricts its operations and financial assistance to Islamic countries or communities. Where the funds are representative of a national body – such as the Abu Dhabi Fund for Arab Economic Development (ADFAED)⋄ – the field of operations is less confined.

ARAB BANK FOR ECONOMIC DEVELOPMENT IN AFRICA (ABEDA) An institution formed by the Arab League⋄ states in 1974 with the intention of promoting the economic development of African countries, specifically those that are not members of the Arab League. The Bank also aims to encourage the participation of other funding agencies from the Arab sector to accelerate the development process in Africa. In addition to using its own funds for projects, the Bank provides the necessary administrative and technical assistance to enable other Arab funding agencies to lend their support.

ARAB FUND FOR ECONOMIC AND SOCIAL DEVELOPMENT (AFESD) The AFESD aims to improve economic and social development in Arab countries through the provision of financial backing for development projects, the promotion of investment in all economic and industrial sectors, and the provision of technical assistance. Established in 1968, with a secretariat in Kuwait, the Fund became operational in 1973. The major beneficiaries have been those Arab League⟳ countries which do not produce their own oil⟳.

ARAB LEAGUE (LEAGUE OF ARAB STATES) An organization formed in 1945 to promote unity and co-operation among all Arab countries, with a secretariat in Cairo. The League originally consisted of those Arab nations that were independent at the time – Egypt, Iraq, Jordan, Lebanon, Saudi Arabia, Syria and North Yemen. Others – Algeria, Bahrain, Djibouti, Kuwait, Libya, Mauritius, Morocco, Oman, Qatar, Somalia, Tunisia, United Arab Emirates and South Yemen – joined on attaining independence. The Palestine Liberation Organization (PLO) is also a member. Although the League's activities have met with some success in the fields of science and culture, it has been beset by difficulties arising as a result of the firm but contradictory political stances adopted by member states. Egypt was expelled from the League in 1979 following the signing of a peace treaty with Israel, but re-admitted in 1990.

ARAL SEA The Aral Sea in the Soviet Union covers 67,000 sq km and is the world's fourth largest inland body of water⟳. A massive, long-standing scheme to irrigate large areas of a desert in Soviet Central Asia has caused the Sea to shrink drastically. For the last thirty years the irrigation⟳ scheme has diverted water from two of the lake's major tributaries, the River Amu and the River Syr Darya. As a result, since 1960 the Aral Sea has lost 70 per cent of its water and 55 per cent of its surface area. It shrank more than 96 km from its original boundaries and the water level dropped some 13 m over a twenty-five-year period, with the result that the Sea is now split into two separate basins of bitter, salty water. The ecology⟳ of the entire surrounding area has suffered, with violent dust storms along the southern and eastern shores. The Sea is now too saline for fish⟳ and too shallow for ships to navigate. In the irrigated desert areas, water in the irrigation channels is now badly polluted, drinking water is of an almost unacceptable quality, and soil erosion⟳ is increasing rapidly.

ARCTIC TREATY In 1990, following the increasing freedoms and liberalization in the Soviet Union, a pact was agreed between Canada, China, Japan and the Soviet Union that committed the signatories to carry out joint environmental protection and scientific research in the Arctic. Amongst other things, the pact covers the monitoring of air pollution⟳, the exchange of technical information and the funding of environmental protection programmes as well as the organization of co-ordinated research activities on timber harvesting, wildlife management and offshore oil⟳ exploration.

ARMS CONTROL Several bilateral arms limitation and reduction treaties have been agreed between the Soviet Union and United States since the first effort to reduce the risk of nuclear war, the Hot Line Agreement, was concluded in 1963. This established direct communication between the two nations in times of crisis. However, most treaties have failed to meet their objective of preventing the proliferation of armaments and halting the development of newer and deadlier forms of nuclear, chemical or biological weapons.

Since 1945, only two multilateral agreements governing actual disarmament have been reached. The Biological Weapons Convention of 1972 not only prohibits the production of biological weapons, it also provides for the destruction of existing stockpiles. The Conventional Forces in Europe Treaty (1990) calls for the Soviet Union to give up its superiority in conventional arms and for the sixteen members of the North Atlantic Treaty Organization (NATO)⟳ and the six Warsaw Pact⟳ nations to destroy large numbers of tanks, combat aircraft and other non-nuclear weapons within a strictly limited geographical area in Europe.

Since the 1920s several other multilateral arms control treaties have been negotiated, each supported by a varying number of countries. They mostly concern nuclear weapons, but although they impose certain restrictions, they do not involve the total destruction of weapons:

Geneva Protocol (1925) – Prohibits the use of asphyxiating, poisonous or toxic gases and all bacteriological forms of warfare. It does not restrict research into or the development and build-up of chemical weapons and so, effectively, bans only the first use of such weapons.

Antarctic Treaty⬦ *(1959)* – Declares Antarctica⬦ to be a demilitarized zone which should be used exclusively for peaceful purposes, rendering the continent the first nuclear-free zone on earth.

Partial Test Ban Treaty (1963) – Bans the testing of nuclear weapons in the atmosphere, outer space and underwater. It does not prevent nuclear testing underground.

Outer Space Treaty (1967) – Prohibits nuclear weapons from being stationed in outer space, on celestial bodies or in earth orbit. It does not legislate against weapons which enter outer space and return to earth.

Latin America Nuclear Free Zone Treaty (1967) – Bans all states and territories in the region from testing, possessing or deploying all forms of nuclear weapons.

Non-Proliferation Treaty⬦ *(1968)* – Prohibits the transfer of nuclear weapons and associated technology outside the five major countries recognized as having nuclear weapons. The Treaty also commits these states to stop further arms production and development.

Seabed Treaty (1971) – Bans the deployment of nuclear weapons on the seabed beyond the limit of territorial waters. There is no ban on mobile underwater weapons systems, or on servicing them from the seabed.

Biological Weapons Convention (1972) – Outlaws the development, production, stockpiling and use of all biological warfare⬦ agents and toxins, and requires the destruction of any existing stocks.

Environmental Modification Convention (1977) – Bans military or any hostile use of technology or substances to change weather patterns, ocean currents or the ozone⬦ layer, or to alter the ecological balance in any way for military purposes. Certain techniques which could be used in tactical military operations are not banned.

Inhumane Weapons Convention (1981) – Places restrictions on the use of weapons which are held to be particularly indiscriminate or injurious (so-called 'wicked weapons'). It bans the use of fragmentation bombs that produce destructive agents which are not detectable in the human body and prohibits the use of mines, booby traps, napalm and incendiary devices against civilians.

South Pacific Nuclear Free Zone Treaty (1985) – Bans testing, manufacture, acquisition and stationing of nuclear weapons in the region. It requests the five major nuclear weapons states to sign a protocol banning the use or threat of nuclear weapons and nuclear testing.

Conventional Forces in Europe Treaty (1990) – Commits members of NATO and the Warsaw Pact to reduce their numbers of tanks, artillery and other offensive weapons to a point where numerical balance is struck between the two alliances. The agreement applies to an area of Europe stretching from the Atlantic coast to the Ural Mountains.

ARMS SPENDING Since 1900, global military expenditure has increased more than thirtyfold. By the mid-1980s the world's annual military expenditure was estimated to be in the region of $860 billion and rising steadily. Five years later, in 1990, the funds devoted each year to military research and development worldwide were approaching the $1 trillion mark. Military defence and armaments programmes consume over 25 per cent of the entire global expenditure on scientific research and development, with over 500,000 of the world's scientists (25 per cent of the total) engaged in the development of new or advanced weaponry. In sum, over 70 million people, including personnel in the regular armed forces, are engaged in military activities worldwide. The United States currently spends $300 billion annually on defence, 6 per cent of the country's Gross National Product (GNP)⬦. The nation's Strategic Defence Initiative (SDI)⬦ research programme, the most ambitious, technologically complex and costly ever devised, cost nearly $4 billion in 1988 alone.

The unit cost of weapons, especially nuclear, is soaring and a single test of a nuclear weapon costs, on average, $12 million. Overall investment in military programmes and expenditure on arms far exceed investment and expenditure in any other field, and this is generally true for developed nations as well as those in the Third World⬦. Developing countries⬦ spend $200 billion annually on arms – four times the economic assistance they receive and more than enough to service their overseas debt⬦. While the national budgets of these countries has been cut in some areas – notably on health, education⬦ and the environment – arms spending levels have tended to be maintained. Since the mid-1970s, the Ethiopian government has been spending $275 million per year on the war against secessionist movements, compared to $50 million a year in environmental improvement projects. Major weapons systems continue to be imported by ninety developing countries, and the traffic in arms accounts for a large proportion of their total overall trade. Most of the arms involved in global trade are consigned to areas of conflict in the Middle East, and account for over half the total of weapons imported into developing nations. More than 90 per cent of all the weapons traded come from six countries: France, the Federal Republic of Germany, Italy, the Soviet Union, the United Kingdom and the United States.

ARTESIAN WELL A well in which pressurized water⬦ rises to the surface. The pressure is formed because the underground aquifer is tilted, so the source of the water is at a higher level than the well.

ARTIFICIAL RAIN Clouds can be induced to produce rain by the addition of certain chemicals. The two most common methods for this so-called 'seeding' of clouds are the aerial spraying of coolants or the injection of reagents such as silver iodide. Both these methods stimulate the formation of ice in clouds, which eventually leads to precipitation. The Soviet Union, the leading nation in the creation of artificial rain, is also working on ground-based systems involving firing rockets into clouds, where they will either release a reagent or create smoke to induce rain. Causing clouds to release the water trapped in them in the form of rain can increase or delay rainfall in any specified area; thus favourable weather conditions can be created to improve agriculture⬦, douse forest fires or clear fog. According to Soviet experts, the artificial seeding of clouds is now responsible for around 15 per cent of seasonal rainfall in the country.

ASBESTOS There are three distinct types of asbestos, a name applied to a group of naturally occurring fibrous mineral silicates. These are crocidolite ('blue' asbestos), chrisotite ('white') and amosite ('brown'). After mining, asbestos fibres are separated and spun into a cloth. Asbestos is highly resistant to heat and a poor conductor of electricity. In the past it has been used for a range of fireproof materials and in the building industry.

Asbestos fibres range in size from 30 cm down to a few thousandths of a centimetre in length. If inhaled they lodge in the tissues of the bronchial tubes and cause a disease of the lung known as asbestosis. Victims become unable to breathe properly. Asbestos is also known to cause cancer⬦ of the lung, the gastrointestinal tract, and the inner lining of the chest cavity.

Since 1982, the use of asbestos has become restricted or prohibited in many industrialized nations. In the United Kingdom, blue and brown asbestos have been banned, but not white. In Switzerland a major manufacturing company has agreed to eliminate asbestos from all its cement products sold within the country and in those sold in the neighbouring Federal Republic of Germany – except for that incorporated in asbestos piping. In the United States, where it is expected that two million people will die as a result of asbestos-related diseases, a total ban on all forms of asbestos has been proposed. Following the restrictions being imposed in industrialized countries, firms manufacturing asbestos have opened new factories in Third World⬦ nations such as India and Mexico or other countries such as South Africa and South Korea where regulations governing production and use are less restrictive.

ASCARIASIS A debilitating disease caused by the common roundworm, *Ascaris lumbricoides*, a widespread parasite in temperate and tropical countries. At least 1 billion people suffer from infection with the worms, cases commonly being reported from 153 countries. The prevalence and intensity of infection are higher in children than in adults.

Adult worms inhabit the human small intestine. Females lay approximately 240,000 eggs per day (65 million during their reproductive lifetime); the eggs are passed out of the human body in excreta. They are extremely resistant to desiccation, low temperatures and chemical attack. They remain in the soil, and larvae develop inside the egg cases within three to four weeks, remaining infective for up to two years. After eggs have been ingested with contaminated food or water⬦, the shell is broken down in the small intestine and the larvae burrow into the mucosal wall and the bloodstream.

Worm infections lead to loss of appetite and faulty food absorption, either causing or exacerbating malnutrition⬦. Twenty adult worms, an average burden for those parasitized, can consume 2.8 g of carbohydrates daily. As well as severely impairing nutrition, the worms cause intestinal obstruction, biliary duct obstruction, hepatic abscesses and other complications requiring surgery.

Ascariasis is common among those living in communities characterized by poverty, poor nutrition, inadequate hygiene and santitation, and scarcity of health services. It is more common in rural than in urban communities, and infection rates are highest where conditions are humid and where night soil is used to fertilize crops and contaminated water for irrigation⬦. Control can be effected through attention to personal hygiene, health education, provision of a potable water supply and medical treatment, usually using broad-spectrum anthelmin- thics at four-monthly intervals. The disease has been eradicated in Israel, Japan and the Republic of Korea by these techniques.

ASIAN CENTRALLY PLANNED ECONO- MIES (ACPE)

A regional grouping of countries formed to promote economic co-operation and development among member states. It originally included Burma (Union of), China, Kampuchea, Korea (Democratic People's Republic of), Lao (People's Democratic Republic), Mongolia, and Vietnam.

ASIAN DEVELOPMENT BANK

A Multi- lateral Development Bank (MDB)⬦ located in the Philippines, which began operating in 1966. Its aims are to promote economic and social develop- ment in the Asian and Pacific region by lending funds and providing technical co-operation to the developing countries⬦ in the area. The Bank has thirty-two regional and fifteen non-regional mem- bers, although membership is open to those states who are members of the United Nations Economic and Social Commission for Asia and the Pacific (ESCAP), as well as to all other member states of the United Nations⬦. The Asian Development Fund was opened in 1974 as the major source of conces- sional lending by the Bank. Other specialized funds, such as the Technical Assistance Special Fund, have also been established to help finance feasibility studies, project implementation reports, sectoral studies, project planning, and any other ventures necessary to promote development programmes.

ASSOCIATION OF SOUTH EAST ASIAN NATIONS (ASEAN)

A regional alliance of countries, comprised of Brunei, Indonesia, Malaysia, the Philippines, Singapore and Thailand. Established in 1967 and based in Jakarta, the Association hopes to foster economic growth, social progress and cultural development in the region, bring about peace and stability, promote collabor- ation and mutual assistance in matters of common interest, and maintain close and beneficial co- operation with existing international and regional organizations with similar aims. ASEAN took over the non-military role of the South-East Asia Treaty Organization (SEATO)⬦ in 1975. Although the Philippines and Thailand maintain close defence links with the United States, the other member states chose to adopt a non-aligned stance.

ASWAN HIGH DAM

The Aswan High Dam in Egypt was planned to allow the level of the River Nile to be maintained throughout the year, prevent flooding, and provide water⬦ for irrigation⬦, domestic and industrial needs as and when required. Construction began in 1960 – building costs were met largely by the Soviet Union – and the dam became operational in 1971. The High Dam is sited 7 km upstream from the smaller Aswan Dam, which was built in 1902 to provide irrigation water. It is 114 m high, and the artificial lake, Lake Nasser, impounded behind the dam is one of the largest in the world, extending over 560 km upriver. The dam's hydroelectric plant contributes significantly to the nation's energy⬦ needs and has helped to re- duce Egypt's fuel import bill.

However, many adverse aspects of this large-scale project have subsequently been identified. Traditional farming relied on the annual flooding for water and silt to improve soil fertility. The build- ing of the dam has forced farming practices to change and markedly increased the use of agricul- tural chemicals. In addition, the dam has caused a drastic increase in waterborne diseases. Schisto- somiasis⬦ has become commonplace in villages around Lake Nasser, with some communities show- ing 100 per cent infection rates. Poorly managed perennial irrigation schemes have led to widespread salinization⬦. Before the High Dam was built, annual flooding washed away salts that had accumu- lated in the soil, but now 35 per cent of the total of irrigated land is affected by salinization.

Downstream, soil fertility has been sharply reduced. The Nile traditionally deposited 100 mil- lion tonnes of organically rich sediment on 10,000 sq km of land. Construction of the High Dam has reduced this total to a few tonnes per year. The resultant loss of fertility must be combated by the addition of expensive chemical fertilizers⬦. More- over, the silt is rapidly building up behind the dam and will curtail electricity generation and eventually prevent it altogether. Further downstream the re- duction of silt flow has also led to coastal erosion and a significant reduction in fishing⬦ potential along the Mediterranean coast. Following the open- ing of the dam, annual catches of sardines in coastal waters fell by as much as 97 per cent, although there is now a 36,000-tonne fishery in the dam's reservoir.

ATLANTIC COMMUNITY DEVELOPMENT GROUP FOR LATIN AMERICA (ADELA)

A private investment trust founded in 1963 by com- mercial enterprises and banks in Canada, Japan, the

United States and Western Europe. It concentrates its activities on development projects and collaborates with national and regional sources of funding and technical assistance, such as the Inter-American Development Bank⌀ and the International Finance Corporation⌀.

ATMOSPHERIC POLLUTION The atmosphere is subject to a wide array of natural pollutants and to those produced by human industry. Natural sources include smoke from forest fires, wind-blown dusts and materials from volcanic activity and natural gaseous emissions. Human actions contribute a vast cocktail of pollutants, including oxides of sulphur and nitrogen, particulate matter, hydrocarbons, lead⌀, and a variety of other chemicals. On a global basis, about 110 million tonnes of sulphur oxides were vented into the atmosphere in 1980, accompanied by 59 million tonnes of particulate matter (mostly soot), 69 million tonnes of nitrogen-based oxides, 194 million tonnes of carbon dioxide⌀ and 40 million tonnes of ammonia⌀. The member nations of the Organization for Economic Co-operation and Development (OECD)⌀ accounted for more than half of this pollution⌀, levels of which are continually rising.

Most air pollutants enter the human body by being breathed in; consequently the respiratory tract is the organ that suffers most damage. In the former German Democratic Republic, which produced 5.5 million tonnes of sulphur dioxide⌀ annually and was recognized, before German unification, as having the highest air-pollution rate in the world, 40 per cent of the population suffered from chronic breathing difficulties.

Air pollution and the need to protect human health is not a new issue and legislation to reduce atmospheric pollution has been introduced ever since the Middle Ages, culminating in the global programme to monitor air quality continuously, introduced by the World Health Organization (WHO)⌀ in 1973.

Many industrialized nations recognized the seriousness of air pollution long ago and have introduced strict clean-air legislation to ease the problems. The political importance of air pollution became evident when it was identified as a trans-boundary problem whose effective control would require international co-operation. In political terms, its significance was re-emphasized when plans to reduce air pollution from motor vehicles by 70 per cent by the year 2010, announced by the Dutch

government in 1989, caused the fall of the incumbent coalition administration. However, despite all the political activity and legislation, more than 1 billion people around the world are now regularly breathing air so polluted that it breaches internationally recognized safety limits. Cities are worst affected. In Athens, on heavily polluted days, death rates increase sixfold. Breathing the air in Bombay is equivalent to smoking ten cigarettes. Capital cities of developed nations, including Brussels, London and Madrid, regularly breach air-pollution safety limits set by the WHO.

In addition to a detrimental effect on human health, air pollution also restricts agricultural productivity and damages forests⌀, water⌀ resources and buildings. Estimates of annual damage resulting from air pollution are difficult to quantify, and calculations are scarce. However, an OECD report in 1985 indicated that damage amounted to 1 per cent of the Gross Domestic Product (GDP) in France and 2 per cent of GDP in the Netherlands.

AUSTERITY MEASURES A name used to describe the rigid and tight controls imposed by the governments of developing countries⌀, usually brought about as a condition of financial assistance and loans provided by developed nations or multilateral agencies such as the International Monetary Fund (IMF)⌀. The measures often lead to a reduction in government spending in areas of social welfare, such as health and education⌀, but not in areas such as arms spending⌀ or importation of goods and services. Increased unemployment and rises in the cost of basic foods often follow, and several developing countries have been faced with riots as the local population has expressed its anger against the results of the measures. Austerity measures in Brazil initiated by the World Bank⌀ and the IMF have not only increased poverty and unemployment but have also resulted in an outflow of money from a nation with an already beleaguered economy. In 1989 alone the World Bank's operations reportedly took $724 million more out of Brazil than the institution put back in.

AUTOMOBILES The Organization for Economic Co-operation and Development (OECD)⌀ reports that the number of motor vehicles in the world increased from 246 million in 1970 to about 427 million in 1980. By 1990 the world was boasting around one motorized vehicle for every ten inhabitants. The average annual growth rate for cars has been 5.9 per cent over the last forty years and, as

100,000 cars roll off the world's production lines daily, estimates for the year 2000 state that the world will boast 650–700 million passenger cars alone – one for every eight people.

Three out of every four of the world's 500 million vehicles are to be found in the industrialized OECD countries, but the number in developing countries⟡ is increasing rapidly – doubling from 53 million in 1975 to about 111 million in 1983. The overall number of automobiles being produced in the world has actually been falling since 1979 as markets in the industrialized world have almost reached saturation point, although those that are produced are tending to have a longer and longer lifespan. In France, Germany and Italy, the average is now one car for every three people, with Britain and Japan not far behind. However, most developing countries are way behind this density of vehicles – Mexico, for example, has one car for every twenty-one people and Brazil has one for every sixteen.

Although most motor vehicles are found in the developed world – notably the United States, which has 41 per cent of the world's vehicles – their impact is global as they are dangerous, inefficient users of energy⟡ and a major source of pollution⟡. Worldwide, around 250,000 people died in officially reported car accidents in 1985 and the number of casualties in the developing world, where reporting is lax, is showing a worryingly high rate of increase. Fatality rates in the Third World⟡ exceed 50 per 100 million vehicles/km, ten times worse than the figure for developed nations.

In terms of the total energy consumption of a car, one third goes in the manufacturing process and 60 per cent in the fuel used to run it. Globally, on average, cars account for around half the oil⟡ used in each country. Fuel efficiency will be greatly improved if cars are made lighter but this will involve the increased use of both aluminium⟡ and plastics, which may not be easily recycled. Modern vehicles use only a maximum of 20 per cent of the potential energy in their fuel; the rest is converted to pollutants and heat. Improving engine performance and exhaust systems with 'catalytic converters'⟡ has been a favoured option in controlling pollution, although the converters can work only when the engine and exhaust are running hot. Moreover, emissions of carbon gases will not be diminished through the use of catalytic converters. Engines which burn 'lean' fuel, which is far less polluting than ordinary fuel, are under development but will not replace the millions of 'dirty' engines currently on the road for decades to come. Most vehicles run on rubber tyres⟡ and there is virtually no environmentally sound way of disposing of them, with the result that 'tyre mountains' are beginning to appear in various countries throughout the world.

Motor vehicles cause a variety of other environmental problems, most notably through the vast amounts of land⟡ taken up by roads and parking places. In densely populated China, where land is scarce, the privately owned automobile is virtually unknown; the bicycle is the preferred method of transport. Conversely, in the United States, providing for the automobile is astronomically costly, environmentally as well as economically. Construction costs for new roads in the USA during 1988 alone totalled $22 billion, and the building cost of a road planned for Manhattan is forecast at about $625 million per kilometre.

B

BABY MILK Dried, powdered milk or any other substitute for breast-milk can have a devastating effect on infant mortality rates in the developing world. A newborn child fed on reconstituted bottle-milk becomes totally dependent on it within a week, because the mother's breast-milk will dry up when the sucking stimulus is removed. In addition, clean water⟡, free of disease-causing agents, to make up the bottle-milk formulas and cash to pay for the milk powder are not readily available. Five million babies die from diarrhoea each year, and the United Nations Children's Fund (UNICEF)⟡ reports that a million lives could be saved if the use of powdered milk was discouraged and the value of breast-feeding promoted.

Nearly $2.4 million worth of powdered baby milk is sold to the Third World⟡ annually, and the transnational company Nestlé has the biggest market share. In 1984 Nestlé agreed to abide by World Health Organization (WHO)◇ recommendations, detailed in the International Code for the Marketing of Breast Milk Substitutes, which were designed to stop the promotion of powdered milk at the expense of breast-feeding. The company took this move partly to end a boycott of the firm's products imposed by consumers in many industrialized countries. However, in 1989 Nestlé reportedly began supplying vast quantities of free powdered milk to hospitals in the Third World. This was in contradiction of a World Health Assembly resolution passed in 1986 which called for such supplies to be ceased and prompted the consumer boycott of Nestlé products to be renewed.

BAGHDAD PACT A treaty between Iran, Iraq, Pakistan, Turkey and the United Kingdom. Signed in 1955 and based in Iraq, its goals were military, economic and social co-operation in the Middle East. When Iraq withdrew in 1959, the group's headquarters moved to Turkey and it was renamed the Central Treaty Organization (CENTO)◇.

BAMBOO There are over 500 species of bamboo, a unique, remarkably versatile plant resource for well over a quarter of the world's population. It can be used for construction, furniture-making and as a source of livestock◇ fodder or human food. Bamboo's hollow stems make it extremely strong and useful for building both as a material and for scaffolding. It is also a valuable source of livestock feed, providing up to four times as much protein as most other grasses. Bamboo is native to parts of Asia, notably India and China, where bamboo forests grow up to 40 m high. In China, species of bamboo provide the sole food source for the giant panda. Because of the unusual life cycle of the bamboo plants, the panda is threatened with extinction◇, for it's food source is in danger of disappearing.

Bamboo flowers and produces seeds once in a lifetime, sometimes after a growth period of 120 years. When bamboo forests◇ flower, all plants flower together, no matter what stage of growth each individual plant has reached. Then the entire forest dies, leaving behind seeds to begin a new growth cycle. Exactly how the the bamboo flowering cycle works remains a mystery, but it is extremely regular: bamboo plants flower every 15, 30, 60 or 120 years. Owing to this unique growth cycle, plant-breeders have been unable to improve bamboo by growing new strains from seeds. In 1990 researchers eventually discovered a means of causing shoots to flower within a few weeks and so ensuring a steady supply of seeds for plant-breeding experiments.

BANANA The term banana is generally applied to all varieties of *Musa* plants which produce edible fruits. It covers both bananas and plantains. The fruit of these plants can be cooked, made into flour or eaten fresh. Most commonly grown varieties of bananas and plantains are cultivated primarily in tropical humid lowlands. Bananas have a high demand for water◇ and are prone to wind damage. In optimal growing conditions, they mature 12–15 months after planting. For maximum yield, bunches of fruit should be left until the fingers are fully mature. Bunch weight increases markedly during the last two weeks, leading to full maturity. Average yields are in the region of 750 bunches per hectare.

Throughout temperate zones, imported bananas are usually eaten raw as a dessert fruit or used in the bakery and confectionery industries. The fruit is sugary and easily digested. Both bananas and plantains contain carbohydrates, iron◇ and vitamins A, B_1, B_2 and C. Plantains contain considerably more starch than bananas, and although they can be consumed raw, they are usually cooked and eaten boiled, fried or roasted. There is virtually no taste for plantains outside the tropics, and they do not enter into export trade.

In the developing countries◇ where they are grown, bananas are produced as a source of export earnings. They are harvested early, in an unripe state, and weigh less than when they are mature. They are ripened in transit under conditions of controlled temperature, humidity and ventilation, usually as they are transported in specially adapted holds of ships at 12°C. The spread of fungal disease and other pests is reduced at this temperature, and the fruits remain in good condition for two to three weeks.

In their natural habitat, banana plants have several other uses. Their large green leaves are widely used as a roofing material, to produce makeshift umbrellas and as wrapping for goods sold at market. Fibres from the stems can be used for making bags and ropes. In Africa beer is made from bananas; it has a low alcohol content but is high in vitamins.

Global production of bananas is of the order of 41 million tonnes; Brazil (5 million tonnes), India (4.6 million tonnes), the Philippines and China (both 2.4 million tonnes) are the major producers. The annual production of plantains is estimated to be 21 million tonnes, all destined for home consumption.

BAND AID   The Band Aid Trust was established at the end of 1984 by the Irish pop singer Bob Geldof⌕ and others, originally as a response to the effects of the famine in Ethiopia. It initially raised $14 million through the release of the record 'Do they know it's Christmas?', sung and played by a mixed group of pop celebrities. In the summer of 1985 Geldof helped to organize Live Aid, a massive concert performed by pop groups simultaneously in Britain and the United States which was broadcast to the world's largest ever television audience. As a result of this, the Band Aid Trust raised a further $160 million. Band Aid has no development programmes of its own; it simply funds projects submitted by other agencies which are considered suitable for support. The Trust was expected to operate until 1990, by which time, it was forecast, the money in the fund would have been exhausted. This reflected Geldof's original conception that Band Aid should provide assistance and reach those in need in the Third World⌕ quickly through a combination of the innovative nature of the Trust's activities and the avoidance of the bureaucratic troubles and slow operational methods that are symptomatic of many established aid⌕ channels.

BANDA, HASTINGS KAMUZU (1905–) Malawian statesman. A student and medical practitioner in Britain and the United States, he returned home to what was previously Nyasaland in 1958 to lead his people's struggle against federation with Rhodesia and to campaign for full independence. only to be arrested in 1959 for belonging to the outlawed Malawi Congress Party. Following his release in 1960, he finally led his country to independence, becoming Prime Minister of Nyasaland from 1963 and first President of Malawi from 1964, eventually taking the title Life President in 1971.

BANDARANAIKE, SIRIMAVO (1916–) Sri Lankan politician who succeeded her husband Solomon Bandaranaike⌕ as leader of the Sri Lankan government. She was the world's first woman ever to hold the position of Prime Minister. She served two separate terms in the post, 1960–65 and 1970–77, but was eventually expelled from Parliament in 1980 for alleged abuse of her powers while in office.

BANDARANAIKE, SOLOMON WEST RIDGEWAY DIAS (1899–1959) Sri Lankan politician. An ardent nationalist, he founded the Sri Lanka Freedom Party (SLFP) in 1951 and eventually

rose to the position of Prime Minister. He pledged himself to carry out a socialist programme and followed a strictly neutral foreign policy. His failure to satisfy the demands of the extremists in the country caused an escalating series of problems and unrest before he was assassinated by a Buddhist monk in 1959. He was succeeded as head of the SLFP by his wife Sirimavo Banderanaike⌕, who later became the world's first female prime minister.

BANDUNG CONFERENCE   In 1955 representatives of twenty-nine African and Asian countries – including China but excluding Japan – met at Bandung in Indonesia to voice their individual and collective opposition to colonialism. It was the first conference of the Afro-Asian nations who have continued to proclaim anti-colonialism and neutrality between East and West. A subsequent conference held in Algiers in 1965 foundered owing to the variety of conflicting interests and concerns of the countries attending, most of which joined the Non-Aligned Movement⌕.

BANK FOR INTERNATIONAL SETTLEMENTS (BIS)   An inter-governmental banking institution located in Switzerland which acts as a reference centre for central banks from developed countries around the world, including twenty-four from Europe and those from Australia, Canada, Japan, South Africa and the United States. The BIS itself is governed by representatives from several of these central banks. It was originally set up in 1930 to co-ordinate financial dealings and reparations arising after the end of the First World War. Most of its functions are now carried out by the International Monetary Fund (IMF)⌕, although the BIS maintains an important role as a trustee. Its main activities now focus on the purchase and sale of gold⌕, foreign currency and bonds for its member central banks.

BASIC NEEDS   Defined as the essential items of private consumption and basic services needed by every individual to maintain a reasonable standard of living. These include adequate food, shelter, clothing and household equipment, together with essential community services such as safe drinking water⌕, sanitation, health services, education⌕, transport and cultural facilities. The term was largely originated by the International Labour Office (ILO)⌕ and is sometimes held to include the right to work.

BATAAN NUCLEAR PLANT A pressurized-water nuclear reactor plant built in the province of Bataan in the Philippines which remains shrouded with controversy. The facility was intended to provide 620 megawatts of electricity. Built by the us-based Westinghouse organization, the project cost $2 billion; much of its funding was provided by the United States. After years of rumours of suspect financial dealings during the administration of President Marcos⟡ and reports that the plant could result in massive environmental damage, the Philippines government formed under President Aquino cancelled the whole project. In 1990 the Philippines launched a court case in the us to seek return of funds invested in the construction and refused to pay any monies due on all outstanding loans relating to the project, saving an estimated $350,000 daily in interest payments.

The Bataan nuclear plant was erected in an area which is recognized to be geologically unstable. It was constructed on land less than 20 km from a major seismic fault, in an area criss-crossed with fault lines and only 16 km from several volcanoes, some of which were thought to be active. The project went ahead despite critical reports from the us Nuclear Regulatory Commission and White House Council on Environmental Quality. Experts concluded that, amongst other problems, strong earth tremors could easily crack tubing, thereby letting water coolant escape. This could lead to a meltdown of the reactor core, as occurred in an identical reactor at Three Mile Island⟡ in the United States. Construction of the Bataan plant was actually halted for a year while the accident at Three Mile Island was investigated, and this was put forward as one reason why the Philippines plant was so expensive. Exactly why the plant costs $2 billion to build, whereas a similar plant built in the us would have cost only $550 million, has never been fully explained.

The Bataan nuclear power⟡ station never became operational, and after the overthrow of President Marcos the new government decided to replace the plant with a conventional coal-fired power station at a cost of $300 million. It also decided to dismantle the nuclear facility and sell off some of the parts to help pay towards the construction costs. The new power station will form part of a new $1 billion energy⟡ programme, a programme that virtually ignores one of the Philippines's greatest natural assets – its enormous geothermal resources. The nation is already the world's second largest user of geothermal energy⟡, and if only 5 per cent of the country's geothermal resources were to be tapped, energy needs would be met until well into the next century.

BAUXITE The chief ore of aluminium⟡, bauxite is a residual deposit, found in surface blankets formed through the weathering of aluminium-rich rocks and clays under tropical conditions. These consist mainly of hydrated aluminium oxide. Aluminium, in the form of alloys, is widely used in aircraft and vehicle manufacture, as foil wrappings and in cans or containers for domestic purposes. Many of the developed countries have now begun to recycle the aluminium in cans and containers, reducing the need for bauxite. Approximately four tonnes of bauxite are needed to produce one tonne of aluminium via a process that requires large inputs of energy⟡. For this reason, aluminium is mainly produced in the developed countries such as Canada, Japan and the United States, rather than in the predominantly developing countries⟡ which produce the bauxite ore.

The annual global production of bauxite is approximately 98 million tonnes; Australia (36 million tonnes), Guinea (17 million tonnes) and Jamaica (7.4 million tonnes) are traditionally the leading producers. In comparison, aluminium production is in the order of 17 million tonnes annually; producing countries are led by the United States (3.9 million tonnes). the Soviet Union (2.4 million tonnes) and Canada (1 million tonnes).

BEGIN, MENACHEM (1913–) Israeli statesman. Born in Poland, he became leader of the extremist Irgun Zvai Leumi organization in Palestine in 1942. He was Prime Minister of Israel from 1977 to 1983, as head of the right-wing Likud Party. In 1978 he shared the Nobel Peace Prize with President Anwar Sadat⟡ of Egypt for their efforts in achieving the Camp David Peace Agreements⟡ and subsequent peace treaty between the two countries.

BELAU Also known as Palau, a small island nation in the northern Pacific Ocean administered for the United Nations (UN)⟡ by the United States, the only remaining remnant of the United Nations Trusteeship system. In 1980 the 12,000 islanders voted to adopt the world's first totally nuclear-free constitution prohibiting the presence of all nuclear materials or activity within its territory. Ninety-two per cent of the population voted in favour and the new constitution can be overturned only by a 75 per cent vote. The US covets the island as the possible site for the location of a fallback base for its naval and nuclear forces currently based in the Philippines, and has persistently campaigned for the new constitution to be overturned. This pressure forced seven elections during the 1980s in the attempt to reverse the anti-nuclear ban and persuade the islanders to adopt a Compact of Free Association which would effectively allow the US to use the island for military purposes. The 1990 vote, which also attempted to replace the requisite 75 per cent voting requirement needed to overturn the constitution with a simple majority verdict, narrowly failed to achieve either goal.

BENELUX   An economic union between Belgium, the Netherlands and Luxembourg, agreed in 1944. Originally designed to cover Customs, tariffs and other trade matters, it developed into a full economic union. It became fully operational in 1960 with headquarters in Brussels, and is looked upon as the precursor of the Common Market.

BERGEN CONFERENCE   A conference in Norway in 1990, held under the auspices of the United Nations⋄, at which representatives of thirty-four governments gathered to discuss the 'sustainable development'⋄ concept formulated in the 1987 Brundtland Report.⋄ Many regarded the Conference as a wasted opportunity, primarily because of the lack of commitment shown in producing decisive measures to ease the environmental burden on the Third World⋄ and reduce global pollution⋄.

Most nations attending favoured the stabilization of carbon dioxide⋄ emissions at 1990 levels by the year 2000, but opposition from Britain, the Soviet Union and the United States meant that the outcome of the Conference was a weak compromise. The wording of the final Bergen declaration on carbon dioxide stated that 'most of the nations will strive for stabilization of emissions by the year 2000'. Similar compromises were reached on assistance to the Third World to deal with the chlorofluorocarbons (CFC)⋄ problem, and with ozone depletion. The principle that 'new and additional' funds on top of World Bank⋄ loans should be given to poorer nations in the South⋄ to enable them to carry on their industrialization process without the use of CFC did not receive support from wealthy nations like the United States. As a result the final declaration was toothless, concluding that

nations must 'strengthen international action to protect the ozone layer, including, for example, through additional resources and technology transfer. . . . It will be necessary to identify new ways and means of providing such resources to developing countries.'

However, the Conference did make some progress: delegates accepted the need to adopt the Precautionary Principle⋄ when dealing with matters of the global environment. The Conference stressed that 'environmental control measures must anticipate, prevent and attack the causes of environmental degradation. Lack of conclusive scientific proof should not be cited as a reason to delay action.'

Perhaps of greatest significance, the Conference created a notable precedent in establishing a dialogue between governments and Non-Governmental Organizations (NGO)⋄. NGOs, ranging from youth groups and trade unions to industry, scientists and environmental activists, were invited to construct an agenda for ministerial delegates and allowed into ministerial drafting sessions.

BERI-BERI   Endemic polyneuritis, more commonly known as beri-beri, is an inflammation of nervous tissue occuring mainly amongst people living in the tropics. The disease is caused by a deficiency of Vitamin B_1 (thiamine) in the diet. It is most prevalent in areas where the staple diet is polished rice⋄, as thiamine occurs mainly in the rice husks, which are discarded. The disease occurs in two forms, wet and dry. Dry beri-beri is accompanied by severe emaciation and affects the peripheral nerves, causing muscular weakness and pain. Wet beri-beri is the result of protein malnutrition⋄ combined with thiamine deficiency, and causes accumulation of fluid and swelling of the limbs. Nervous degeneration occurs in both forms, and death from heart failure is often the outcome. Treatment with vitamin supplements achieves total recovery.

BERNE CONVENTION   Common name for the Convention on the Conservation of European Wildlife and Natural Habitats⋄, which is designed to protect all forms of endangered wildlife on parts of the European continent.

BETEL NUT   The seed of the palm, *Areca catechu*, commonly chewed in several countries in the tropics, mainly in Asia and almost exclusively by men. In India, chewing betel quid is a regular habit for more than 20 million people, and betel-chewers outnumber smokers. More than twenty brands of betel quid, or pan masala, are commercially available.

Betel is usually chewed together with the leaves of the betel, pepper or lime, and a variety of spices. The nut contains a mild narcotic and red dye. Studies have shown that continued chewing of betel can cause oral cancer⌀.

BHOPAL One of the world's worst industrial accidents, which re-emphasized the differences between human rights⌀ in the developed and developing worlds. As a result of a leak of deadly methyl-isocyanate gas at the Union Carbide factory in Bhopal in India in 1984, thousands of people died. Exact figures are impossible to obtain, for the bodies of many victims were cremated and others died later, having left the area. Four years after the accident, the Indian government estimated that at least one person per day in the area surrounding the factory was dying as a result of injuries or illness caused by the leak, and the death toll was reported to exceed 3,150. Unofficial figures suggest that as many as 500,000 could be injured, disabled or physically or mentally disturbed as a result of the accident.

In 1989 studies showed that more than 70 per cent of people in the worst-affected areas around the factory were either ill or disabled. Diseases of the eyes and gastrointestinal tract were commonplace. In addition to the gas-affected population in the worst-hit locations, 40 per cent of those living in nearby areas that had received only mild exposure to the gas cloud were also found to be ill.

Following lengthy legal battles, the Indian government and Union Carbide finally agreed on a $470 million settlement of all civil and criminal proceedings arising from the disaster, although Union Carbide denied liability. The Indian government originally claimed $3 billion, far less than would have been claimed if the victims had been American. Its attempt to have the case heard in the US courts failed, Union Carbide claiming that the leak was caused deliberately by a disgruntled employee. India's Supreme Court controversially upheld the decision to accept $470 million in final settlement from the US-based transnational. The settlement was partially based on the Indian government's own figures, which estimated the number of people with permanent total or partial disability to be 30,000, with a further 20,000 temporary cases. The number assessed as having suffered minor injuries was put at a mere 50,000.

The Directorate of Claims set up in India to deal with requests for compensation has been slow in processing the 300,000 applications which have already been lodged, and by early 1990 only 123,000 had been assessed. Claimants are expected to have proof of hospital examination or treatment, something which was denied to the majority of slum-dwellers exposed to the gas. Little credence is given to patients suffering from indirect or long-term effects, such as impaired immune systems, psychiatric problems or menstrual problems in women. The 1989 survey of those in areas affected by the gas cloud found that around 50 per cent of women were experiencing menstrual problems and the number of spontaneous abortions, stillbirths, and offspring with genetic defects had risen markedly. Post-traumatic stress disorder, a deterioration in mental health commonly associated with wars and disasters, was also discovered in 57 per cent of those examined. The national and local governments in India are claiming a total of $67 million in expenses, which means that the 500,000 Bhopal residents who are finally expected to seek compensation are likely to receive less than $1,000 each.

BIKO, STEVE (1946–1977) A black South African, a leader in the 'black consciousness' movement in the struggle against apartheid⌀, who founded the South African Students' Organization in 1968. He was arrested in 1977 and died of injuries received while in detention.

BILHARZIA Common name for schistosomiasis⌀, a tropical disease found in the Americas, Asia and Oceania, although it is most prevalent in Africa. At least 200 million people are infected, and a total of 600 million in seventy-six countries are are at risk of contracting it.

BIO CONTROL *See* Biological control

BIODIVERSITY (BIOLOGICAL DIVERSITY) A measure of species richness and natural genetic variation which can apply either within or between species of wildlife. Species diversity means the variety of differing wildlife species, while genetic diversity refers to the mixture and range of genes.

Diversity within ecosystems⌀ has been equated with the stability of the system. It is vital to the proper functioning of ecosystems and is the basis of biological wealth and adaptability. The human race has long taken advantage of natural genetic variation, selecting out plants or animals with characteristics that they prefer. Genetic diversity has had its most noticeable impact in agriculture⚬. Between 1930 and 1975, over half of the yield increases in cereal crops, sugar cane, cotton⚬ and groundnuts⚬ could be attributed to the contribution of genetic resources. Improved genetics also accounted for 25 per cent of increased milk yields in cows over the same period. Human activities – bringing about loss of habitats, overexploitation of natural resources⚬, and competition and predation through introduced species – are rapidly destroying or restricting the range of both wild species and genetic resources. The loss of habitats which show the greatest diversity, such as tropical forests⚬ and coral⚬ reefs, is depleting the world's pool of genetic material, weakening the ability of the system to evolve and adapt to any changes, and reducing the global genetic pool from which a selection can be made.

BIOGAS   A methane⚬-rich gas produced by the fermentation of animal dung, human sewage or crop residues. There are several advantages in converting agricultural residues or human wastes into biogas. In addition to the gas produced, the slurry formed is rich in nutrients; it can be used as an organic fertilizer⚬ and is particularly useful for feeding fish-farm ponds. The gas can be used as a fuel or converted to fertilizer for developing countries⚬. Methane has a high octane rating and burns cleanly. Moreover, it can be used in diesel engines⚬ as well as spark-ignition engines, and can replace up to 80 per cent of diesel used. However, the methane produced by treatment of wet biomass wastes, such as manure or sewage sludge, is not an ideal fuel for vehicles. Even highly compressed – a process which necessitates the use of heavy, pressurized cylinders – it has a relatively low energy⚬ density which limits the range of vehicles.

BIOLOGICAL CONTROL (BIOCONTROL) A term used to describe the control of pestilential organisms such as insects and fungi through biological means rather than the application of man-made chemicals. This can include breeding resistant crop strains, inducing fertility in the pest species, disruption of breeding patterns through the release of sterilized animals or spraying juvenile hormones to interrupt life cycles, breeding viruses that attack the pests, or the introduction or encouragement of natural or exotic predators to control pest outbreaks.

In nature, all pests are controlled by natural forces and the methods used in biocontrol programmes attempt to re-create or introduce these forces in an invaded area. When they are successful, a permanent equilibrium between pest and predator is established. Biological control methods therefore tend to be self-regulating: predators of a pest will reduce in number when the pest population falls to a low enough level, or increase in line with pest numbers. The process obviates the need for the use of persistent, toxic chemicals⚬ which pose a marked hazard to human health.

One of the earliest forms of biocontrol occurred in the 1890s when the cottony cushion scale, *Icerya purchasi*, was accidentally introduced in California and, in the absence of natural predators, quickly decimated citrus orchards. A natural predator of the scale insect, the beetle *Rodalia cardialis*, was discovered in Australia, introduced into California, and brought the scale insect under control within two years. Another well-known example of biocontrol concerned the use of the myxamotosis virus, which is native to America, to control rabbit populations in Australia and Britain. Within three years of its introduction into Australia in 1950, over 99 per cent of the nation's hundred-million-strong rabbit population had been killed. A similar destruction rate was seen in rabbit numbers in Britain following introduction of the virus in 1953. However, as a result, the vegetation of the countryside changed as the grazing pressure imposed by rabbits was removed, thus allowing the unchecked growth of many plants previously eaten by these rodents.

Modern-day examples of major biocontrol programmes include those aimed at combating pests of cassava⚬ in Africa, the screwworm⚬ fly in the United States and North Africa, and pine aphids in Africa. All these pests had been accidentally introduced by humans.

Living systems are very complex, and substantial research must be carried out to ensure the safety of biological control programmes – several of which have, in the past, created secondary problems of varying severity in terms of environmental impact.

BIOLOGICAL DIVERSITY *See* Biodiversity

BIOLOGICAL EFFECTS OF IONIZING RADIATION (BEIR) A specialized Committee on the Biological Effects of Ionizing Radiation – the BEIR Committee – was formed by the United States National Academy of Sciences to investigate all aspects of the way ionizing radiation affects the human body, and to propose safety limits based on its findings. In 1972 it produced a report stating that there is a risk in radiation of any dosage, and therefore no threshold below which exposure to radiation can be regarded as safe. In 1980 the BEIR Committee produced another report that contradicted its previous findings. Uncertainty in the global scientific community continues, because there is relatively little information about the effects of long-term exposure to small doses of radiation. A substantial amount of the epidemiological data available is based on observations of survivors of the atomic bomb blasts in Japan during the Second World War.

BIOLOGICAL OXYGEN DEMAND (BOD)

The BOD is the most widely accepted means of measuring how polluting an effluent is when it is discharged into a living watercourse. Most aquatic animals need oxygen to respire. Organic matter normally present in streams, rivers and lakes, all of which have a natural level of oygen dissolved in the water⊂, gives a background BOD of a few parts per million. When an effluent – such as sewage – containing a great deal of organic matter is discharged into the water, it encourages bacteria and micro-organisms naturally present in the watercourse to reproduce and multiply. In the process, all – or the greater part – of the oxygen in the water may be used up. In these circumstances, the effluent is said to have a high BOD. If an effluent has a high BOD and the watercourse is unable to dilute it, oxygen levels will fall to a point where fish⊂ and other aquatic animals will be unable to respire. One of the main functions of sewage-treatment works is to reduce the BOD of the liquid that is discharged. The BOD concept can be applied to pollutants such as pesticides⊂, nitrates from farmland, or leachates from chemical dumps or industrial sources only if they raise the oxygen demand of the watercourse.

BIOLOGICAL WARFARE

The use of living micro-organisms, or infectious material derived from them, to bring about death or disease in humans, animals or plants. During the First World War the Germans infected cavalry horses of the Allied Powers with bacteria. Several other bacteria – such as *Bacillus anthracis*, which causes anthrax, and *Pasteurella pestis*, which causes plague⊂ – have been evaluated as possible biological weapons. The Japanese were reported to have used plague bacteria against the Chinese army during the 1930s. Biological warfare was officially condemned by the Geneva Convention⊂ (1925), to which the United Nations⊂ has urged all member states to adhere.

Organisms for use as weapons are required to be highly virulent, but not necessarily lethal. They may be used in an aerosol form, in bombs, or simply added to water⊂ supplies. Although biological warfare is now officially banned under the 1975 Biological Weapons Convention, which prohibits the development, production and stockpiling of biological and toxin weapons, research probably continues into developing new strains of hazardous organisms, partly because they are relatively cheap and easy to produce and partly because many of the organisms involved are naturally occurring and need to be studied either in order to counteract their effects as and when they may appear in epidemic form, or for their potential medicinal qualities. Botulin, the toxin produced by the bacterium *Clostridium botulinum*, is the most poisonous substance in the world. It is still being produced regularly at establishments such as Porton Down, the former germ warfare centre in the United Kingdom, ostensibly for medical purposes. Dystonia, a rare neurological disorder, is incurable, but treatment with minute amounts of botulin has proved successful for some forms of this condition

BIOMASS

A scientific term used as a means of describing the amount of living material that exists. It may be used to refer to the gross weight of organisms present in a limited area, or applied on a global scale. It may be specified for one species, for a family, or for all species. Usually expressed in tonnes per hectare, biomass is commonly used to describe only dry animal and plant matter. Wet matter is usually three to four times heavier than dry, owing to the water⊂ content. Biomass is composed of plant material (phytomass) and animal matter (zoomass), although 99 per cent of the world's biomass is plant material.

BIOSPHERE

The thin covering of the planet, encompassing the atmosphere, soil⊂ and water⊂, that contains and sustains life and in which all living organisms exist together with their inanimate (abiotic) environment. Thirty kilometres above the sea the air is far too thin and cold to support life, while 100 km below the sea there is no light and temperatures are in the region of 3,000°C. Most living organisms exist in an area extending 6,000 m above to about 10,000 m below sea level. Although energy⊂ is continually being added from the sun, the earth represents a closed, finite system with regard to the natural resources needed to sustain life within the biosphere.

The biosphere concept was first introduced by the French naturalist Jean Lamarck, and developed by the Russian scientist Vladimir Vernadsky. The idea of the biosphere has recently been taken up and refined by the British scientist James Lovelock, who proposes that the global ecosystem⋄ – which he calls Gaia⋄ – is able to maintain its own equilibrium creatively by responding to disruptive changes through a system of natural feedback mechanisms.

BIOSPHERE RESERVES An international network of protected areas set up in selected parts of the world under the Man and the Biosphere (MAB)⋄ programme. Long-term research is being carried out within these areas on the structure, function and dynamics of various ecosystems⋄ with a view to improving human understanding of natural systems and so improving conservation⋄ of natural resources⋄ and maintaining genetic diversity. Biosphere reserves are selected for their particular ecosystem type, not for their uniqueness, and all the various ecosystems and biogeographical provinces of the world, each with distinctive flora and fauna communities, are being included in the programme. Conservation is a major objective of the reserves, reflecting the integrated concept of the MAB programme. The reserves are made up of natural 'core' areas, and surrounding 'buffer' zones which have been modified by human activity, where practical and experimental work is undertaken. The creation of a biosphere reserve does not affect the sovereignty of the host nation. By 1988, 269 biosphere reserves had been established in seventy countries around the world, encompassing a total area of 143 million hectares.

BIOTECHNOLOGY The application of biological organisms, systems, or processes to industrial processes. Living organisms can be manipulated to produce food, drugs or other products, and have been exploited for centuries. Micro-organisms have been used to make bread, convert milk to cheese and brew alcoholic beverages. Common substances such as vinegar, antibiotics and vitamins are manufactured through manipulation of microbial organisms. Over the past fifteen years, genetic engineering has allowed an unprecedented degree of flexibility and control to be exerted in these industrial processes, with the ability now to alter or insert single genes.

Genetic engineering is a form of biotechnology in which single-celled organisms have their DNA modified so that they can be used to produce useful substances such as insulin⋄, new vaccines and diagnostic tools. The ability of genetically altered micro-organisms to produce chemical compounds will lead to applications in the food-processing, chemicals, and energy⋄ industry as well as in medicine and agriculture⋄. Modern industrial processes using engineered micro-organisms include fermentation to produce methane⋄ and alcohol from sugar⋄ or waste food products for use as a fuel, the use of bacteria to remove heavy metals from polluted areas, the production of proteins from oil⋄ residues, and the separation of precious metals from mining waste.

Fermentation of bacteria or the cross-breeding of plants appear relatively harmless, but the production and release into the environment of unique, genetically engineered organisms could lead to a potentially dangerous situation if any such releases are not strictly controlled and monitored.

BIRTH CONTROL The limiting of human reproduction, usually carried out with a view to restricting population⋄ increase through either reducing the number of offspring or lengthening the interval between pregnancies. Originally coined in 1924 by the United States reformer Margaret Sanger, the term has many synonyms such as planned parenthood, family planning and fertility regulation.

Birth-control techniques have been in evidence for thousands of years, with controls being carried out mainly as a means to improve family living standards. Control may be effected through a range of chemical, physical or cultural means, including contraception, sexual abstinence, induced abortion, sterilization and promotion of normal breast feeding⋄.

Most world leaders agree that there is a pressing need to regulate human fertility, although the means through which this should be achieved is an extremely contentious issue, especially amongst religious groups and on ethical grounds. In the past, infanticide – which is still practised in some traditional societies around the world – was a common form of regulating family numbers. However, there has been a general global trend to move from curbing actions after birth (infanticide) – first to action before birth (abortion) and then to before conception (contraception), the pace and scope of these trends varying from region to region.

By 1900 all the commonly used methods of birth control, apart from oral contraceptives, were understood, if not widely available. Although hormonal contraceptives, based on synthesized derivatives of natural steroid hormones, were known in the 1920s, they did not become available until the mid-1950s, and then only in North America and Europe. Following a more or less global expansion in their availability, today over 100 million women around the globe currently use oral contraceptives or have done so at some time. Although the Pill is remarkably effective – fewer than one woman in a hundred per year of use will conceive – its use, especially over a lengthy period, can seriously jeopardize a woman's health; users are more likely to suffer from heart attacks, strokes and embolisms than non-users.

For a variety of reasons, research on chemical contraceptives for men has progressed slowly and has tended to be based on naturally occurring compounds such as gossypol, which is derived from cotton▷. Not until the end of 1990 was a highly effective male contraceptive developed. This contraceptive, which is reversible and causes minimal side-effects, makes use of a hormonal drug, testosterone enanthate, administered in the form of weekly injections.

Despite advances in birth control, techniques and materials take a long time to become available in poor rural areas of the developing world. In these areas, women are usually frustrated in their desire to control their fertility in the effort to improve their own welfare and that of their family. In 1980 use of contraceptives in developed countries was estimated to be 70 per cent compared with only 25 per cent in developing nations. Part of the problem in the developing world was the lack of co-ordinated family-planning programmes, resulting in poor access to modern contraceptives: in 1988, only half the population in 40 per cent of developing countries▷ had easy access to modern methods. Yet the example of China shows that family planning can be carried out successfully, even on a massive nationwide scale and in resource-poor rural areas, if the commitment and political will are there.

China, the most populous country on earth, adopted a sustained national birth-control programme during the early 1970s. The government urged all couples to stop at two children and, in 1979, launched a one-child-family programme with the aim of limiting the national population to 1.2 billion by the year 2000. The plan aimed to ensure that living standards were raised, particularly for those who complied with the policy, and the programme was supported by extensive and readily available family-planning services. Social acceptance of the programme was difficult in a society where families are traditionally large, and especially so in rural areas where there is a strong preference for sons. Despite all the constraints the programme attained a fair degree of success, although at the end of 1990 the Chinese authorities admitted that it would not be possible to reach their year-2000 target and had modified the goal to a limit of 1.25 billion – 50 million more than the original figure.

BIRTH RATE   The number of live births in any given year divided by the mid-year population▷ of the country in question, gives the birth rate. It is used for comparative purposes and is usually multiplied by 1,000 and expressed as the Crude Birth Rate (births per thousand population). Since 1965, the global Crude Birth Rate has shown a general decline, falling from 33·9 in 1965–70 to 26 during 1985–90. As living standards improve, so the Crude Birth Rate tends to decline. Regionally, Africa, at 45·2, exhibits the highest rate, and Europe (13·5) the lowest.

BLACK DEATH   A name used to describe plague▷, principally bubonic but also pneumonic and septicaemic, pandemic in Europe and Asia during the Middle Ages. The name is derived from the black buboes, or swellings, that erupted on the bodies of those infected. Originating in the Far East, the plague – which is still widespread in the world today and especially so in developing regions – was spread throughout Europe by rats and fleas, with three outbreaks occurring during the fourteenth century. Mortality from the disease – whose transmission cycle was unknown at the time – ranged from 20 to 50 per cent. The outbreak in Europe killed around 25 million people, at least a quarter of the population, and had such a drastic effect on rural society and the economy as a whole that it took the continent 200 years to recover fully. In total, 50 million people were believed to have died from the Black Death.

BLACK SEA   Large inland sea bounded by Bulgaria, Romania, the Soviet Union and Turkey, the site of a unique coastal energy▷ project. The Black Sea is the world's largest depository of several mineral gases: it is estimated to contain about 75 billion tonnes of high-calorie fuels such as hydrogen sulphide, ammonia◇, methane◇ and ethane. Hydrogen sulphide, naturally produced from decayed animal or vegetable matter, is toxic to all living organisms. A layer of debris containing this deadly gas has been rising steadily over the past 7,000 years, and it is thought that this layer may reach the surface within the next fifty years.

A power station is to be constructed to harness the potential offered by the Black Sea's deposits of mineral gas and generate electrical power. The plant will generate an estimated 80 billion kilowatt/hours of energy per year. It will also help to clean up the high levels of pollution◇ caused by seashore factories and waste dumping.

In the proposed power plant, a series of pumps will siphon water◇ saturated with the gases. Once pressure from deep-sea water is removed, the combustible gases are released. These gases will be separated from the water, and sulphide will be recovered to manufacture fertilizers◇ or sulphuric acid for industrial purposes. The energy costs of pumping and purifying the sea water will be about 20 per cent of the envisaged power to be generated. The clean water will be returned to the Sea.

BLACKWATER FEVER   A rare but extremely serious complication of malaria◇ in which the malarial parasite, *Plasmodium falciparum*, causes widespread destruction of red blood cells, leading to the excretion of blood pigment in the urine. The dark-brown urine formed gives rise to the disease's name. Sufferers from this complaint develop a high fever and jaundice, and a reduced flow of urine caused by a blockage of the kidney tubules. If the latter occurs, the results are fatal. The condition can also be brought on by mismanaged treatment with quinine, a drug which used to be the sole treatment for malaria. Blackwater fever can be treated successfully through regular blood transfusions, coupled with prolonged rest, administration of intravenous glucose and alkaline fluids, and careful nursing.

BORAN CATTLE   A specialized breed of cattle developed from animals originating in the region of the Kenya–Ethiopia border, which may help to reduce environmental degradation in livestock-rearing areas of Africa. Most cattle are unable to tolerate drinking either too much or too fast, but Boran cattle have the ability to take in considerable quantities of water◇ at a rapid rate – up to 106 litres in six minutes. Being highly adapted to arid environments, Boran cattle can restore their water deficit

and have an intake sufficient for forthcoming days; this gives their owners the added benefit of being able to utilize pasture much further from waterholes than would otherwise be the case. Under normal conditions, livestock◇ are watered daily, but this is not possible in water-shortage areas of Africa. Instead, livestock have to adapt to watering regimes which vary from up to every three days for cattle and donkeys, five days for sheep and goats, and seven to fifteen days for camels. Horses need to be watered daily. The shorter the time spent at the waterhole and the longer the time between drinking bouts – the outstanding features of Boran cattle – the less the physical degradation around the well and overgrazing◇ of surrounding vegetation.

BORLAUG, NORMAN (1914–)   American plant-breeder who was instrumental in developing the new high-yield varieties (HYV) of wheat◇, maize◇ and rice◇ that formed the basis of the 'Green Revolution'◇ in underdeveloped countries. He has been involved in work on improving cereal crops in Mexico since 1944 and received the Nobel Peace Prize in 1970 for his role in stimulating the 'Green Revolution'.

BOTSWANA BEEF EXPORT PROJECT   Botswana is the driest independent country in southern Africa: the Kalahari desert constitutes almost 75 per cent of its land◇ surface. An $18 million livestock◇ development project, funded by the World Bank◇, aims to increase beef production for export by 50 per cent in Botswana, a country which is already facing severe desertification◇ and overgrazing◇ problems. There are as many as three head of cattle for each inhabitant in the country, and massive fences have had to be erected to protect the fragile rangelands and contain the beef herds. The fences have also had the effect of channelling hundreds of thousands of wildebeest, following their traditional northward migration routes in search of water◇, towards the now dried-up Lake Xau. The beef project will also be operating in the ecologically unique Okavango Delta, one of the remaining untouched wildlife areas in the region. Beef exports have proved of little benefit to the majority of the country's population in the past, and sacrificing scarce rangeland to cattle production, at the expense of local game, will worsen their plight, for wild animals provide up to 60 per cent of the Botswanan diet.

BOTTOM-UP DEVELOPMENT A theory of development that calls for funds and projects to be administered directly to the rural poor and targeted on the reduction of poverty and efforts to meet basic human needs.⌂ It is the opposite to trickle-down development⌂, the prevalent system of donating aid⌂ until 1970.

Until the late 1960s, overseas aid⌂ was generally allocated to specific projects, mostly large-scale industrial schemes. It was assumed that improved industrial productivity would provide economic benefits which would eventually filter through to those most in need (so-called trickle-down development⌂), a situation that was rarely achieved. In reality the aid benefited only a wealthy minority in urban areas, although developed nations continued to provide it in this form because it allowed them to sell advanced machinery and equipment and to charge fees for providing the essential technical assistance needed to operate it. Moreover, donors could maintain a degree of control over the way in which the aid was spent. Since the early 1970s there has been an increasing swing towards bottom-up development, with the emphasis firmly on rural development and promotion of small-scale rather than large-scale projects. Exactly how the process should best be effected and through what channels assistance should perferably be directed remains to be determined.

BOVINE   SPONGIFORM   ENCEPHALO-PATHY (BSE)   BSE, also referred to as mad cow disease, is a disease of cattle that came to prominence during the late 1980s following a major outbreak in the UK. Attention was focused on the disease amid concern that it could spread to humans through the consumption of contaminated meat. Cattle with the disease lose the ability to balance and stumble about as if intoxicated. The disease causes cavities to appear in their brains, the tissue of which becomes spongy. First discovered in 1986, BSE is believed to related to scrapie, a disorder − first identified over two hundred years ago − that affects sheep and has similar symptoms. It has confounded scientists ever since its discovery, as the causative agent is thought to be a simple protein, containing no nucleic acid and therefore with no genetic code.

The agent causing BSE may be related to or derived from scrapie, having been fed to cattle in protein-rich food supplements produced from sheep remains. Another disease, Creutzfeldt-Jakob Disease, which has similar symptoms to those of BSE and scrapie, is already known to afflict humans,

although it is rare, affecting fewer than one in a million. By 1990 there was no evidence that BSE could pass to humans, although several other animals, including pigs, mice, cats and antelopes were found to have developed a similar disease. Although tests for identifying the disease had been developed, they could be used only on dead animals. Tests for detecting BSE in live animals are not expected to be available until 1993.

BRANDT COMMISSION   The common name of the Independent Commission on International Development Issues (ICIDI)⌂ established in 1977 under the chairmanship of Willy Brandt⌂, ex-Chancellor of the Federal Republic of Germany. The eighteen-member Commission was given the task of examining the grave global issues arising from the economic and social disparities within the global community, and to suggest solutions and ways in which they could be implemented. Its findings were published in 1980 in a report, *North–South; A Programme for Survival* (more commonly known as the 'Brandt Report'). In essence, it called for urgent action to improve relations between the rich North⌂ and the poor South⌂ and made specific recommendations in four main areas:

Aid⌂ − The Commission called for at least $4 billion of extra aid to be given annually to the least-developed countries, on grant or concessional terms, for a period of twenty years. This level of funding would require all donating governments in the North to increase their official development aid to 1 per cent of Gross National Product (GNP)⌂ by the year 2000. The Commission also called for a new international system of taxation, and proposed not only that developing countries⌂ should be allowed greater access to the funds available through global financial institutions, but also that these funds should be given on preferential terms. In addition, it was proposed that those countries classified as 'Least Developed' should be allowed more say in the world's multilateral financial institutions such as the World Bank⌂ and International Monetary Fund (IMF)⌂, and that a World Development Fund should be established which would have universal membership.

Food − The Commission recommended that a further $8 billion should be spent every year until 2000 to improve agricultural development and specific aid should be given to developing countries to assist with land-reform programmes. A further annual fund of $200 million should be set aside to ensure both the availability of cereal stocks and the provision of emergency food supplies as and when required. International agreements should also be concluded to stabilize world food supplies and prices and ensure the dismantling and removal of all protectionist barriers erected by the North to safeguard their agricultural markets.

Trade – More equitable trading regimes would be produced if the North removed all protectionist measures which discriminated against processed exports emanating from the South. The Commission also called for the provision of funding and resources to allow the South to process more of its own raw materials and facilitate improved marketing of these processed goods. Commodity prices should be stabilized through a system of international agreements, and a Common Fund⟲ was proposed to finance buffer stocks and other measures through which commodity prices could be guaranteed.

Energy⟲ – The Commission urged that a long-term financing facility should be made available which could be used to promote energy resources in the South. Meanwhile, oil-producing nations should maintain production levels and consumer countries should do their utmost to meet strict energy-conservation targets.

Despite making a number of specific recommendations on how development in poorer countries could be improved, the report stimulated disappointingly little international action. A second report, *Common Crisis*, was published in 1983, after which the Commission was disbanded.

BRANDT, WILLY (1913–) German statesman, an anti-Nazi active in the Resistance movement during the Second World War. He became mayor of West Berlin in 1957 and was appointed Minister of Foreign Affairs in 1966. Three years later he became Federal Chancellor, serving until his resignation in 1974. His 'Ostpolitik' led to treaties with the Soviet Union and Poland and a 'Basic Treaty' between 'East' and 'West' Germany. He was actually forced out of office when one of his aides was found to be spying for the German Democratic Republic ('East' Germany). He was awarded the Nobel Peace Prize in 1971, and became chairman of the Independent Commission on International Development Issues⟲ in 1977. He served as a Member of the European Parliament (MEP) from 1979 to 1983.

BREAST FEEDING After the birth of a child, a yellowish fluid called colostrum is produced in the mother's breast. This fluid, which is passed on to an infant in mother's milk, contains protective proteins and antibodies which provide babies with increased resistance to a variety of infections, parasites and diseases. Most infants require no other food or liquid other than breast-milk during the first four to six months of life. Breast-feeding also inhibits menstruation and, worldwide, suckling is estimated to prevent more births every year than all other forms of birth control⟲ put together. Women in developing countries⟲ who breast-feed have, on average, four fewer children than if they had not breast-fed. Breast-feeding also increases the intervals between births, thus giving lower-birth-weight children a better chance of survival. Transnational⟲ companies which produce powdered milk are trying to persuade millions of women, particularly those in the Third World⟲, to cease breast-feeding and use artificial milk instead. In the developing countries around 10 million cases of infant malnutrition⟲ and diarrhoea occur each year as a direct result of mothers stopping breast-feeding. Babies fed on bottled milk are twice as likely to die as those who are breast-fed.

In the developed countries in Europe and North America the trend away from breast-feeding reached a peak in 1970, when only 30 per cent of mothers breast-fed their children. Since then, a growing realization of the value of breast-feeding has seen the figure rise again to reach 90 per cent in some countries.

BRETTON WOODS CONFERENCE A conference held in 1944 at Bretton Woods, New Hampshire in the United States, at which the governments of Canada, the United Kingdom and the United States established a system of international financial rules which ultimately led to the foundation of the International Monetary Fund (IMF)⟲ and the International Bank for Reconstruction and Development (IBRD)⟲. The chief features of the new financial system were an obligation for each country to maintain the exchange rate of its currency within 1 per cent of a value fixed in terms of gold⟲, and the provision by the IMF of finance to enable nations to overcome temporary payment imbalances. The increasing strain placed on the system eventually caused the exchange-rate component to collapse in 1971, mostly as a result of the US government's suspension of convertibility from dollars to gold. Proposals put forward at the conference to originate an International Trade Organization could not be finalized, but were instrumental in leading to the formation of the General Agreement on Tariffs and Trade (GATT)⟡ in 1948.

BROWNS FERRY A nuclear power⟡ station in the US state of Alabama which was the site of a potentially catastrophic accident in 1975. The power station, at the time the world's largest, consisted of two 1,065-megawatt boiling-water reactors. Foam insulation was accidentally ignited by technicians, and the resultant fire spread to the reactor building. Both reactors were shut down and the fire burned for seven hours, the station remaining out of commission for eighteen months. The nature and potential severity of the incident coincided with the turning point for interest in nuclear power in the United States. The last order for a nuclear plant that was not subsequently cancelled was placed in 1975, although 1990 witnessed a renewal of interest in the possibilities of nuclear power in the country.

BRUNDTLAND COMMISSION Popular name for the World Commission on Environment and Development (WCED)⟡, which was chaired by the Norwegian Prime Minister, Mrs Gro Harlem Brundtland⟡.

BRUNDTLAND, GRO HARLEM (1939–) Norwegian politician. She originally trained as a doctor, specializing in public health, before entering Parliament in 1977. She served as Minister of Environment in the 1974–9 administration. Elected as leader of the nation's Labour Party in 1981, she became Prime Minister in the same year and Chairman of the World Commission on Environment and Development (WCED)⟡ in 1984.

BSE *See* Bovine Spongiform Encephalopathy

BUSHMEN The name given to the aboriginal people of Southern Africa, also known as San people; they call themselves !Kung. They were once widespread across the savannah region of southern and eastern parts of Africa, but there are estimated to be only 50,000 surviving in the Kalahari region in Botswana and Namibia. Of this total, only 5,000 truly nomadic groups of bushmen still exist, restricted to parts of the Kalahari Desert, where they lead a traditional hunting-and-gathering lifestyle. Bushmen are typically of small stature with dark, leathery, yellowish skin. They communicate in a language which is based on the same 'click' sounds as found in several other languages in southern Africa.

BUTHELEZI, CHIEF GATSHA (1929–) Zulu statesman, Chief Minister of the semi-independent KwaZulu in South Africa from 1970. He is strongly opposed to KwaZulu becoming a nominally independent Black Nationalist State (or Homeland⟡) but envisages a confederation of the black areas, with eventual black majority rule over all of South Africa under a one-party socialist system. He is founder-president of the Inkatha movement, formed in 1975, a non-violent cultural-political organization set up to fight for a non-racial democratic political system. (The name Inkatha comes from the word for the grass coil used by Zulu women to support loads being carried on their heads; the many strands give the coil its strength.)

C

CADMIUM A soft, dense, fairly uncommon metal which occurs naturally as the mineral greenockite (CdS) and, more significantly, in zinc⟡, copper⟡ and lead⟡ sulphide ores. All commercial cadmium is recovered as a by-product of the smelting of copper, zinc and lead ores. Chemically similar to lead, it is used in the control rods of nuclear reactors, as well as in light meters, television tubes, solder, low-friction alloys for bearings, and rechargeable batteries and solar cells⟡. Cadmium is also an important component of phosphate⟡ fertilizers, and is discharged when they are refined. Cadmium is released into the air as a result of the incineration or disposal of cadmium-containing products such as rubber tyres⟡ and plastic containers, and through the refining of other metals, primarily zinc.

Cadmium and all its compounds are poisonous. Normally, they are irritants and have an emetic action, usually resulting in their rapid expulsion from the human body. However, if large quantities of dust or vapours are inhaled, pulmonary oedema and death can result. Chronic poisoning causes damage to the kidneys and heart. Prolonged exposure causes a loss of calcium from the bones, which become brittle and break easily. Through the widespread use of fertilizer⊘, cadmium occurs in high levels in some industrial effluents and in sewage sludge. Restrictions on its use have already been introduced in Britain, the Federal Republic of Germany and Sweden, and further measures are being contemplated by the European Community⊘. Around 18,000 tonnes of cadmium are produced each year; the world's leading producers are the Soviet Union (2,900 tonnes), Japan (2,400 tonnes) and the United States (1,700 tonnes).

CAMP DAVID AGREEMENTS Camp David is the official country retreat of the President of the United States. During President Carter's⊘ term of office, in 1978, Egypt's Anwar Sadat⊘ and Israel's Menachem Begin⊘ met at Camp David for discussions, through which they agreed on a framework for establishing peace in the Middle East. Two official agreements, mediated by President Carter, arising from that meeting laid the foundations for the peace treaty which was signed between Israel and Egypt in 1979, and led to Egypt's phased withdrawal from the Sinai, which was completed in 1982. The agreements also called for the election of a 'self-governing authority' by the Palestinians of the West Bank and Gaza Strip – something which is yet to be realized.

CANCER Cancer is a breakdown of the normal process of cell growth and differentiation. It commonly begins with a change in a single cell involving a mutation in the genetic apparatus. Transformed cells then divide without restraint and differentiate into other forms, eventually giving rise to billions of aberrant cells. These cancerous cells can invade and destroy nearby tissues, or be transported in the blood or lymph to other parts of the body. Cancers of epithelial tissue (such as skin, breast and lung) are called carcinomas; those in connective tissue (bone, muscle and cartilage) are known as sarcomas; and cancers in cells of the blood system are leukaemias.

Cancer can be caused by a wide variety of factors, including exposure to chemicals, dietary content and viruses. But socioeconomic factors, lifestyles, behaviour and environmental components appear to be a major influence in determining the varieties of cancer that commonly occur in any country or region. It is now widely believed that 85 per cent of all cancers are caused by environmental factors; the remainder have a hereditary basis or arise from spontaneous metabolic disturbances.

An estimated 6.4 million new cases of cancer worldwide are reported each year, with well over half of these arising in the developing countries⊘. The annual death toll from cancer is put at 4.8 million, representing around 8 per cent of total deaths from known causes.

CANCUN SUMMIT Cancun, a coastal resort in Mexico, was the site of a North⊘–South⊘ Summit meeting held in 1981 following proposals for such a meeting put forward by the Brandt Commission⊘ in 1980. It was one of the largest Summit meetings ever held; the eight industrialized and fourteen developing countries⊘ attending discussed the widening gap between the developed countries and the Third World⊘ and examined ways in which it could be reduced. The meeting provided no new initiatives or agreements, and overall results were disappointing. In its second report in 1983, the Brandt Commission called for a second North–South Summit, but none was forthcoming.

CAPITAL FLIGHT Every year the International Monetary Fund (IMF)⊘ calculates the total value of goods and services officially bought around the world, and all interest payments made internationally. This is then compared with a similar calculation made for the goods and services officially sold and the interest received. Between 1982 and 1990 there has been a regular discrepancy in these accounts which shows that, on a global scale, between $88 and $100 billion is unaccounted for annually. Much of this money is lost through what is called capital flight – the transfer of large sums of money out of a country, often illegally.

The removal from power of President Ferdinand Marcos⊘ of the Philippines and 'Baby Doc' Duvalier in Haiti served to focus attention on the enormous wealth that is often stashed abroad illegally. Vast sums of money are transferred by wealthy individuals or those in power in developing countries⊘ to secret bank accounts in the developed countries for purely personal gain. The practice often involves monies appropriated from funds allocated as international aid⊘. Such sums seriously damage the economies of many indebted Third World⭤ nations. At the end of 1982 the foreign debt of Zaïre stood at $4.2 billion, yet some reports indicated that certain individuals in the country have between $4 and $6 billion invested in bank accounts and real estate in Switzerland.

Financial experts calculate that Mexico suffers worst from capital flight, with over $53 billion passing out of the country between 1976 and 1985. During the same period, Venezuela 'lost' $30 billion, Argentina $26 billion, South Africa $17 billion, India and Malaysia $12 billion, and Brazil, Nigeria and the Philippines $10 billion each.

CARBON CYCLE   Carbon is an integral component of the molecules that make up all living cells, and an essential element for all organisms. It is a fundamental constituent of chemicals such as proteins, carbohydrates and hormones, without which life could not go on. The process by which carbon circulates through the natural world is known as the carbon cycle. In the inorganic component of the cycle, carbon gases – principally carbon dioxide⭤, methane⭤ and carbon monoxide – naturally present in the atmosphere are washed out in the form of carbonic acid, which interacts with rocks to form bicarbonates. These then flow into rivers, lakes or the oceans. Carbon dioxide from the atmosphere also dissolves directly in any area of open water.

Living organisms add another component to the cycle by changing the speed at which carbon dioxide interchanges between land, sea and air. Elemental carbon, derived from atmospheric carbon dioxide, is taken up by plants during photosynthesis. All vegetative matter, and other organisms such as marine plankton, are capable of carrying out the photosynthetic process during which carbon is formed into long-chain molecules which are used to make part of the growing organism. The carbon they accumulate is released back into circulation when the organisms die and decay, or when they are eaten. The burning of fossil fuels⭤ and vegetation, either by human action or as a natural occurrence, also releases carbon back into the atmosphere in gaseous form.

The natural carbon cycle is increasingly being compromised by expanding human consumption of fossil fuels, coupled with destruction of vast tracts of tropical forests⭤ and other vegetation which plays an important role in absorbing carbon dioxide from the atmosphere. The resultant build up of this gas in the atmosphere is contributing to the greenhouse effect⭤ and warming the planet.

CARBON DIOXIDE   A colourless, odourless, non-toxic, non-combustible gas which is naturally present in air and is a vital part of the carbon cycle⭤. It is produced when animals respire and when any material containing carbon is burned. It is also released by natural combustion processes such as volcanic eruptions. The atmospheric concentration of the gas has risen markedly over the past hundred years: from 280 parts per million (ppm) in 1850 to 350 ppm today. In comparison, during the last Ice Age atmospheric concentration was estimated to be 200 ppm, and this had risen to around 270 ppm just before the Industrial Revolution. The significant increase in atmospheric concentration seen over the last century is almost solely due to the burning of organic carbon-based fuels such as coal⭤, oil⭤, natural gas⭤ and wood⭤. It is known that carbon dioxide (CO_2) contributes more than any other gas to the greenhouse effect⭤; nevertheless, the rate of accumulation in the air is still increasing, rising 10 per cent since 1983.

Most experts accept that concentrations of atmospheric CO_2 will at least double in the long term owing to continuing use of carbon-based fuels. The burning of fossil fuels⭤ currently adds 5.6 billion tonnes of CO_2 into the air annually, and the effects of deforestation⭤ release another 2–4 billion tonnes each year. In total, the air today is receiving a billion tonnes more carbon from the destruction of vegetation than it did fifteen years ago, and a similar amount extra from the burning of fossil fuels.

It is forecast that global emissions of CO_2 will grow by 2.7 per cent annually over the next forty years. Of the current total, the United States contributes 23 per cent, followed by the Soviet Union (18 per cent) and China (11 per cent). The pollution from these countries is still increasing, whereas other nations, such as France and the Federal Republic of Germany, have managed to reduce their emissions.

Paradoxically, although there is a growing international effort to curtail production of the gas, carbon dioxide is actually manufactured commercially from chalk or limestone. It has a number of industrial uses: as a coolant in nuclear reactors, as a refrigerant, in fire extinguishers, and in carbonated or 'fizzy' drinks.

The Toronto Conference on the Changing Atmosphere, held in 1988 to consider the CO_2 problem, suggested that emissions should be cut some 20 per cent by 2005 and 50 per cent by 2020.

This is highly unlikely to occur. In 1989, a Japanese study predicted that global emissions of carbon dioxide in 2030 would be 18 billion tonnes – over three times what they were in 1986. Grandiose attempts to reduce atmospheric CO_2 concentrations are also likely to have little effect. US administration plans to plant 100 million trees each year for the next decade to soak up some of the CO_2 and reverse the effects of deforestation reflect similar projects being implemented around the world. Each tree locks up around 12 kg of CO_2 and the contribution of tree-planting schemes to carbon dioxide reduction will be negligible.

CARIBBEAN COMMUNITY (CARICOM)
CARICOM came into existence in 1973, basically comprising ex-British colonies in the Caribbean, and superseded the Caribbean Free Trade Association (CARIFTA). The main aim of CARICOM is to promote economic integration among its member states, largely through the introduction of a viable common trading market to which all members will belong. As part of this initiative, it is hoped that tariff and non-tariff barriers in the region will be removed by 1991. The free movement of individuals within member countries was planned to start in 1991, and a regional stock market is scheduled to begin operating in 1993. CARICOM also facilitates the cohesive operation of common services such as shipping and metereological systems, and allows member countries to co-ordinate their national policies and maintain a unified position with regard to extra-regional affairs. Antigua and Barbuda, Bahamas, Barbados, Belize, Dominica, Grenada, Guyana, Jamaica, Montserrat, St Kitts and Nevis, St Lucia, St Vincent, and Trinidad and Tobago make up the Community, which is based in Guyana and controlled by a Heads of Government Conference. In 1979, a coup in Grenada led to the setting up of a progressive regional subgroup.

CARIBBEAN DEVELOPMENT BANK (CDB)
A regional development bank, located in Barbados, which came into existence in 1970. The Bank aims to play a major part in stimulating and improving the economic growth and development of member countries in the Caribbean, and to promote their economic co-operation and integration with particular regard to the needs of the less-developed nations in the region. Membership is open to all states and territories in the Caribbean Basin and non-regional countries which are members of the United Nations⬦. In addition to the member

nations of CARICOM⬦, the Bank has seven regional members: Anguilla, British Virgin Islands, Cayman Islands, Colombia, Mexico, the Turks and Caicos Islands, and Venezuela. There are also three non-regional contributing governments: Canada, the United Kingdom and the United States. Three of the regional members – Colombia, Mexico and Venezuela – are non-borrowing members. The CDB has established a Special Development Fund which will allow lending at concessionary rates to countries in the Caribbean. Canada, Colombia, the United Kingdom, the United States and Venezuela have made contributions to this fund.

CARIBBEAN DEVELOPMENT FACILITY (CDF)
The first ever meeting of the Caribbean Group for Co-operation in Economic Development⬦ established the Caribbean Development Facility (CDF) which is organized by the World Bank⬦. The CDF has no institutional structure and is not a centrally managed fund; rather, it is an accounting mechanism to record the sum of separate financing arrangements concerning projects and operations undertaken jointly by individual donors and recipients. It is looked upon as a means of directing to the Caribbean additional assistance which is needed to help finance essential imports, and of encouraging supplementary financing for development programmes.

CARIBBEAN GROUP FOR CO-OPERATION IN ECONOMIC DEVELOPMENT (CGCED)
The Group is composed of all donor countries, multilateral donor organizations and the recipient nations in the Caribbean. It was established to promote flows of quick-disbursing aid to the Caribbean, to encourage regional co-operation, and to provide a forum for discussion of matters of common interest. The initial meeting of the CGCED was held in 1978 under World Bank⬦ sponsorship.

CARRYING CAPACITY
In ecology, the carrying capacity of the environment refers to the maximum number of plants or animals that a particular area can support. It can be applied to communities as well as individual species. When the carrying capacity is exceeded, there are insufficient natural resources⬦ to sustain the population⬦, which may then be reduced by emigration, reproductive failure, or death through starvation, thirst or disease. Most species oscillate in number, fluctuating just above and just below the carrying capacity of the environment. Some, such as lemmings and snowshoe hares,

show a massive increase in numbers, far in excess of the carrying capacity, followed by a catastrophic reduction caused by mass starvation or the adoption of behaviour patterns which result in mass fatalities. This takes the number of animals down to well below the carrying capacity, from which point they begin to increase and the cycle is repeated. As a relatively new animal species on earth, the human race has yet to go through a complete population cycle, so which of these two models – if either – will be followed by humans has yet to be determined.

The concept of carrying capacity is not yet refined enough to determine the actual number of people that the earth could support. Certainly, human activities such as intensive agricultural production can exceed the carrying capacity of the land⇨. Where it is exceeded, land becomes prone to soil erosion⇨, desertification⇨ and depletion of nutrients. Under these circumstances, agricultural production is not sustainable and can be maintained only artificially through the addition of massive amounts of expensive, energy-intensive agricultural chemicals. In 1984, the Food and Agriculture Organization (FAO)⇨ attempted to estimate the carrying capacity of 117 developing nations. Assuming that no agricultural chemicals were used, no improved crop varieties were introduced, no land-conservation measures were adopted and traditional crop-mix and rotation patterns were maintained, the results indicated that the developing world should be able to support twice its 1975 population of 4.1 billion. However, as with most models, general assumptions were made which would be unrealistic in the real world, and the method is too unreliable to be used as a policy-forming instrument.

Between 1960 and 1980, as the world's population moved from 3 billion to over 4 billion, human demands began to outstrip the sustainable supply of natural resources. The global *per capita* production of several basic commodities peaked during this period – wood⇨ in 1967, fish⇨ in 1970, oil⇨ in 1973, beef in 1976 and grains in 1978 – and all have been declining ever since.

One complicating factor in determining the earth's carrying capacity with regard to humans is the standard of living. According to ecologist Arthur Westing, the world population would have to be at least halved if an affluent standard of living, deemed to be that seen in the richer nations which have twice the world average Gross National Product (GNP)⇨ *per capita*, were to be extended to everyone. In comparison, around 70 per cent of the world's current population could be supported at the standard of living prevailing in nations with average global GNP *per capita*.

CARSON, RACHEL LOUISE (1907–64)
American science writer who, between 1936 and 1952, worked as a genetic biologist before becoming an editor with the United States Fish and Wildlife Service. Her books – most notably *The Sea Around Us*, published in 1952, and *Silent Spring*, published in 1962 – greatly increased public awareness of the threats to the natural environment, warning of the dangers of pollution⇨ and the need for environmental monitoring and protection.

CARTEGNA GROUP
A grouping of eleven Latin American countries formed in 1984 to facilitate discussion of their problems of heavy foreign debts⇨, which had primarily been caused by over-ambitious borrowing from banks drawing on the the huge sums amassed by members of the Organization of Petroleum Exporting Countries (OPEC)⇨ after the 1974 oil price rise. At the time, Brazil and Mexico each owed in excess of $100 billion, and two years earlier they had become unable to keep up with payments due on these debts.

CARTER, JAMES EARL (JIMMY) (1924–)
American statesman, Democratic President of the United States of America from 1977 to 1981. He served in the navy, studied nuclear physics and managed his family's peanut farm before entering politics in 1953. He was governor of his home state, Georgia, from 1970 to 1974, and his original presidential campaign was based on promises for ambitious social and economic reforms. During his presidency he was instrumental in concluding the Panama Treaty and the Camp David Agreements⇨ and introduced measures to limit energy consumption in the US. He was also in office during the time when US hostages were being held in Iran. His failure to resolve this crisis contributed significantly to his defeat in the 1980 presidential election, won by Ronald Reagan⇨. Following his defeat he became associated with GLOBAL 2000⇨, an independent voluntary organization involved with a wide variety of global development projects, including the attempt to rid the world of Guinea worm⇨.

CASH CROPS
Cash crops are crops which generate revenue for the producer through their sale or export value, as opposed to crops grown to provide

food for a farmer or to sustain the local population. Major examples are coffee⟁, tea⟁, oil palm⟁, cocoa⟁ and non-food crops such as rubber⟁, tobacco⟁, jute⟁ and cotton⟁. Cash crops are predominantly grown in the developing world and sold to satisfy demand in the industrialized nations. Consequently they are a major source of foreign exchange, particularly for those tropical nations with no oil⟁ or mineral wealth. Both production and prices are fickle, open to disruption by natural forces or manipulation of commodity markets, and results can be catastrophic. Honduras depends for 70 per cent of its export earnings on bananas⟁ and Dominica also earns more than half its foreign earnings from bananas, crops which were devastated by hurricanes in 1979 and 1980. Unfortunately, even when production is good and prices are favourable, the purchasing power provided to developing countries⟁ by agricultural exports rarely keeps pace with the price of the manufactured goods or foodstuffs that they need to import.

From the mid-1950s to mid-1960s the growth rate of cash crops increased twice as fast as the total agricultural growth rate in developing countries. Then, as now, every aspect from production through to point of sale was dominated by transnational⟁ business enterprises. Critics argue that production of cash crops uses up land⟁ that could otherwise be used to grow food; cash crops are therefore a major contributory factor to world hunger⟁, as shown in Brazil, where efforts to grow soya beans for export displaced cultivation of black beans. Black beans, the staple diet of the poor, became scarce, food riots followed, and Brazil was forced to import black beans from Chile.

Poor countries find themselves growing crops and introducing industrial agriculture⟁, usually causing peasant farmers to be thrown or priced off the land, in order to obtain foreign exchange. Throughout the Third World⟁, peasant farmers are forced to use marginal land to grow food for their families, the best land is devoted to cash crop production. In Mali, more land is given over to growing cotton and groundnuts⟁ for export than is used to produce food for home consumption. During the Sahel droughts⟁ and localized famines of the 1970s and 1980s, production of cash crops for export continued to increase.

Globally, cash crops now take up almost 750,000 sq km of land. Approximately 5 per cent of the world's cultivated land is being used to grow cotton, much of it land which once provided a livelihood for peasants. Cotton is prone to attack by a variety of pests and has an extremely high demand for nitrogen, quickly exhausting soils unless nitrogen is added – this is why, in developing countries, it is often grown in rotation with a leguminous crop such as groundnuts.

The negative aspects of cash crops are manifold. Small-scale farms, the most productive agricultural units, have virtually no incentive to produce cash crops. Smallholders rarely benefit from increased agricultural export earnings. The United Nations⟁ reported that between 1968 and 1973 coffee prices rose by 58 per cent, but producer prices in Rwanda over that period remained fixed. Similarly, in 1968–9, when the world price for groundnuts rose, the Senegalese government's price to farmers actually fell. In addition many cash crops, like cotton, are particularly damaging to the soil. Land used to grow groundnuts requires a minimum of six years' fallow to recover fertility. In the fickle growing conditions faced by most farmers in the developing world, it is unwise to give priority to cash crops which tend to require one or two harvests per year, rather than follow a pattern of diversified cropping that produces food year-round. Even where production constraints are limited, cash crop producer countries not only become vulnerable to the fluctuations of the world's commodity markets, which are primarily manipulated by the richer, consumer nations; they also face having production curtailed by natural disasters such as droughts⟁ or floods⟁, or a basic lack of the infrastructure needed to get the crops from the fields to the marketplace.

CASHEW Cashew is an evergreen, nut-bearing tree with a wide-spreading root system; the roots spread out far wider than the canopy. The tree belongs to the family *Anacardiaceae* and originated in tropical America. It was one of the first fruit trees of the New World to be introduced throughout the tropics – brought to India in the sixteenth century by the Portuguese, primarily to make use of the tree's extensive rooting system to stabilize the shifting, sandy soil on the country's east coast. Cashew trees need no fertilizer⟁ and grow with a minimum of maintenance, reaching maturity at eight years. A tree yields, on average, 1,130 kg of nuts per hectare and the trees continue to bear fruit for between 25 and 40 years. The nuts contain an edible kernel with a distinctive taste and are used extensively in the confectionery and baking industries. They also contain a caustic black liquid which is very corrosive and blisters the skin, making processing complicated.

There are three processing techniques, but in all three the nuts are first dried, cleaned, moistened and kept for twelve hours to soften the shell. They can then be roasted, treated with solvents or exposed to superheated steam in order to remove the caustic oil content. The liquid collected as a by-product contains 90 per cent anacardic acid and 10 per cent cardol. These are both used in the varnish, plastic and synthetic resin industries. The oil also has uses as a lubricant, an insecticide⋄, and in the manufacture of cement, tiles and brake linings. The fruit from the tree can be used to make an alcoholic gin-like beverage, sap from the bark is used to make indelible ink, and gum from the trunk is extremely adhesive and can be used to make glue. India is now the largest producer of cashew nuts in the world, accounting for 30 per cent of the total global production of 510,000 tonnes and providing 95 per cent of total exports.

CASSAVA   A shrubby flowering plant, *Manihot esculentus*, also known as manioc, native to tropical America. Many varieties are now extensively cultivated throughout the tropics for their edible starchy, tuberous roots. These can be processed into tapioca, ground to produce manioc or cassava meal, used as animal fodder, or cooked and eaten as a vegetable. Cassava is a staple food crop in many areas in the tropics, especially for poor people, and it is capable of providing high yields under conditions of poor soil fertility and low rainfall. It produces more calories per hectare than most other staple crops and is the principal source of carbohydrate in Latin America and parts of Oceania. Worldwide, around 100 million tonnes are eaten by the 400 million people whose daily diet is based on cassava.

Cassava tubers, like other similar tuber crops such as taro, potatoes and yams, consist of almost pure starch, However, leaves of the plant contain about 17 per cent protein. Cassava tubers contain a poison, hydrocyanic acid, which must be removed before the tubers are eaten by humans or animals. The outer peel and core of the tuber contain the most poison, so these must be discarded. Drying, roasting or boiling the tubers removes the remainder. Cassava is increasingly being used as a livestock⋄ feed and for the production of industrial starches, the bulk of the massively expanded production in Thailand being imported into Europe as cheap cattle fodder to replace cereal-based concentrates which had become too expensive. Global production is in the order of 137 million tonnes per year; Brazil (25 million tonnes), Thailand (19 million tonnes) and Zaïre (16 million tonnes) are the major producing nations.

CASTRO RUZ, FIDEL (1927–)   Cuban revolutionary and statesman, son of a wealthy sugar⋄ planter, imprisoned between 1953 and 1955 for his political activities. He later led an invasion of Cuba launched from Mexico in 1956 and, commanding a guerrilla group of 5,000 supporters, he and his followers waged a continual war against the Cuban dictator, Fulgencio Batista, until they emerged victorious in 1959. He was elected Prime Minister later in the year and established a socialist government which the United States has continually attempted to subvert. He became President of Cuba in 1976 and was elected President of the Non-Aligned Movement⋄ in 1979.

CATALYTIC CONVERTER   A device fitted to a motor vehicle to destroy some of the harmful gases produced by the internal combustion engine. Expensive metals, notably platinum⋄, are used to facilitate the process, which takes place in two stages. Oxides of nitrogen are reduced to a mixture of nitrogen and carbon monoxide, while unburned hydrocarbons are oxidized to produce harmless products. Although toxic fumes are reduced by 90 per cent using the latest three-way regulated catalytic converters, levels of carbon gas emissions are not curtailed. To function properly, the proportion of air in the exhaust must be carefully controlled, and this is done electronically. Other converters, known as oxidation catalysts, do not cope with nitrogen oxides⋄, so supplementary measures, such as the use of lean burn engines which produce cleaner exhausts, are required to bring this about. Catalytic converters are rendered inoperative by lead⋄ and are ineffective when an engine is run on leaded fuel. They are also fairly expensive – early versions of the converters raised vehicle prices by an average of $500 – and must be maintained regularly if they are to remain effective. Catalytic converters must be fitted by law to all new cars in Japan and the United States, and the European Community⋄ has adopted similar legislation which comes into force after 1992.

CATHOLIC   FUND   FOR   OVERSEAS DEVELOPMENT (CAFOD)   Based in London, CAFOD is the official aid⋄ agency of the Catholic bishops of England and Wales. It aims to promote world development by co-operating in self-help projects and programmes overseas, through the provision of aid and relief and by a programme of education⋄ directed to improve awareness in the United Kingdom.

CENTRAL AMERICAN BANK FOR ECONOMIC INTEGRATION (CABEI) A regional development bank established in 1960 which began operating in 1961, with its headquarters in Honduras. The Bank aimed to promote the complementary economic development of the five members of the Central American Common Market⋄: Costa Rica, El Salvador, Guatemala. Honduras and Nicaragua. It also established a special integration fund for financing infrastructure-improvement projects in the region.

CENTRAL AMERICAN COMMON MARKET (ODECA) The *Organanización de Estados Centro-americanos* (ODECA) was established in 1960 to promote economic co-operation and development between member states. Founding members included Costa Rica, El Salvador, Guatemala, Nicaragua and Honduras, the latter seceding in 1970.

CENTRAL ARIZONA PROJECT (CAP) The Central Arizona Project is one of the most ambitious and expensive water supply projects in the world. In the state of Arizona in the south-western United States, the towns of Phoenix, Scottsdale, Tempe and Mesa make up a sprawling metropolitan area of a million people. The towns regularly play host to thousands of winter visitors in search of the sun and are actively encouraging new residents. The city of Tucson also houses a further 500,000 people. All these cities exist in a desert where rainfall is minimal. Phoenix and Tucson were originally founded on rivers, both of which have been completely dry for decades. The state capital, Phoenix, now barely satisfies its water needs from reservoirs which depend on the winter-snow melt from mountains much farther north. Tucson is probably the largest city in the world which depends totally on wells for its water supply. However, overuse of aquifers has seen the water table in southern parts of Arizona drop by up to 90 m in the last fifty years.

To offset the shortage, the CAP will eventually bring water⋄ from the Colorado River to central and eastern Arizona. These areas are a long way from the river, and much higher. A complex system of pumps, canals and aqueducts, well over 500 km in extent, costing billions of dollars, is therefore being built to raise and transport the water. Construction began in 1973 and was due to be completed by 1990. Lake Havasu is impounded behind the Parker Dam on the Colorado and water is to be pumped from it up 365 m to the reservoirs in Phoenix. A 150-km pipeline will carry water on to Tucson. It is eventually planned to supply 790 million cubic metres (cu m) of water annually to the seventy-one towns and cities in the centre and south of the state. A further 380 million cu m will be supplied to Indian reservations in the region.

CENTRAL TREATY ORGANIZATION (CENTO) A mutual defence alliance between Iran, Pakistan, Turkey and the United Kingdom which became operational in 1959. It succeeded the Baghdad Pact⋄ and was intended to promote co-operation and economic development of its Middle Eastern members. The UK – together with the United States, which subsequently became an associate member – treated CENTO primarily as a military alliance against the Soviet bloc and neglected the economic development of the member states in the Middle East. As a consequence, Iran, Pakistan and Turkey formed a separate regional co-operation organization in 1964. CENTO eventually ceased to exist after the official withdrawal of Iran and Pakistan in 1979.

CENTRE FOR INDUSTRIAL DEVELOPMENT (CDI) A co-ordinating institution established under the first Lomé Convention⋄. Based in Brussels, the Centre is responsible for gathering and disseminating information on conditions and opportunities for industrial co-operation between parties to the convention, namely the ACP countries⋄ and members of the European Community⋄. It is empowered to carry out, at the request of the EC or ACP countries, feasibility studies on possible industrial development opportunities in the latter, to facilitate contacts between industrial policy-makers, firms and financial institutions, to provide specific industrial information and support services, and to help identify opportunities for industrial training and applied research geared to the needs of the ACP countries.

CENTRE FOR OUR COMMON FUTURE An independent foundation, based in Geneva, established in 1988 to mark the first anniversary of the publication of the report by the World Commission on Environment and Development (WCED)⋄. The Centre acts as a focal point for all follow-up activities arising from the publication and recommendations of the WCED report *Our Common Future*. Working in conjunction with a variety of international and national Non-Governmental Organizations (NGOS)⟲, academic and research institutions, and networks, the Centre aims to ensure that the report forms an integral part of national and international policies and actions in the movement towards promoting and achieving development that is both economically and environmentally sustainable.

CENTRE FOR WORLD DEVELOPMENT EDUCATION (CWDE)   Based in London, the CDWE is an independent, voluntary educational organization whose main objective is to increase knowledge, awareness and understanding in the United Kingdom of all facets of world development and the developing countries⟲ and the interdependence between the UK and the Third World⟲. Producing and disseminating a wide range of informative and educational materials, it is active in both the formal and informal education⟲ sectors. The Centre has also played a pioneering role in bridging the gap between the business community and those concerned with environment and world development matters, as evidenced by the World Development Awards for Business, created by CWDE and presented to commercial enterprises which have exemplified the positive impact that businesses can have on economic and social development in the Third World.

CFCs   *See* Chlorofluorocarbons

CHAGAS DISEASE   A disease caused by infection with a protozoan of the genus *Trypanosoma*, endemic in Central and South America, although transfusions with blood from infected migrants have led to cases in the northern United States and Canada. In endemic areas 17 million people are infected, and 90 million people are at risk. Named after a Brazilian physician, it is a form of trypanosomiasis⟲ transmitted to humans when *Trypanosoma cruzi*, present in the faeces of nocturnal bloodsucking bugs, come into contact with wounds on the skin or the delicate tissue of the nose and mouth.

There are two stages of the disease. The acute stage develops within three weeks of infection and may last up to two months. Symptoms consist of fever, lymph node swelling and enlargement of the spleen and liver. The chronic stage can last for years, and is manifested by invasion of the parasites into the vital internal organs. Once they are in the muscles of the heart and in the central nervous system, serious – potentially fatal – inflammation and lesions can occur. The disease is mostly limited to poor rural areas of South and Central America. It is especially prevalent in children, and there is no effective cure.

CHARTER OF ECONOMIC RIGHTS   The Charter of Economic Rights and Duties of States, prepared by the United Nations Conference on Trade and Development (UNCTAD)⟲, was adopted by the United Nations (UN)⟲ in 1974. The Charter sought to expand the participation of Third World⟲ countries in the decision-making processes which govern the global economy, and allow them more say in both global and national development programmes. The voting was 120 to 6, with 10 abstentions. Belgium, Denmark, the Federal Republic of Germany, Luxembourg, the United Kingdom and the United States were the dissenting votes. Most of the remaining Western European nations, Canada and Japan abstained.

CHEMICAL WARFARE   The use of toxic substances – gas, liquid or solid – to kill or disable personnel, pollute food, poison water⟲ supplies, or any other military use of chemicals. The first large-scale use of chemical weapons took place during the First World War, when chlorine, phosgene and mustard gas were all used. These gases killed around 100,000 soldiers and injured 1.3 million. They were rendered ineffective through the development of gas masks and were not used in the Second World War. Toxins have now been developed which penetrate through the skin, necessitating the use of protective clothing which covers the whole body.

Binary toxic weapons have also been produced in which two harmless chemicals, in liquid or solid form, are fired in the same artillery shell. On explosion, these mix to provide a lethal aerosol. It is alleged that these types of chemical weapons were used during the Iran–Iraq conflict in the 1980s. Many nations have developed the research and technology needed for chemical weapons programmes because most of the chemicals used are similar to, or derivatives of, common pesticides⟲. The Soviet Union and the United States have considerable stockpiles of chemical weapons and France is also reputed to maintain supplies. The United Kingdom reportedly destroyed its stockpiles in the 1960s but undertakes a significant amount of research which is an integral part of the US chemical-warfare effort. In

1990, following the apparent end of the Cold War, the United States began openly to remove stocks of its chemical weapons from Europe, transporting them by road and sea to Johnson Atoll in the Pacific for disposal. The use of chemical weapons is prohibited under the Geneva Protocol of 1925, but this does not ban the research into, development, production or stockpiling of chemical weapons.

The subject of chemical weapons has been under constant review with the goal of producing regulations to effect a total prohibition of such weapons. A forty-member United Nations⊃ Conference on Disarmament has been meeting in Geneva since 1980 with the intention of negotiating a complete ban. Since 1986, a group of experts gathered in Berne has also been discussing means to curb the proliferation of chemical weapons. The main stumbling block to progress has been the problem of verification. The West has argued for compulsory inspection of chemical plants, the Soviet Union has steadfastly maintained that any inspection system should be voluntary. In 1988 representatives from the international chemical industry declared their unequivocal opposition to chemical weapons, an action which led to two American companies refusing to supply the US government with thionyl chloride (used in the production of a nerve gas, sarin), thereby laying themselves open to prosecution.

CHERNOBYL   A nuclear power⊃ station in the Soviet Union which in 1986 suffered the world's worst nuclear accident. The accident, which occurred as the result of an experiment involving the switching off of safety systems, caused the deaths of 31 people. At least 129 others suffered acute radiation sickness and many thousands are expected to die prematurely as a consequence of exposure to radiation released from the stricken plant.

Conservative estimates of the cost of the disaster have already reached the $340 billion mark and are constantly being revised upwards. An estimated $13 billion has already been spent including the cost of encasing the ruptured reactor in steel and concrete. Clean-up and reparation costs will also include the transportation of the 116,000 people evacuated from the area immediately around the devastated power station and their resettlement in safe areas, the building of 13,500 new dwellings, compensation for the loss of property and crops, and the cost of long-term medical treatment for those suffering from the effects of radiation sickness. Efforts to reduce high radiation levels in the vicinity of the Chernobyl plant failed to meet original expectations,

and at least another 100,000 people may have to be resettled in the future.

Medicines to counteract radioactive fallout were not available in some regions until two months after the accident, despite the fact that for every eight hours that pass before the medication is taken, its effectiveness is halved. By 1990, the incidence of tuberculosis⊃ in the vicinity of Chernobyl had risen 14 per cent and the incidence of common illnesses in the worst-affected areas had increased by 70 per cent due to a weakening of immune systems caused by exposure to the radiation given off by the accident. Up to 250,000 people may now be living in areas too polluted for normal life.

In the south-east of Mogilev province in the Soviet Union, radiation was found to be higher than in the 30-km zone around Chernobyl itself, with levels up to three times higher than the safe level, 15 curie per sq km. Around a quarter of the region's children have exhibited thyroid deficiencies, and the cancer⊃ rate is rising. The high level of radioactivity⊃ which fell to earth in this region was due to the rain produced when the Soviet government decided to 'seed' the strontium-caesium radioactive clouds produced by the Chernobyl disaster to prevent them drifting towards Moscow.

Radiation released into the atmosphere from Chernobyl drifted across much of Europe polluting vast areas of countryside and numerous animals. Restrictions placed on the consumption of meat and vegetables for fear that they might be contaminated were still in place in 1990 as far away as the United Kingdom. Estimates suggest that a minimum of 2,000 people in the European Community⊃ are likely to develop cancers over the next fifty years as a direct result of this radioactive fallout. Upper estimates put the figure at around 1 million people.

CHILD LABOUR   Official figures produced by the International Labour Office (ILO)⊃ say that at least 100 million young children around the world are leading a working life, with the actual total likely to be almost twice as high. In India alone there are an estimated 16 million child labourers, and the ILO reports that around the world, 75 million children between the ages of eight and fifteen are actually being forced to work for a living. Many are compelled to work to pay off their family's debt⊃. Wages are frequently paid in advance, so the children are in effective 'debt bondage' to their employers. Most child workers cannot go to school, but as – in many places in the developing world, at least – education⊃ has no significant bearing on

employment opportunities, there is thus no real
incentive to go.

Children are generally docile, fast, agile, and
above all cheap and dispensable. Consequently, millions, mostly in developing countries⌀, are toiling
long hours for little reward, with no fringe benefits,
insurance or security. Working children are more
likely to suffer occupational injuries because of
unsafe working conditions, inexperience, fatigue
and the fact that most workplaces and machinery
have been designed for use by adults.

Many countries have adopted legislation stipulat-
ing minimum ages below which children cannot
legally be employed and specifying conditions under
which they can work. The ILO Minimum Age
Convention of 1973 stipulates twelve is the lowest
age, and then for light work only.

CHIPKO MOVEMENT A grass-roots,
community-level movement opposing government
or official deforestation⌀ programmes. It originated in the Himalayan foothills of the Indian state
of Uttar Pradesh in the early 1970s. 'Chipko' means
'to embrace' in Hindi, and the movement is so called
because of the initial efforts of women who
embraced trees to prevent them from being felled.
The movement owes a great deal to the philosophy
of Mahatma Gandhi in terms of its commitment to
non-violent resistance and its village-orientated
approach to economic development. The Chipko
campaign in India resulted in a fifteen-year ban on
commercial green felling in the forests⌀ where the
movement originated, and its success has spawned
similar movements elsewhere in the world.

From its outset, the Chipko Movement was concerned with the just allocation of rights to exploit
forest resources. It is now recognized that forests are
fundamental resources ensuring the viability of
human communities and wildlife. They prevent soil
erosion⌀, preserve water⌀ and agricultural land⌀,
and are a crucial source of food and fuel. In view of
this, the Chipko Movement has since matured into a
fully fledged environmental-improvement and con-
servation⌀ movement.

CHLORINATED HYDROCARBONS Org-
anic chemicals containing only carbon and hydrogen
are known as hydrocarbons. When one (or more) of
the hydrogen atoms in the molecule is replaced with
an atom of chlorine, the result is a chlorinated
hydrocarbon. These chemicals are almost entirely
synthetic and do not occur in nature. As a result, the
micro-organisms responsible for natural decay and

recycling in nature cannot easily break down the
compounds, which consequently persist in the
environment.

The best-known chlorinated hydrocarbon is the
insecticide⌀ DDT⌀. Extensive and ill-managed use
of DDT, together with similar pesticides⌀ such as
dieldrin⌀, has brought about widespread pollu-
tion⌀ and massive loss of wildlife caused by these
toxic chemicals⌀ concentrating as they pass up
natural food chains. Chlorinated hydrocarbons tend
to be insoluble in water⌀ but dissolve in fat. They
thus become stored in fatty tissue or fluids in the
body and minute amounts, consumed over a period
of time, continually accumulate in animals, causing a
variety of physical ailments. In addition to the
potential of causing direct physiological damage,
chemicals such as DDT are passed on to newborn
babies in breast-milk.

Chlorinated hydrocarbons pose a variety of
dangers. Some, manufactured as industrial solvents,
are both toxic and carcinogenic. When fluorine is
added to the hydrocarbon in addition to chlorine,
chlorofluorocarbons (CFC)⌀ are produced which,
when released into the upper atmosphere, cause the
breakdown of the protective ozone⌀ layer. If chlorinated hydrocarbon compounds – or products
which contain them, such as plastics – are disposed
of by incineration, and the temperatures reached are
relatively low – around 500°C – other highly toxic
compounds, mainly dioxins⌀, are formed.

Owing to the enormous risks posed by chlorinated hydrocarbons, the production or use of
insecticides and other products containing them
have been banned in many industrialized countries.
However, they are still manufactured in bulk and
sold for use in the Third World⌀, with many companies simply transferring production to countries
in the developing world where regulations governing their manufacture and use are lax.

CHLOROFLUOROCARBONS (CFCs)
Chemical compounds in which hydrogen atoms in a
hydrocarbon are replaced by atoms of chlorine and
fluorine. Synthetically produced, the chemicals are
used as aerosol propellants, refrigerants, coolants,
sterilants, solvents in the electronics industry, and in
the production of insulating foam for packing.
Although there are several such compounds, two,
trichlorofluoromethane (CFC-11) and dichlorodi-
fluoromethane (CFC-12), account for 80 per cent of
global CFC production.

As there are no natural systems for breaking down
these man-made chemicals, they build up and persist

in the environment. Atmospheric concentrations of CFC-11 and CFC-12 were building up at a rate of 5 per cent annually during the 1970s – the former last for 65 years in the air, the latter for 110 years. They are broken down only by the action of ultraviolet radiation when they have risen into the upper atmosphere.

CFCs are greenhouse gases⋄, allowing incoming solar radiation to pass but blocking outgoing radiation reflected from the earth's surface. This effectively heats up the planet, adding to the greenhouse effect⋄. CFCs are amongst the worst of the greenhouse gases, and are thought to contribute between 14 and 25 per cent of global warming⋄. CFC-12 is 5,700 times as effective as a Greenhouse gas than carbon dioxide⋄. CFCs are also responsible for the depletion of the stratospheric ozone layer, as the chlorine, released when CFCs break down, reacts with ozone⋄, destroying it in the process.

Concern over the action of CFCs during the early 1980s led to a ban on their use in certain countries and to the Vienna Convention⋄ of 1985, which took the first steps towards a global ban on their production and use. If production of CFCs were to continue at the level seen in the mid-1980s, it would bring about a 3 per cent steady-state reduction in total global ozone over the next sixty to seventy years. It is estimated that to prevent any further rise in the concentration of atmospheric CFC levels, releases of the gases would have to be cut immediately by 85 per cent. The Vienna Convention, however, called for production of the gases to be frozen in 1990 to 1986 levels, the eventual aim being to bring about a 50 per cent reduction by the end of the century. The subsequent Montreal Protocol⋄, provided an international agreement to cut CFC production and consumption, including provisions for technical assistance to achieve this end.

Over eighty governments are now committed to phasing out CFC by 2000. Wealthy governments, such as the United Kingdom and the United States, originally refused to contribute to a fund to help developing nations such as China and India to curtail their CFC production and develop suitable alternatives. However, at a conference in London in 1990 the nations represented, including the UK and the USA, agreed on means to provide funding and transfer technology to the developing countries to help them phase out CFC production and find suitable alternatives. The fund proposed to help phase out ozone-depleting CFCs and similar gases could cost up to $6 billion by 2000, with $240 million to be spent over the first three years.

Some of the alternative compounds originally developed as replacements, notably the hydrofluorocarbons (HFCS) and hydrochlorofluorocarbons (HCFCS), also proved to attack ozone. The HFCS are also greenhouse gases. HFC-134, for example, is 3,000 times more potent than carbon dioxide in its impact on increasing global warming.

CHOLERA   An acute intestinal infection caused by the bacterium *Vibrio cholerae*, which is transmitted in drinking water⋄ contaminated by faeces from an infected person or via contaminated food. Formerly causing a high death rate, the disease is now less dangerous, although the mortality rate in untreated cases is over 50 per cent. Cholera is still prevalent in tropical countries and epidemics occur in areas where sanitation is poor. After an incubation period of one to five days, symptoms commence suddenly: severe vomiting and diarrhoea which, if untreated, leads to potentially fatal dehydration. In extreme cases, imbalance in the concentration of body fluids can cause death within twenty-four hours. Treatment consists of replacing lost body fluids and salts. Vaccinations against the disease can provide temporary immunity, usually lasting four to six months.

CHRISTIAN AID   A voluntary British organization founded in 1949 by the British Council of Churches to co-ordinate and direct the use of donations made by the public and church groups to assist in improving living standards in the Third World⋄. Christian Aid makes no distinction as to race, religion, or politics and supports development programmes in agriculture⋄, health, and education⋄, as well as giving relief aid in emergencies.

CHROMIUM   A hard, grey metal that occurs in nature primarily as chromite. It is widely used as an electroplated layer on steel and metal objects, but is mostly used to make ferrous alloys. When it is incorporated in alloys with iron⋄, nickel, cobalt⋄ and tungsten, exceptionally hard steels can be manufactured. Chromium salts are also used in the leather and dye industries. The major producers of chromium are South Africa (3 million tonnes), the Soviet Union (2.3 million tonnes) and Albania (1 million tonnes).

CLEAN-UP COSTS   The cost of cleaning up after environmental pollution⋄ and the destruction of ecological equilibria is much greater than the cost of prevention. The costs of cleansing the environment after oil spills⟲ are thought be around $1,000 per barrel spilled. In the United States, measures to effect small reductions of sulphur dioxide⟲ would cost around $200 per tonne, but as the proportion of the pollutant removed from exhaust gases is increased, so is the cost – up to $1,200 per tonne removed. The cost of reducing sulphur dioxide by 12 million tonnes by the year 2000 – almost 50 per cent of 1980 levels – would be $4 to $7 billion per year, representing a cost of $500 per tonne. For nitrogen oxides⟲ the costs of cleaning up exhausts from vehicles may be considerably more, reaching as high as $10,000 per tonne of pollutant removed.

Costs of cleaning up after environmental disasters connected with the nuclear industry are even higher. The costs arising from the accident at the Three Mile Island⟲ nuclear power plant in the United States exceed $2 billion, a sum that will be dwarfed when the full cost of the Chernobyl⟲ disaster in the Soviet Union, already over $340 billion, is calculated. Many experts forecast that in future there will be an increase in the incidence of environmental disasters, despite the availability of economic and practical methods to reduce them.

In general, pollution damage in developed countries is estimated to cost 3.5 per cent of Gross National Product (GNP)⟲. The cost of air pollution damage in the United States alone ranges between $2 and $35 billion each year. By the early 1980s, member nations of the Organization for Economic Co-operation and Development (OECD)⟲ were spending an estimated $75 billion annually on pollution control and technology to protect or clean up the environment. In the developed countries as a whole, the cost of environmental policies has been estimated at between 1 and 2 per cent of GNP – well below the cost of damage – with most funds being allocated to pollution abatement and natural resource protection. A report in 1990 predicted that the costs of cleaning up Western Europe could amount to around $200 billion annually by the year 2005, cutting output by 2.5 per cent.

In Eastern European nations such as Hungary, investment in environmental protection amounts to considerably less than 1 per cent of GNP – far below the cost of damage, estimated to be in the 2–5% range. Rectifying the major pollution problems in Poland alone will cost an estimated $20 billion.

Spending in the Third World⟲ is far lower, and is usually directed towards drinking water⟲ and sanitation improvement projects. Yet despite the growing awareness, especially within the industrialized world, of the need for environmental protection, public and private spending on environmental protection in the ten major Western industrial countries between 1980 and 1988 did not rise in line with economic growth. Over the period, France, Japan, Sweden and the United Kingdom permitted their spending to stagnate or drop, while Canada, Denmark, the Federal Republic of Germany, Italy, the Netherlands and the United States increased their outlay by between 25 and 35 per cent. In reality, several OECD countries allowed spending on measures to improve or protect the environment to fall from well over 1 per cent of GNP to well under. Overall, spending has fallen sharply as defined as a proportion of the GNP, the $75 billion spent representing 0.7 per cent of the collective GNPs.

COAL   A carbon-containing mineral deposit, widely used as a fuel and a raw material for the plastics and chemical industries. Coal is formed through the prolonged compaction and heating of partially decomposed fossil vegetable matter, a process which takes thousands of years. There are several types of coal, classified according to carbon content.

In its simplest form, peat⟲, a lot of water⟲ and volatile compounds are bound together and the carbon content is relatively low. Lignite (brown coal) contains more carbon, followed by 'hard' coal and anthracite, which has the highest carbon content. When burned, one tonne of hard coal produces the same energy as three tonnes of lignite, these two being the most commonly used forms. Coal began to be widely used as a fuel during the Industrial Revolution in about 1800. By the 1920s it accounted for 80 per cent of the world's total commercial energy⟲ consumption. Natural gas⟲ and petrochemicals gradually replaced coal as a preferred energy source until the major rise in oil⟲ prices in the 1970s. Coal still accounts for around 30 per cent of commercial energy production and its use increased in both 1988 and 1989 by almost 4 per cent. Owing to increasing demand, world coal trade, which expanded by 40 per cent during the 1980s, is expected to increase by another 40 per cent during the 1990s. Worldwide, over 4 billion tonnes of coal are mined each year, China (950 million tonnes), the United States (830 million tonnes) and the Soviet Union (687 million tonnes) are the major producer and user countries.

Annually, over 3 billion tonnes of coal are burnt around the world. Known reserves of bituminous/ anthracite coal should last for 193 years, while supplies of sub-bituminous coal/lignite are calculated to be sufficient to meet current demand for another 370 years. As a measure of coal's abundance, known reserves in the United States alone are thought to be 43 per cent greater than the world's total oil and natural gas reserves combined.

The burning of coal produces considerable amounts of air pollution⬦ through the release of soot and other particulate matter and the production of carbon dioxide⬦ and other gases which add significantly to the greenhouse effect⬦, coal-fired power stations worldwide producing around 10 per cent of all the greenhouse gases⬦. Concern has also been growing about emissions of sulphur dioxide⬦ and nitrogen oxides⬦ which are emitted when coal is burnt, as both are involved in the production of acid rain⬦. Emissions of both can be reduced by the use of scrubbers and by fluid-bed technologies to ensure more complete combustion, but due to the cost of implementation these techniques are not being applied to many coal-fired power stations.

As a reaction to the growing environmental concerns about sulphur dioxide and carbon dioxide emissions, alternatives to coal are actively being investigated. Oil is now too expensive. Adoption of Combined Heat and Power (CHP)⬦ systems would greatly reduce the amount of coal needed to generate a given amount of energy, and utilizing waste to generate heat and power would also help. Replacing the world's coal-fired power stations with nuclear ones would create huge and virtually unmanageable stockpiles of radioactive waste, and the projected $151 billion costs of the building programme would be prohibitive. The use of renewable energy sources such as solar power⬦, wind power⬦ and wave power⬦ may prove to be the best and most ecologically sound alternative.

COASTAL DEGRADATION   The world's oceans help to determine the climate and are a vital source of food for half the population of the developing world, as well as supplying a large part of the diet of developed countries. Over 90 per cent of the living harvest of the sea is from coastal waters, yet most human waste and industrial effluent ends up in these very waters and remains trapped in water⬦ and sediment close to shore – poisoning all forms of marine life. This leads not only to a massive loss of food production but also to a loss of two of the major buffers preventing coastal erosion.

The destruction of coral⬦ reefs – as rich a source of wildlife species as tropical forests⬦, home to one-third of the world's variety of fish⬦ and a major barrier to tidal erosion throughout the tropics – is increasing rapidly as a result of both pollution⬦ and overexploitation.

Vast tracts of mangrove trees, which have a similar coastal protection function throughout the tropics, are similarly being lost through human activities. Massive swamp-draining schemes and felling of mangroves⬦ is occurring throughout the tropics, notably in India, Madagascar, Sabah, Sarawak, and Venezuela. In Asia the trees are used mainly for export to Japan as wood chips and in Madagascar for fuelwood⬦, tannin and lumber.

COBALT   Cobalt is produced as a by-product during the processing of copper⬦, nickel and silver ores, in which it appears in low concentrations. It is a hard, magnetic metal used in the production of magnets and cutting tools. Alloyed with other metals it provides resistance to corrosion and oxidation and is particularly important in the manufacture of gas turbines and jet engines. Major producers are Zaïre (12,000 tonnes), Australia (3,000 tonnes) and New Caledonia (2,500 tonnes).

COCAINE   The leaves of the coca bush (*Erythroxylum coca*), which commonly grows in the Andes Mountains in South America, have traditionally been chewed by local inhabitants for centuries. The leaves are rich in nutrients such as calcium, phosphorus and vitamins A and B_2. They also dull the appetite and act as a stimulant, enabling people to work for long hours on small meals. Coca is also used in hospitals as a local anaesthetic, and in some areas it is the only anaesthetic available. The leaves can also be processed into a white powder, cocaine hydrochloride, which is much more potent, and it is this which is traded to meet the global demand for the drug. Cocaine stimulates the nervous system, producing a feeling of euphoria which is very short-lived, but the drug itself is not physically addictive. Use, possession and trade of cocaine are illegal in most nations, but the cocaine trade offers huge rewards and is growing despite internationally co-ordinated efforts to prevent it.

The United Nations Fund for Drug Abuse Control (UNFDAC)⬦ is co-operating with governments in the Latin American countries on programmes to prevent production of the drug. Millions of dollars are being spent in campaigns to persuade farmers to grow alternative crops such as coffee⬦, fruit and vegetables. Governments in the donor nations, where demand for the drug is fuelling the cocaine trade, are also trying to curtail production but are having little success in curbing demand.

The economies of indebted Latin American nations indirectly benefit from the activities of those involved in the cocaine trade, although only a relatively few corrupt individuals make the greatest gains. Certain members of Colombian society are making an estimated $3 billion annually from illegal drug activities and drugs have become the nation's most profitable export, worth twice as much as coffee. At least $2.5 billion of illegally earned income is flowing into Bolivia, almost four times the total value of all the country's legal export earnings. In Peru too, a $600 million income from illegal drug exports makes it the nation's biggest export earner; the coca leaf is the largest cultivated crop in the Peruvian Amazon.

Wherever it is now grown for illicit trade, coca causes severe pollution⬦, deforestation⬦ and soil erosion⬦. Paraffin, sulphuric acid, acetone, toluene, lime and carbide are all used in the process which produces cocaine hydrochloride. All are dumped or discharged into watercourses by those engaged in the illegal production of powdered cocaine, who are also responsible for cutting down and clearing large sections of forests⬦ for coca cultivation. Moreover, governments and local authorities are responsible for subjecting vast tracts of land and local inhabitants to aerial spraying with defoliants and other hazardous chemicals. In 1989 a caterpillar, *Eloria noyesi*, was found that feeds exclusively on the leaves of the coca plant, and the US government is examining plans to use the insect in biological-control⬦ programmes.

COCKROACH   Cockroaches, insects of the order Blattaria, are one of the most primitive of all insects. There are several species, most of which live in tropical forests⬦, although many are now considerable pests in both temperate and tropical climes. They invade and thrive in virtually all human habitats. With a predilection for dark, humid places, they are most often found in kitchens, bakeries, breweries and hospitals. They avoid light, generally hiding by day and venturing forth at night to feed. They eat everything that humans do, as well as virtually anything else derived from natural vegetative or animal-based material. Cockroaches manage to spoil more food than they eat. They regurgitate digestive juices on to potential food and feed readily on human

faeces or soiled hospital waste, crawling over everything and fouling it with their droppings and unpleasant odour. Cockroaches are a major health hazard and are known to carry more than thirty types of harmful bacteria, viruses, fungal pathogens and protozoan parasites, including the organisms that cause polio⬦, typhoid, leprosy⬦, infectious jaundice and salmenollosis.

If conditions are favourable, cockroaches can breed prolifically all year round, and if all the offspring of one pair of insects survived and bred, the total number of insects would reach 20,000 in one year. In the United States, cockroaches are regarded as the nation's greatest pest. In the state of Georgia alone, estimated losses due to cockroaches are over $16 million annually. At least 17 million Americans, or 7 per cent of the total population, are allergic to either cockroach excretions or material that has been shed when the insects moult. In the battle against the roaches, US homeowners spend $340 million a year on control sprays. Official expenditure is even greater with the authorities in New York City alone allocating $2.6 million annually for cockroach control; the insects are a common pest in low-income and substandard housing. Worldwide, billions of dollars have been spent on cockroach control – mostly to no avail, as the insects breed so quickly and rapidly develop resistance to the majority of pesticides⬦ used against them. Cockroach eggs are also impervious to all commonly used chemical pesticides.

COCOA   Cocoa is obtained from the bean-shaped seeds of a tree, *Theobroam cacao*, which grows in warm, humid lowland areas throughout the tropics. The crop requires a high rainfall – over 1,200 mm annually – distributed evenly throughout the year, with no dry season. Commercial production also needs average temperatures of 30°C. Cocoa is a very suitable crop for small farmers and can be grown under an umbrella crop such as coconut⬦. Seedlings must be raised under a shade crop, as intense sunlight damages the leaves. Trees begin to bear fruit three to five years after planting, harvesting taking place at two-to-three-week intervals. Harvested pods are split open and the beans are then fermented and dried until the two halves separate easily; this process takes seven to ten days.

Cocoa has become one of the world's major primary commodities⬦ and is an important cash crop⬦, particularly in West Africa, earning valuable foreign currency. Cocoa has a high food value, containing 10 per cent protein and over 50 per cent fat (cocoa butter). It is used commercially to produce chocolate, cocoa drinks and as a flavouring. World production is around 2.2 million tonnes; the Ivory Coast (720,000 tonnes), Brazil (347,000 tonnes) and Ghana (255,000 tonnes) are the leading producers.

COCONUT The coconut-palm tree, *Cocos nucifera*, is distributed throughout the wet tropical lowlands, where it is grown at low elevations since it does not thrive at high altitude. The fruit of the coconut palm is one of the most important of tropical crops. Coconuts have a thick, fibrous husk surrounding a single-seeded nut. The hollow core contains coconut milk. The white kernel can be eaten raw or dried to produce copra, from which coconut oil is extracted. Coconuts are used primarily for oil production; the dried flesh (copra) is the richest source of vegetable oil, with an oil yield of 64 per cent. Coconut oil is used for making soaps, creams, synthetic rubbers and confectionery, and for cooking. The residual coconut cake is used as a livestock◇ feed, and the coarse husk fibre (known as coir) is used to make matting, floor coverings, brushes, furniture and water-resistant ropes.

Two varieties of coconut palm, tall and dwarf, are cultivated commercially. Tall trees, reaching a height of 30 m, take longer to mature and do not begin to yield until six or seven years after planting, although the copra from tall varieties is of a better quality than copra from dwarf trees. Nuts usually require twelve months to reach maturity. Harvesting takes place about six to eight times a year and only mature nuts are harvested, as immature nuts have low-quality copra. Harvesting is done by skilled climbers, although trained monkeys are sometimes used. Knives attached to long bamboo poles may also be used. The average annual yield is about 50 nuts per tree, which will produce around 10 kg of copra.

Freshly harvested nuts are husked and split into two halves, which are then dried in the sun or in kilns. The meat or kernel becomes detached from the shell after two to three days. The copra is collected and stored. In addition to their commercial uses, the shells of coconuts are used as a fuel in drying kilns. Toddy, the sap of the palm obtained from the inflorescence before the flower matures is used to make wine, and the trunks of older trees may be cut down and used for house construction or boat-building.

CODE OF ENVIRONMENTAL PRACTICE
A code assembled by delegates attending the sixth Economic Summit Nations Conference on Bioethics, held in 1989. No common set of rules and regulations governing the environment and development practices currently exists, and the Code has been submitted to the G7◇ Heads of State – the leaders of the nations which normally attend the annual Economic Summit meeting – for their approval.

The Code aims to promote 'stewardship of the living and non-living systems of the earth in order to maintain their sustainability for present and future generations, allowing development with equity'. It contains both guidelines and obligations for those nations who choose to adopt it. The Code has been drawn up using principles embodied in earlier initiatives to protect the environment and conserve natural resources◇, such as the World Conservation Strategy◇ (1980, revised in 1991) and the report of the World Commission on Environment and Development (WCED)◇ published in 1987.

CODEX ALIMENTARIUS COMMISSION
An intergovernmental body established jointly in 1962 by the United Nations◇ Food and Agriculture Organization (FAO)◇ and the World Health Organization (WHO)◇ with the purpose of ensuring fair practices in international trade and protecting the health of consumers by setting guidelines for the control of food quality and safety. The Commission has developed over 200 individual commodity standards, 35 codes of hygienic and technological practice, and approximately 2,000 Maximum Residue Limits concerning the safe level of pesticides◇ in food. Codex standards for raw and processed food commodities◇ moving in international trade cover many aspects, including uniform labelling requirements, use of additives, the presence of contaminants, sanitary requirements, and composition and analysis of any foodstuff. Although the Commission is an intergovernmental body, representatives of the food industry, consumers' associations, and those from the marketing and distribution sectors participate in meetings. An international conference was held in 1991 to promote and co-ordinate widespread international agreement on food standards, chemicals in food, and quality and safety procedures for food exports and imports.

COFFEE Coffee is one of the most important export crops for many developing countries◇. *Coffea* is a genus of shrubs originating in the Old World tropics, especially Africa. Two varieties are now extensively grown for commercial purposes, *Coffea arabica* and *C. robusta*. The arabica variety produces a mild, much-sought-after coffee, whereas robusta, although easier to grow, produces coffee of a lower quality. The plants need average temperatures of 20–28°C and rainfall of around 1,700 mm, so they grow best in equatorial regions. Coffee must be picked when it is ripe and the surface of the fruit is red; harvesting is undertaken every twelve days or so during the season. Once the pulp has been removed from the fruit, the coffee beans are dried and stored.

Coffee was 'discovered' by Europeans, who began to grow it throughout distant tropical colonies. It was introduced into Ceylon (Sri Lanka) only to be decimated by a disease, coffee rust, as a result of which the land⊝ was given over to growing tea⊝ as a substitute crop. Coffee was introduced into Brazil only in the 1870s but quickly became the nation's leading export. The country rapidly became the world's leading producer and exporter.

Coffee is the most important internationally-traded commodity after oil⊝, with approximately 75 per cent of the world's crop entering into international trade. In 1978, close to the high point for coffee prices, Brazil produced almost one-quarter of the world's coffee. A decade later, trade was still worth $8.5 billion and over fifty developing countries were growing coffee beans for export. The United States, importing well over one million tonnes annually – accounting for one third of the global crop – is the world's largest coffee consumer, followed by the Federal Republic of Germany.

Annual production of coffee is in the order of 5.6 million tonnes, with Brazil (1.3 million tonnes) maintaining its position as the leading producer nation followed by Colombia (780,000 tonnes) and Indonesia (350,000 tonnes). Over the years, producers and consumers have made several efforts to safeguard and stabilize coffee trade. The most comprehensive, the International Coffee Agreement (ICA), failed in 1989. The ICA, which specified quotas that each country was allowed to export, collapsed mainly because the USA wanted the pact's quotas changed so that more top-quality arabica coffee could enter the world market. Brazil, which held a quota amounting to 30 per cent of global exports and produced the cheaper robusta coffee, would not agree to this change.

COLD CHAIN SUPPORT   Cold Chain Support was the name given by the World Health Organization (WHO)⊝ in the mid-1970s to a comprehensive programme designed to ensure the widespread availability of viable vaccines. The programme encompassed the provision of cooling equipment, teams of workers trained to order, distribute, dispense and administer the vaccines, and development of the necessary technology for packing and processing. To maintain quality and viability, vaccines must be refrigerated from the time of production: through storage, processing, transport and distribution. Refrigeration units had to be designed that could protect vaccines during power failures, a common occurrence in rural areas in the Third World⊝. Equipment was developed in which the cooling energy is stored in the icepack bank, maintaining vaccine storage temperatures even when the unit receives no power. Chemical indicators accompany vaccines to show that they have been kept below critical temperatures at all times. As equipment needs to be checked thoroughly twice a day, thermometers for this and a detailed management plan for each unit are included, together with a comprehensive training programme.

COLOMBO PLAN   Signed in Colombo (Ceylon/Sri Lanka) in 1951, the Plan was designed to foster co-operative economic development in the countries of South and South-East Asia. Conceived as a result of a meeting of Commonwealth⊝ Foreign Ministers the previous year, it sought to improve living standards of the people in the region by reviewing development plans and co-ordinating all manner of development assistance. There is an annual meeting of a Consultative Committee, and financial arrangements are negotiated and administered bilaterally rather than through a central fund. Membership comprises twenty-one developing countries⊝ in the region, including Afghanistan, Bangladesh, Bhutan, Burma, Fiji, India, Indonesia, Iran, Kampuchea, Republic of Korea, Laos, Malaysia, Maldives, Nepal, Pakistan, Papua New Guinea, the Philippines, Singapore, Sri Lanka and Thailand. There are six non-regional members: Australia, Britain, Canada, Japan, New Zealand and the United States.

COMBINED HEAT AND POWER (CHP)   CHP stations generate electricity less efficiently than most conventional power stations, but because the heat they produce is also exploited, efficiency may be as high as 70–80 per cent – at least twice the level found in conventional power stations that generate electricity only. In Britain alone, if the energy⊝

wasted in the form of heat discharged to the atmosphere as a by-product of power generation could be harnessed, through currently available CHP technology, it could provide 30 per cent of the nation's space-heating and hot water needs and save the equivalent of 30 million tonnes of coal⊙ annually. Large CHP district-heating systems are also capable of burning virtually any type of fuel, including municipal refuse. CHP plants are well established in many of the industrialized countries – major district-heating schemes operate in Austria, Denmark, Finland, France, Hungary, Italy, the Netherlands, Norway, Poland, the Soviet Union, Sweden, the United Kingdom and the United States.

COMECON *See* Council for Mutual Economic Assistance.

COMMITTEE ON FOOD AID POLICIES AND PROGRAMMES (CFA) A body instituted jointly by the United Nations⊙ General Assembly and the Food and Agriculture Organization (FAO)⊙, following a recommendation from the World Food Conference⊙ to re-establish an intergovernmental body to review and orchestrate the operations of the World Food Programme⊙. In addition to its functions as the governing body of the World Food Programme, the CFA is mandated to help evolve short- and long-term food aid⊙ policies.

COMMITTEE FOR INDUSTRIAL CO-OPERATION (CIC) Founded as a joint European Community⊙/ACP⊙-group body under the Lomé Convention⊙, the CIC, located in Brussels, is responsible for examining problems in the field of industrial co-operation. Grievances can be submitted by the ACP countries and/or by the Community, and suggestions for appropriate solutions and co-operative ventures and development projects may also be proposed. The CIC also oversees the work of the Centre for Industrial Development⊙.

COMMITTEE OF INTERNATIONAL DEVELOPMENT INSTITUTIONS ON THE ENVIRONMENT (CIDIE) An organization of financial institutions set up in 1981 to examine the environmental aspects of all forms of economic development and ensure that all member states of the United Nations⊙ take environmental considerations into account when formulating their economic development policies. Membership is drawn

from a number of the major multilateral donor organizations – the African Development Bank⊙, Arab Bank for Economic Development in Africa⊙, Asian Development Bank⊙, Caribbean Development Bank⊙, Commission of the European Community, European Investment Bank⊙, Inter-American Development Bank⊙, Organization of American States⊙, United Nations Development Programme (UNDP)⊙, United Nations Environment Programme (UNEP)⊙ – which provides the Secretariat – and the World Bank⊙.

COMMODITIES Essentially, a commodity is anything that is produced for sale. Commodities may be *consumer goods*, such as radios, or *producer goods*, such as copper⊙ bars or cereals. However, the term is commonly used to refer to the primary products most widely traded internationally. These include food and agricultural crops, such as bananas⊙, cassava⊙, cocoa⊙, coffee⊙, maize⊙, wheat⊙ and other cereals, sugar⊙, and tea⊙; fibre crops such as cotton⊙, sisal, wool, and jute⊙; base metals, including gold⊙, silver, copper⊙, tin⊙, lead⊙ and aluminium⊙; oilseeds, such as soya, olive and groundnuts⊙; fuels, basically coal⊙, oil⊙ and natural gas⊙; and certain cash crops⊙ such as tobacco⊙ and rubber⊙.

In real terms, the price of most commodities exhibited a downward trend throughout the latter half of the 1980s. Taking the year 1980 as 100, the World Bank⊙ index of 33 non-fuel export commodities averaged only 70.1 in 1989, emphasizing how badly many developing nations had fared during the 1980s.

Third World⊙ commodity exporters were hundreds of millions of dollars worse off during the decade because of low prices. Protectionist measures implemented by the United States, the European Community⊙ and other importing countries exacerbated the situation. The decade started with high hopes that producer and consumer co-operation would lead to policies which would stabilize the global prices for most commodities and reduce fluctuation. Instead, most of the old treaties governing trade in commodities had collapsed by 1990, causing prices to fall with them. Pacts on coffee, cocoa, wheat, tin and sugar all failed, and by 1990 only one, covering natural rubber, was still operating.

In 1990, the wholesale prices of tea, most vegetable oils, rubber, grains and base metals fell, some reaching their lowest point since the early 1980s. Price increases during the 1990s for most of the major commodities are, at best, expected to be slow and gradual, keeping below the average global inflation rate, with the World Bank projecting small annual falls over the early part of the 1990s for its index of non-fuel export commodities, after adjusting for inflation. This means that developing countries⟡, many of which depend heavily on only one or two commodities for their export earnings, could become even worse off. Fourteen countries in Asia, nineteen in Latin America and thirty-seven in Africa rely heavily on a single commodity for at least 20 per cent of their export earnings.

COMMON AGRICULTURAL POLICY (CAP)

A far-reaching policy, adopted by members of the European Community⟡, to safeguard agricultural production and farm prices in EC member nations. Throughout the 1970s and 1980s the CAP paid out massive incentives to encourage the production of surplus food. These were first stored in expensive and notorious 'food mountains', which were then thrown away, fed to livestock⟡ or dumped on the world market, depressing global prices for agricultural produce in the process. The subsidies also led to smaller areas of land⟡ being farmed more intensively, often to the point of exhausting the soil. The bulk of the huge CAP budget, some $26 billion in 1991, is still being spent on subsidies and incentives to keep farm prices high. Despite favourable economic conditions around the world, it still cost the EC taxpayer close to $9 billion in 1989 to maintain the international competitiveness of the Community's overpriced agricultural produce.

Despite limited improvements during the late 1980s, climaxing in a major Brussels Summit agreement in 1988, the CAP programmes continue to inflict widespread environmental damage throughout Europe, even though the overall policy now includes incentives to curtail agricultural production. The CAP also funds Environmentally Sensitive Areas (ESA) in which farmers are paid for farming in ways that benefit the environment and its wildlife – some 2.3 million hectares of ESAs had been designated by 1990.

In 1990, the Worldwide Fund for Nature (WWF)⟡ reported that even the semi-natural habitat of farmlands that now dominate from 40 per cent to 80 per cent of land in EC countries is under continuing threat from the CAP. The Council for the Protection of Rural England (CPRE) concluded that in Britain over 90 per cent of agricultural spending promoted by the CAP is potentally damaging to the environment. The intensification of farming promoted by the CAP is thought to be a crucial factor in causing many of Europe's amphibians, fish⟡ and reptiles to join the list of threatened species. In the Federal Republic of Germany, 173 species of plants are declining because of land drainage, 89 owing to the use of herbicides, and 56 because of excess nutrients in surface water⟡ arising from overuse of fertilizer⟡. Grassland habitats throughout Europe are now being threatened, as the CAP promotes deliberate afforestation. Up to 5 million hectares of marginal land may be afforested if current CAP restraints on agricultural production continue. Wetlands⟡ also continue to be under threat: both the Netherlands and Germany lost half of their wetlands between 1950 and 1985.

Future prospects are brighter, as by 1992 some 16 million hectares of marginal agricultural land may have gone out of production as a result of farmers becoming uncompetitive. A quarter or more of CAP funds are currently directed to assist poorer farmers working 'environmentally sensitive' land, and other measures are being taken to reverse the destructive policies encompassed within the CAP. In addition, the deadlock during the final stages of the Uruguay Round of the General Agreement on Tariffs and Trade (GATT)⟡ at the end of 1990 was primarily caused by EC governments' refusal to reduce farming subsidies, paid under the CAP, by as much as the United States wanted. The threatened collapse of the talks, brought about by the failure of the USA and the EC to reach a negotiated settlement over CAP subsidies, and the prospect of the trade wars that would follow such a collapse, stimulated the EC to undertake a comprehensive review of the CAP and its policies.

COMMON FUND

At the fourth United Nations Conference on Trade and Development (UNCTAD)⟡, delegates decided to establish a Common Fund to finance measures introduced to maintain equitable trading in eighteen basic commodities.⟡ Agreement was reached in 1979 on the Fund's basic structure and mode of operation. Through its First Window it will provide support for International Commodity Agreements⟡, lending money to existing organizations to strengthen existing agreements and stabilizing commodity prices through the creation of international buffer stocks. The Fund's Second Window will provide finance for measures to improve commodity production, quality and trading volume, such as research and development programmes and market promotion, concentrating on the provision of support for agricultural commodities which are not suitable for stockpiling.

Following lengthy negotiations lasting fifteen years from initial proposals, by 1990 enough countries had ratified the agreement for the Fund to be launched. However, the starting overall capital of $750 million was well below the $6 billion target suggested by the UN. Consequently, the Fund's impact will be limited. Contributions will mostly come from UN member governments, partly assessed on a country's proportion of world trade and part voluntary contribution.

COMMONWEALTH
A loose association of forty-nine independent nations which were once subject to the imperial government of the United Kingdom. The Commonwealth of Nations was established in 1931, and the name was modified after the Second World War. The total population⟳ of its member states comprise nearly 25 per cent of the world's population. Full members include Antigua and Barbuda, Australia, the Bahamas, Bangladesh, Barbados, Belize, Botswana, Canada, Cyprus, Dominica, Fiji, The Gambia, Ghana, Grenada, Guyana, India, Jamaica, Kenya, Kiribati, Lesotho, Malawi, Malaysia, Malta, Mauritius, New Zealand, Nigeria, Papua New Guinea, St Kitts-Nevis, St Lucia, Seychelles, Sierra Leone, Singapore, Solomon Islands, Sri Lanka, Swaziland, Tanzania, Trinidad and Tobago, Tonga, Uganda, Vanuatu, Western Samoa, Zambia, Zimbabwe and the United Kingdom. In addition, Maldives, Nauru, Tuvalu and St Vincent and the Grenadines are special members not usually represented at the meetings of heads of government, the executive body.

Commonwealth government leaders meet every two years but Finance Ministers meet annually. The organization is serviced by a permanent secretariat based in London. Throughout its existence, the Commonwealth has seen only three withdrawals. Ireland left in 1949, South Africa withdrew in 1961 and Pakistan in 1971.

COMMONWEALTH AGRICULTURAL BUREAU INTERNATIONAL (CABI)
The CABI is a co-operative organization owned, administered and financed by member countries of the Commonwealth⟳, established in 1929 under the name Imperial Agricultural Bureaux. Based in the United Kingdom, the CABI controls a series of international academic and research centres concerned with all aspects of agriculture⟳. It acts as a clearing house for the collection, collation and dissemination of information, aiming to provide a global information service for agricultural scientists and other professionals working in agriculture or closely related fields. This covers pest and disease identification as well as biological control⟳. The CABI has four institutes and ten specialized bureaux under its umbrella; the CABI itself is governed by an executive council on which twenty-four Commonwealth countries and several dependent territories are represented, all of which contribute to the organization's funding.

COMMONWEALTH FOUNDATION
An intergovernmental agency originally established in 1966 following a decision by the Commonwealth⟳ heads of government to promote closer professional integration and working within the organization. The charitable status of the initial body was changed in 1983 when the Foundation was reconstituted as an international institution. Although the objectives of autonomous character, organizational arrangements and its original mandate were maintained, the Foundation's scope was widened to include co-operation with Non-Governmental Organizations (NGOs)⟳.

The Foundation provides grants for attending conferences, short-term study visits and training attachments within Commonwealth countries. It also provides financial support to professional associations and professional centres within the Commonwealth. Financial support is also available for short-term fellowship schemes as well as for visits by senior professional practitioners in selected areas of specialization.

COMMONWEALTH FUND FOR TECHNICAL CO-OPERATION (CFTC)
A voluntary scheme to promote economic development, subscribed to by all Commonwealth⟳ governments and administered by the Commonwealth Secretariat⟳. Its purpose is to provide technical co-operation to meet the priority needs of the developing countries⟳ within the organization. The CFTC operates through three main programmes: the Technical Co-operation programme, the Export Market Development programme, and the Education and Training programme. A feature of the Fund is its regular use of the expertise and instructional facilities and capabilities found in the developing countries themselves, and the pioneering use of innovative educational tools such as distance learning.

COMMONWEALTH SCIENCE COUNCIL
(CSC) Established in 1946, the CSC aims to promote collaborative scientific research and development involving member countries of the Commonwealth⌀, and to increase individual nations' capabilities to use all opportunities provided by science and technology to promote their economic, social and environmental development. Membership is open to all Commonwealth nations.

COMMONWEALTH SECRETARIAT
Established in 1965 by the Heads of Government Conference and based in London. It serves the Commonwealth⌀ collectively, providing the central administration for joint consultation and co-operation in many fields. It services the Commonwealth Prime Ministers' meetings, and where appropriate other offical meetings. The Secretariat collects and disseminates information for use by all Commonwealth countries and co-ordinates technical co-operation for economic and social development through various Commonwealth organizations and institutions.

COMPENSATORY FINANCING FACILITY
A customized International Monetary Fund (IMF)⌀ facility through which nations can obtain compensatory cover to overcome a temporary shortfall in export receipts. It is specifically designed for situations where the deficit has arisen due to circumstances largely beyond the control of the country involved, notably natural disasters or climatic disturbances. It can also be used to cover the costs of temporary excesses in cereal or other food imports – again brought about as a result of circumstances beyond a country's control – such as crop losses caused by drought⌀, floods⌀ or hurricanes.

COMPEX *See* System of Compensation for Export Receipts

CONABLE, BARBER S.(1922–)
Before his appointment as the seventh president of the World Bank⌀ in 1986, Barber Conable had served for twenty years in the United States House of Representatives. From 1965 to 1985 he was a member of the Joint Economic Committee, the House Budget and Ethics Committee, served as chairman of the Republican Policy and Research Committee, and rose to become Ranking Minority Member of the House Ways and Means Committee. On his retirement from Congress he served on the boards of several transnational⌀ corporations as well as the board of the New York Stock Exchange before being elected Distinguished Teaching Professor at the University of Rochester. One of the most striking events of his tenure as World Bank president was the change of the Bank's thinking that resulted in environmental matters becoming of apparently greater importance in influencing its activities. This became manifest in the decision to allocate $400 million annually for environmental improvement projects in middle-income nations⌀ and the commitment to a major global facility to help protect the global environment.

However, as a proponent of orthodox economic theory, Conable maintained the Bank's policy of reducing poverty around the globe through growth stimulation, support for free trade and encouragement of profit-driven investment. There appeared to be only minimal movement in the Bank's activities to support Conable's pro-environment rhetoric and pronouncements. By 1990 many saw that the Bank's newly formed Environment Department was of marginal relevance, having precious little input into the Bank's operating systems where projects are prepared and loan portfolios built up. In a major internal row, members of the environmental staff were reported to have circulated papers arguing for more attention to be paid to environmental matters and proposing that the bank should not support any projects that involved logging in tropical forests⌀. Emphasizing their disquiet, they further suggested that no more forestry loans should be granted until a clear forestry policy had been determined. In late 1990 the Environment Department underwent a radical restructuring, carried out by senior members of the Bank's staff.

CONFERENCE ON INTERNATIONAL ECONOMIC CO-OPERATION (CIEC)
A wide-ranging conference, including countries from both North⌀ and South⌀, which was held in Paris between 1975 and 1977. The twenty-seven participating nations from the South included seven from the Organization of Petroleum Exporting Countries (OPEC)⌀. The North was represented by eight developed market economies and the European Community⌀. Delegates held discussions on all matters of development, economics, trade, energy⌀ and finance, but the Conference ended with no positive results or agreements.

CONFERENCE ON SECURITY AND CO-OPERATION IN EUROPE (CSCE)
In 1975, at

the thirty-five-nation CSCE, all the European governments – with the exception of Albania – signed the Helsinki Declaration⋄. This incorporated pledges to promote freedom of contract and movement between their countries, including exchanges between East and West. The Soviet bloc's repeated failure to keep these promises prevented Europe from achieving the security and co-operation that were the CSCE's goals. In 1989, when the Communist governments in Eastern Europe began to be replaced, the CSCE started to take on an increasingly important role.

At a CSCE meeting in Paris in 1990, it was proposed that the Conference should become an instrument of collective security, eventually to replace both the North Atlantic Treaty Organization (NATO)⋄ and the Warsaw Pact⋄. Leaders of thirty-four Western and Eastern European nations, plus the United States and Canada, met to decide a charter for a New Europe. The 1990 CSCE Summit was held partly in recognition of the end of the Cold War, partly as a celebration of the unification of Germany, and partly to discuss the security implications for the changing face of Europe. The countries attending attempted to adapt the 1975 Helsinki Declaration to cope with the new developments in Europe and the economic and political ramifications. Even Albania, which had previously had nothing to do with the CSCE, attended as an observer. The proposed Charter for a New Europe embodied commitments to individual liberties, the rule of law and a property-owning pluralistic democracy.

Several fundamental agreements were reached, including a decision to 'institutionalize' the CSCE by establishing a small secretariat. In future, Summits are to be held every two years, with Foreign Ministers meeting on a more frequent basis. In addition, an 'Observation Unit' will be established to ensure free and fair elections. The Conference agreed that in future 'Emergency' meetings could be convened, but only by consensus. Decisions on other proposals, such as a 'Conflict Prevention Centre' and the formation of a parliamentary body – an Assembly of Europe comprising members of national parliaments – were not finalized.

CONSERVATION 'Conservation' describes the process through which natural resources⋄ are managed to allow partial or total exploitation, for individual, community or commercial use, without in any way jeopardizing the long-term viability of the resource base or inflicting undue or excessive environmental damage. It is held to encompass full consideration of the varying requirements of the local human population⋄, together with those of the wildlife species or habitat to be conserved, including an appreciation of the ability of each to adapt to any changes. It is distinguished from 'preservation', which is considered to be the maintaining of the pristine state of nature as it is or might have been before the intervention of either anthropogenic or natural forces.

CONSULTATIVE GROUP ON FOOD PRODUCTION AND INVESTMENT IN DEVELOPING COUNTRIES (CGFPI) An international group established following a recommendation by the 1974 World Food Conference⋄. The CGFPI is sponsored and staffed by the World Bank⋄, the United Nations Development Programme (UNDP)⋄ and the Food and Agriculture Organization (FAO)⋄. Its purpose is to increase, coordinate and improve the efficiency of technical and financial co-operation for boosting agricultural production in developing countries⋄. It is in reality more than a consultative group in that it can pledge aid⋄ and assistance as it sees fit.

CONSULTATIVE GROUP ON INTERNATIONAL AGRICULTURAL RESEARCH (CGIAR) A highly specialized body devoted to guiding research, on an international scale, concerning a wide range of topics associated with agriculture⋄ and farming systems⋄. Founded in 1971, it is an informal association of governments, international and regional organizations, and private foundations. Government members are mostly the donor countries (about forty) and ten developing countries⋄ elected by the regional caucuses of the Food and Agriculture Organization (FAO)⋄.

There are thirteen international agricultural research centres in the CGIAR system. Their function is to act as 'centres of excellence', their research being mainly focused on specific food crops, the development of appropriate farming systems, or investigation of ways and means to improve livestock⋄ production. The centres essentially work on a regional or global basis, although they have close relations with the needs and environmental conditions prevailing in the individual nations in which they are sited.

The centres are:

International Maize and Wheat Improvement Centre (CIMMYT) – The origins of the international agricultural research system lie in the collaborative research efforts undertaken by the Rockefeller Foundation and the government of Mexico, started in 1941 and designed to raise crop productivity. CIMMYT, located in Mexico, originally started operations in 1943 and aims to improve varieties and production of maize⬦, wheat⬦, barley and triticale⬦.

International Rice Research Institute (IRRI) – Established in 1960, based in the Philippines. Mandated to develop improved rice⬦ varieties and rice-growing systems and maintain a germ plasm bank.

International Centre of Tropical Agriculture (CIAT) – Established in 1966, based in Colombia. Research is directed to improving the production of beans, cassava⬦, rice⬦ and beef, with particular emphasis on production systems and conditions in the Western hemisphere.

International Institute of Tropical Agriculture (IITA) – Established in 1967, based in Nigeria. Charged with improving varieties and production of cowpea, yam, cocoyam, sweet potato, cassava,⬦ maize⬦ and beans, on a worldwide basis but paying close attention to the needs and conditions prevailing in Africa.

International Potato Centre (CIP) – Founded in 1971, based in Peru. Mandated to develop improved varieties of potato⬦ grown in high altitude conditions in the South American Andes Mountains as well as improving lower-altitude varieties.

International Crop Research Institute for the Semi-Arid Tropics (ICRISAT) – Established in 1972, based in India. Undertakes a diverse research programme aimed at improving the quality and reliability of food production in semi-arid tropical regions.

International Laboratory for Research on Animal Diseases (ILRAD) – Established in 1974, based in Kenya. Directed to develop preventive and treatment systems to control African trypanosomiasis⬦ (ngana), theileriosis and other livestock diseases, together with means to overcome other factors restricting cattle production in Africa.

International Livestock Centre for Africa (ILCA) – Established in 1974, based in Ethiopia. Charged with conducting research and development to improve livestock production and marketing systems, train livestock specialists, and gather and disseminate data on animal husbandry.

International Centre for Agricultural Research in Dry Areas (ICARDA) – Established in 1977, based in Syria. Mandated to research and develop rainfed agricultural systems in arid and semi-arid regions, particularly under conditions pertaining to northern Africa and western Asia.

International Service for National Agricultural Research (ISNAR) – Established in 1980, based in the Netherlands. An organization, set up under CGIAR auspices in response to requests from developing countries⬦, to provide guidance, information and expertise to help strengthen individual national agricultural research programmes,

International Food Policy Research Institute (IFPRI) – Established in 1974, based in the United States. Formed to identify, examine and evaluate issues arising from governmental and international agency intervention in national, global and regional food problems.

West African Development Association (WADA) – Established in 1971, based in Liberia, originally founded as the West African Rice Development Association (WARDA). The institute's work programme was directed at the promotion of self-sufficiency in rice in western Africa and the development of improved rice varieties suitable for the area's agroclimate and socioeconomic conditions. The research programme came in for a great deal of criticism, eventually resulting in a complete reorganization of the institute, its research emphasis and name.

International Board for Plant Genetic Resources (IBPGR) – Established in 1974, based in Italy. Originally set up to found and co-ordinate the operation of an international network of vegetative germ plasm banks. The IBPGR was housed at the headquarters of the Food and Agriculture Organization (FAO) in Rome, before international disagreements and controversy caused the decision to relocate it to Scandinavia. This decision was later revoked and the Board is likely to remain in Rome, with support from the Italian government and maintaining close links with the FAO.

The CGIAR supports and influences all these individual centres, the bulk of the Group's funds come from wealthy donor governments and institutions. Finance is mainly derived from the World Bank⬦, other development banks, the United Nations Development Programme (UNDP)⬦, Food and Agriculture Organization (FAO) and donor governments, including Canada, Federal Republic of Germany, Sweden, the United Kingdom and the United States. Substantial contributions are also received from private bodies, notably the US-based Rockerfeller and Ford foundations.

CONTADORA GROUP A regional grouping of Central and South American countries formed to promote economic, social and political development and co-operation between member states, and reduce conflicts among Central American states. The Group, named after an island in the Gulf of Panama, consists of Colombia, Mexico, Panama and Venezuela.

CONTOUR FARMING To prevent soil erosion◇ and loss of precious topsoil, farmers can adopt a variety of methods to protect their land◇. If the land is almost level, few protective measures will be required. As the steepness of the land under cultivation increases, so does the need for protection against erosion. Many commonly used, low-cost protective measures make use of the natural contours of the land. There are five basic techniques, although the names of each vary from country to country:

Bench terraces – Terracing has been practised in Asia for centuries, and the use of bench terraces is the oldest and most widespread form of mechanical protection employed by farmers. Building terraces is particularly labour-intensive because it requires sloping land to be converted into a series of large, almost level steps. The vertical retaining walls for each step are usually made of stone or earth covered with clinging vegetation to improve stability. The terrace bed may actually be built sloping slightly forwards or backwards into the hill; this is particularly common in irrigated terraces which are used to grow rice◇. A lip built on the front of the terrace allows it to be flooded as and when required.

Platform or Orchard terraces – Discontinuous forms of terraces which require far less work. Semicircular retaining walls are used to build small platforms on the hillside, which are then planted with single trees. The terraces are constructed along the contour, and the slopes between them must be covered with a soil-binding vegetation.

Diversion drains – Drainage channels used to separate higher, steeper non-arable land from flatter cultivated land. The drains run along the contour and intercept water runoff coming down from higher ground which would otherwise erode the unprotected arable land.

Bunds – Small earthen or stone banks, built along contours to impede the downhill flow of water. On arable land they are best used below diversion drains to channel off water that actually falls on the fields being protected.

Grass waterways – A system of drains and channels running down sloping ground, used to carry away the runoff from diversion drains and bunds. They must be fairly deep and covered in thick, hardy vegetation in order to be able to cope with storm water.

CONVENTION AGAINST ILLICIT TRAFFIC IN NARCOTIC DRUGS AND PSYCHOTROPIC SUBSTANCES Drafted by 106 nations and signed in 1988 by 43 of them. The Convention was officially due to come into force ninety days after being ratified by twenty nations, but the authorities in several participating nations began to adhere to it immediately. Officially, it came into force in November 1990. It covers the seizure and confiscation of profits and other assets obtained through drug crimes. The annual turnover of profits from illegal drug activities is estimated to be $500 billion, making it second in worth only to the world's armaments trade. But the revenue is used for the benefit of only a relatively few corrupt individuals; at the same time, the illegal drugs industry generates violence and corruption, diverts huge amounts of economic resources, disrupts agricultural production and creates public health problems.

The Convention also covers the extradition of criminals and mutual legal assistance in prosecuting offenders. The regulations focus on drug trafficking but also embrace supply and demand, whereas earlier international treaties concentrated primarily on the eradication of production, paying scant attention to traffic and demand. Actions to curb demand will no longer be limited to anti-drug educational campaigns, but will be extended to cover anything that is deemed to be encouraging demand, including films and journals that promote consumption, as well as trade in drug paraphernalia.

CONVENTION ON THE CONSERVATION OF EUROPEAN WILDLIFE AND NATURAL HABITATS (BERNE CONVENTION) A regional convention, signed in Switzerland by thirteen European nations and the European Community◇ and designed to protect the wildlife of Europe. Originally proposed in 1979 by the Council of Europe and in force since 1982, the Convention contains detailed annexes listing endangered species of animals and plants and their natural habitats which are to be strictly protected by all signatory nations.

CONVENTION ON THE CONTROL OF TRANSBOUNDARY MOVEMENTS OF HAZARDOUS WASTE
Adopted in 1989 after discussions involving a hundred countries plus the European Community⌀ at a meeting in Basel. Following eighteen months of negotiation only thirty-five countries signed it, over half of which were developing countries⌀. The Convention attempts to limit and control international transport and disposal of hazardous waste⌀. It does not ban all transboundary movements, allowing shipment of such wastes between signatory states following the 'prior informed consent' principle.⌀ Under the new regulations, all waste exports which a recipient state has not authorized in writing will be forbidden. Waste exports will be allowed only if the country of origin does not have the technical means or suitable sites for disposing of them and the recipient nation has the facilities to dispose of them in an 'environmentally sound and efficient manner'. Where illegal movements are detected, the country of origin will be liable to reclaim the waste.

Significantly, the Convention does not define hazardous waste, but lists more than forty classes of materials which have to be controlled. Although the treaty covers the shipping and disposal of toxic waste – the exact definition of which is also unclear – it does not apply to nuclear waste⌀. Waste intended for recycling is not deemed 'hazardous'.

Industrialization leads to the production of vast amounts of hazardous waste which must be disposed of. It is estimated that the industrialized countries belonging to the Organization for Economic Co-operation and Development (OECD)⌀ produce 300 million tonnes of hazardous waste annually, 10 per cent of which passes across international borders for disposal. Around 20 million tonnes of hazardous waste are exported to the Third World⌀ each year from Europe alone. Officially over 600,000 tonnes of toxic waste – that which is known to be poisonous – are exported annually from OECD nations. Although 80 per cent is reputed to go to other developed nations, 20 per cent is shipped to the Third World for disposal, usually in untreated form.

Over forty developing countries have imposed a total ban on the importation of any wastes and continue to campaign to prohibit outright any trade in waste products. Most African states are in favour of a total ban, and in 1988, after the appearance of

so-called 'leper ships', sixteen West African states agreed to set up a 'dumpwatch' body to monitor dumping, with all declaring that dumping was to become a criminal offence.

Four 'leper ships' surfaced during the late 1980s. These vessels were loaded with hazardous waste which no country would accept. Two ships, the *Karin B* and the *Deep Sea Carrier*, contained unidentified toxic waste which Italian companies had dumped in Nigeria. These contents – together with that of the *Zanoobia*, which had been sailing around the world for over twelve months in search of a resting place for its cargo – were eventually returned to Italy. The dubious record for 'leper ships' is held by the *Khian Sea*, which spent over two years sailing the high seas in search of a somewhere to offload its cargo of 11,000 tonnes of contaminated incinerator ash.

CONVENTION ON THE REGULATION OF ANTARCTIC MINERAL RESOURCE ACTIVITIES (CRAMRA)
Designed to control exploration and mining in Antarctica⌀, CRAMRA faced mounting opposition led by two of the nations most vital to its success, Australia and France. The Convention, the third amendment to the 1959 Antarctic Treaty⌀, was initiated by New Zealand and adopted in 1988. Under its terms, exploration for minerals⌀ and oil⌀ in Antarctica would have been allowed, but only if strict safeguards were adhered to. The complicated system devised to evaluate, license and police exploration and mining would have permitted the major developed nations a great deal of control.

Ratification of CRAMRA required the signature of the seven countries with pertinent claims under the Antarctic Treaty, and Australia's decision not to ratify led to the virtual demise of the Treaty before it had come into existence. Australia's decision not to comply is based on the desire to try and achieve even stronger protection for the continent and have it declared a World Park, a move later backed by France and other nations. The ploy may backfire, as Australia, which claims 42 per cent of Antarctica, may lose this territorial claim through refusal to ratify CRAMRA. As a result, Antarctica may be opened up to widespread and unregulated exploration.

Thirty-nine delegates met during 1989 and 1990 to discuss the Convention and the proposal to replace it by an agreement that would establish Antarctica as a huge nature reserve. The United Kingdom and United States initially insisted on allowing mining to go ahead in controlled conditions, a stance originally supported by New Zealand. Austria, Belgium, and Greece favoured the Franco-Australian proposal. At a special meeting held in 1990 – organized with the proviso that another should be held at a later date to decide ways of assigning liability for any damage that might occur if the option to allow mining was taken – the UK had become virtually isolated in its opposition to a complete ban on oil drilling and mineral exploration.

CONVENTION ON THE CONSERVATION OF MIGRATORY SPECIES OF WILD ANIMALS (BONN CONVENTION) A specialized convention, originating from a conference in the Federal Republic of Germany in 1979, which lists endangered species of migratory vertebrates and insects and their habitats.

The Convention applies to the entire population of any species or subspecies of wild animals, a significant proportion of whose members cyclically and predictably cross one or more national jurisdictional boundaries. The thirty-four signatory states agree to protect both wildlife and the habitats used by these animals, even if they are merely transient.

CONVENTION ON INTERNATIONAL TRADE IN ENDANGERED SPECIES OF WILD FAUNA AND FLORA (CITES) CITES was agreed in 1973 and became operational in 1975. Also known as the Washington Convention, CITES was originated to control the global trade in endangered species⊖ of wildlife and any products derived from them. It operates through a system of import, export and re-export permits. The Convention has three appendices. Appendix I forbids trade in all animals listed on it or any products derived from them; Appendix II restricts trade in separately listed species considered to be at risk of becoming endangered; Appendix III allows individual nations to announce their own domestically endangered species and trading regulations.

Parties to CITES also examine ways to conserve and exploit wild animals and occasionally remove species from the appendices – for example trade in vicuna wool was reintroduced following a recovery in populations. Similarly, since CITES began operating, the populations of some species, such as the American alligator, have recovered and are now out of danger, and their listing has changed accordingly. Although contracting parties agree not to trade in species on Appendix I and to record all trade in those plants and animals on Appendix II, any signatory nation can announce a 'reservation' which provides it with exemption from the CITES trade restrictions.

Data compiled by the Wildlife Trade Monitoring Unit of the World Conservation Union⊖ (IUCN) indicate that trade is prohibited for about 680 species on Appendix I and regulated for a total approaching 30,000 on Appendix II. Parties to the Convention are required to submit annual reports and records of trade to the CITES Secretariat in Switzerland.

No estimates of total trading volumes are available, but illegal trade surpasses legal trade in several instances, although the CITES Secretariat estimates that the enforcement of the treaty is 60 per cent effective. Illegal trade is monitored by the Trade Records Analysis of Flora and Fauna in Commerce (TRAFFIC), a network of offices with eleven branches, affiliated to the Worldwide Fund for Nature (WWF)⊖ and the IUCN.

With 103 members, CITES is one of the most successful and widely supported international conventions, although the protection of plants lags well behind that of animals. The Convention's worth is also undermined by the behaviour of non-participating nations who are effectively free to trade in all endangered species. The inability to police and enforce the regulations, coupled with the fact that participating nations are allowed to register 'reservations' according to their national priorities, also hampers its effectiveness.

CONVENTION ON LONG-RANGE TRANS-BOUNDARY AIR POLLUTION Agreed in 1979 in Geneva within the framework of the United Nations Economic Commission for Europe (ECE)⊖, the Convention was initially designed to control the atmospheric pollutants that contribute to acid rain⊖. A ministerial Conference on Acidification of the Environment held in Stockholm in 1982 hastened ratification of the Convention, which came into force in 1983. The political ramifications of air pollution⊖ and the economic and legal complexities of acid rain damage have compromised acceptance of the Convention. Twenty-six nations and the European Community⊖ originally signed.

In 1987, a protocol to the Convention which called for a reduction of sulphur emissions or transboundary fluxes by at least 30 per cent came into force after being ratified by sixteen countries. Contracting parties agreed to reduce, by 1993, national sulphur emissions or their transboundary flows to a level at least 30 per cent below 1980 levels. At least twenty-nine countries have ratified the protocol, including most of the industrialized countries, although three of the world's largest emitters of sulphur – Poland, the United Kingdom and the United States – have not.

The protocol concerning sulphur dioxide⟡ followed an earlier one which estbalished a network of monitoring stations. Two more protocols – one calling for member nations to freeze their emissions of nitrogen oxides⟡ at 1987 levels by 1994 and another to limit volatile organic compounds, or hydrocarbons – have also been negotiated.

COPPER A reddish metallic element that is essential to all living organisms and also has extensive industrial uses. Most commercially worked copper deposits are low-grade ore, and at least ten different copper-containing ores are mined. Copper is used widely in the electrical industry, for telegraphic cables, wire and pipes. It is also important in alloys such as bronze, brass and gunmetal, as well as being an essential component of anti-fouling paints, algicides and wood⟡ preservatives.

As copper is a naturally occurring element, part of the biological cycle, 18,000 tonnes are emitted into the atmosphere each year, to which anthropogenic sources add a further 56,000 tonnes. Copper dust and fumes are unhealthy and can lead to respiratory diseases, such as metal-fume fever and vineyard-sprayer's lung. Some 10.7 million tonnes of copper are mined annually; the United States (1.5 million tonnes), the Soviet Union (1.5 million tonnes) and Japan (955,000 tonnes) are now the major producers, overtaking Chile, the long-time leader.

CORAL Coral are small, sedentary marine animals belonging to a class of coelenterates. Corals occur in dense colonies which are found in all oceans, usually in warm, shallow water⟡. The stony corals, of which there are about 1,000 species, secrete a rigid external skeleton made of calcium carbonate (limestone). Coral reefs are formed by succeeding generations of stony corals, the hard skeleton persisting after the individual corals die. The main reef-building occurs at depths of less than 50 m and in waters where the temperature is around 20°C. Within this zone, symbiotic algae are present in the tissue of the corals; these stimulate the secretion of limestone, so accelerating the growth of the reefs. The reefs play a major role in protecting coastal land

from being eroded by the sea, by lessening the force of tides and waves.

The reefs also act as a sink for carbon and will play a major role in slowing the global warming⟡ process by absorbing excess atmospheric carbon. Coral reefs currently form a sink for 111 million tonnes of carbon annually, equivalent to around 2 per cent of present carbon dioxide⟡ emissions. The largest reef in the world, the Great Barrier Reef in Australia, adds 50 million tonnes of calcium carbonate each year. Assuming that the commonly predicted sea-level rise of 20–140 cm occurs, the overall rate of calcium carbonate deposition on the Great Barrier Reef will rise 40 per cent. Ocean atolls will produce a further 160 million tonnes.

Coral reefs are one of the most diverse and productive ecosystems⟡ in the world, rivalling tropical rainforests⟡ as a respository of biodiversity⟡. The broad range of corals secrete a vast number of differing chemicals which are used by humans for a multitude of purposes, most significantly in the field of medicine. The potential yield of fish⟡ from coral reef waters is estimated to be in the region of 9 million tonnes – about 11 per cent of the current global marine fish catch. Yet coral reefs around the world are under threat of destruction.

Overfishing, and the use of dynamite for fishing⟡ purposes, is destroying fish stocks and the reefs themselves. Reefs elsewhere are threatened by pollution⟡ that either kills the coral directly or enriches the water, causing algal blooms which cut off light. Similarly, silt runoff caused by human activities such as deforestation⟡ and ill-considered agricultural practices on land⟡ lead to clouding of coastal waters, and coral must have light to allow their symbiotoic algae to photosynthesize. Increasing amounts of human sewage, produced in coastal regions throughout the tropics by the booming tourist industry, is also posing a serious threat to the existence of reefs.

In Sri Lanka, coral reefs are being quarried for their limestone, which is used in the building industry. In the Maldives and Comoros, large chunks of coral are taken for use as building stone or mortar. Reef damage is worst in the Philippines, where most of the coral is collected and exported to satisfy the huge United States market. In 1989, the Filipino coral curio and souvenir trade was estimated to be running at about 2,000 tonnes annually, with nearly 75 per cent of this shipped to the United States. The previous year, the USA imported 1,456 tonnes of coral, in contravention of the Lacey Act which prohibits the importation of wildlife illegally collected or exported from its country of origin. Belgium, France, Japan and the United Kingdom have all imported significant quantities of stony corals from the Philippines, despite a law passed by the Philippines government in 1977 which banned the collection and export of coral.

CORAL BLEACHING A new and unexplained phenomenon which threatens to destroy coral◇ reefs worldwide and may be an indicator of global warming◇. Widespread bleaching was seen throughout the Caribbean in 1987 and 1989. In the Bahamas, Florida and Jamaica, no bleaching of any sort had been witnessed for over forty years.

Corals are bleached when the colourful symbiotic algae they house are lost. The algae can re-enter their hosts if conditions are favourable, and bleached reefs have recovered, taking up to three months to do so. When the algae are absent for any length of time, the coral dies. The extent of bleaching varies with depth: the shallower the water◇, the worse the bleaching. During the 1980s, reefs throughout the Pacific Ocean and the Caribbean exhibited signs of bleaching. Scientists have hypothesized that the cause of the bleaching is stress, brought on by unusually warm water, changes in salinity, excessive exposure to ultraviolet radiation or extreme climatic changes. Most corals thrive when the water is between 25 and 29°C, and it is believed that algae die when water temperatures exceed the upper limit.

CORN *See* Maize

COTO DOÑANA NATIONAL PARK Spain's Coto Doñana National Park, established in the wetland◇ region in the south-west of the country, is Europe's largest bird sanctuary. It is visited by over a million birds each year, many of them migratory. The ecological stability of the 49,000-hectare park is under threat owing to a major tourist complex and local agricultural development projects which, although outside the park's perimeter, are depleting water◇ supplies. Proposals under discussion in 1990 to construct a second leisure complex, designed to accommodate a further 32,000 people – and which would draw yet more water to satisfy the needs of the tourists – could eventually lead to large areas of the park drying out. Overextraction of water from underground aquifers and the intrusion of brine as the water table drops will irrevocably damage the ecology of the park, which is a key staging post for birds migrating between Europe and Africa and supports a wide range of fauna and flora, including several endangered species◇ such as the Iberian lynx and Spanish imperial eagle, two of the world's rarest animals.

COTTON Cotton is a small, drought-tolerant tropical and subtropical shrub which grows to a height of 1–2 m. It produces fibres around its seeds and is the most significant vegetable fibre grown in the world in commercial terms. The fibres are principally used to produce textiles. Wild species of cotton, *Gossypium spp*, are native to the arid and semi-arid zones in Australia, South America, northern and southern Africa and desert regions in Arabia and South-East Asia. Several varieties are grown commercially. The plant is a perennial but is usually replanted each year to avoid the danger of pest build up. Cotton is attacked by a wide range of pests, some of which have few alternative hosts. Commercial production therefore depends heavily on the use of intensive pest-control methods.

It is common practice to intercrop cotton with a cereal crop or use it in rotation with leguminous crops which help to improve soil tilth. Cotton usually matures five to eight months after sowing. The fruits of the cotton plant are called bolls. These contain seeds, around ten in each boll, each with soft fibres (or lint) extruding from the seed coat; these can be spun into yarn for cloth. During the height of the dry season the bolls burst open, exposing the lint, which can then be harvested, usually by hand. After picking, the seeds are dried and the lint is separated in a ginning mill. Yields of 330 kg per hectare of seed cotton are regarded as average.

About one third of weight is lint; this, once extracted, is made into bales and exported to countries where it is spun into thread for manufacturing textiles. Seeds may also be exported, as they have a high content of edible oil which is used to produce margarine, salad oils and soaps. The by-products from the oil-extraction process also provide a valuable livestock◇ feedstuff. The pigment, gossypol, derived from cotton also has good potential for use as a safe and effective male contraceptive.

Global production of cotton is around 41 million tonnes annually; India is by far the largest producer (3.6 million tonnes), followed by Pakistan (1.9 million tonnes) and Brazil (1.7 million tonnes). Lint production is in the order of 18 million tonnes annually, with China (4.2 million tonnes) producing the most.

COUNCIL FOR MUTUAL ECONOMIC ASSISTANCE (CMEA) Commonly known as

COMECON, the CMEA was an economic association of Communist countries founded in 1949 in opposition to the Marshall⬦ Plan. It sought to promote economic co-operation and development between member states through the creation of a common market, but had no real central organization and no effective free trade between members. The Council, based in Moscow, linked the Soviet Union with Bulgaria, Czechoslovakia, Hungary, Poland and Romania. In 1950 the German Democratic Republic (GDR) joined, followed by Mongolia (1962), Cuba (1972) and Vietnam (1978). Yugoslavia retained associate membership. Albania withdrew from COMECON and the Warsaw Pact⬦ in 1961. The organization was basically under the control of the Soviet Union; opposition to policy decisions were voiced primarily by Romania and, occasionally, by the more developed GDR and Hungary.

In 1990, the collapse of the Communist regimes in Eastern Europe⬦ and the democratization process, coupled with movements towards free-market economies, led to the rapid demise of the CMEA. Its traditional trade and payments system, which was based on bilateral trade agreements between member nations, negotiated prices for goods, and settlement in 'transferable' roubles quickly disintegrated. The problems of how to settle outstanding claims and liabilities in the 'transferable' roubles system, which was abandoned, caused many of the countries involved to place direct controls on their exports and imports, thus creating a virtual trade war. The demise of the CMEA rocked the already fragile economies of many Eastern European countries which were heavily dependent on intra-CMEA trade, and in some cases threatened total collapse. In 1989, intra-CMEA trade accounted for some 40–80 per cent of the member states' total trade.

The CMEA was officially disbanded in early 1991 and replaced by the Organization for International Economic Co-operation⬦, a body incorporating the ex-Warsaw Pact countries, Cuba, Mongolia and Vietnam with Germany acting in an observer role.

COUNCIL OF EUROPE   An association of European states founded by ten countries in 1949, based in Strasbourg, pledged to improve living conditions, uphold the principles of parliamentary democracy and promote the economic and social progress of its members. From 1949 onwards, most of the democratic states in Europe sent parliamentary representatives to the Council. It is organized to provide a framework for intergovernmental co-operation in the areas of culture, education⬦, health, social welfare, crime prevention, harmonization of legislation, youth affairs and relations between developed and developing countries⬦.

The two organs of the Council are the Committee of Ministers, comprising the Foreign Ministers of the member states and a Parliamentary Assembly, consisting of 177 representatives appointed by national parliaments. A mainly consultative body, the Council has negotiated a number of conventions – significantly one in 1950 on human rights⬦, which resulted in the formation of the European Court of Human Rights in 1959.

Before the sudden upheavals in Eastern Europe⬦ during the late 1980s, the Council had expanded to twenty-three members: Austria, Belgium, Cyprus, Denmark, Finland, France, Federal Republic of Germany, Greece, Iceland, Republic of Ireland, Italy, Lichtenstein, Luxembourg, Malta, the Netherlands. Norway, Portugal, San Marino, Spain, Sweden, Switzerland, Turkey and the United Kingdom. Hungary was admitted in 1990. Poland's application for membership had also been accepted, pending the holding of free elections; and applications from Czechoslovakia and Yugoslavia were nearing acceptance. Other former communist bloc countries might apply for membership as they move along the path to democracy.

D

DAMS   Dams are used to restrict or divert the normal flow in watercourses for a variety of purposes including raising the water⬦ level for navigation, storing and providing water for irrigation⬦, industrial use, or water control, and to produce a high-pressure source of water for hydroelectric⬦ purposes. There are many types and sizes of dam; two forms are regularly built for major, large-scale water-resource development projects. Gravity dams depend on their sheer weight to hold back water and have a flat, vertical face. Arch dams consist of curved concrete structures with a convex face upstream. They are less bulky than gravity dams and therefore cheaper to build.

Worldwide, there are over 13,000 'large' dams – dams that are more than 15 m high. The highest, on the Vaksh River in the Soviet Union, reaches 300 m. Large dams are essentially a recent phenomenon – all but seven of the world's hundred largest dams have been built in the last fifty years. The costs of dam projects have risen in line with size. Several dams under construction in 1990 were expected to cost over $1 billion to complete. The estimated cost of the Yacreta Dam being built by Argentina and Paraguay had already soared to almost $6 billion, three times the original prediction.

Many dams are built to produce hydroelectricity, a relatively cheap, non-polluting, renewable source of energy◇. Hydroelectric dams can cause significant problems of their own, including loss of fertile land, wildlife and cultural sites, plus damage to public health and other factors which are usually not considered in cost–effect calculations of their economic viability. However, any energy source that produces no direct carbon dioxide◇, and thus does not contribute to global warming◇, has a significant advantage over the fossil fuels◇. Operation of a 10-kilowatt hydroelectric plant is reputed to save the equivalent of 21 tonnes of petroleum annually, leading to a 70-tonne reduction in carbon dioxide emissions. Dams already produce 20 per cent of the world's electricity, but it is estimated that less than 10 per cent of the potential hydropower is currently being tapped; the remainder could be exploited with little technical difficulty.

Vast areas of land◇ have been flooded as a consequence of major dams being built. In 1985, dams under construction were planned to flood a further 33,000 sq km of land in various parts of the world, adding to the 308,000 sq km already underwater. In the inundation process when the dam's reservoir is filled, valuable and often unique wildlife and habitats are lost, and millions of people are displaced. Downstream ecosystems◇ are disrupted. Silt and valuable nutrients are trapped behind the dam. Fisheries are also severely depleted. Reduced silt deposits, traditionally used to maintain soil tilth in fields downstream, have to be remedied by the application of fertilizer◇. Coastal erosion also increases as a result of silt-load loss. Large dams also contribute to earthquakes because of the weight of the enormous quantities of water which build up behind them, as well as producing perfect habitats for a variety of parasites, diseases and the insect vectors that transmit them.

A further problem with large dams that may arise in the future is the possibility of their full or partial collapse. Most of the world's largest dams have yet to stand the test of time, but between 1970 and 1983 three major dam bursts each were reported from Colombia, India and the United States, with Argentina, Liberia, Mozambique and Nepal also recording dam breaches during this period.

DANUBE DAM The Gabcikovo–Nagymaros River Barrage System, a $5.8 billion hydrolectric power◇ development project on the River Danube, was envisaged to become the largest water-diversion scheme in Europe. The project was planned to re-route 30 km of the Danube into a concrete canal to provide a two-basin, three-dam system with water◇ to generate electricity for the three sponsoring nations, Austria, Czechoslovakia and Hungary.

The project aroused fierce and widespread opposition even in Hungary where, following liberalization, opinion polls suggested that 60 per cent of the nation was against the project. Opponents in all three countries concerned claimed that the scheme would destroy the local ecosystem◇, flood historic sites and wipe out unique plants and wildlife. Hungary's drinking water and a vast underground reservoir could be contaminated if the project went ahead.

The Hungarian government had the option of preventing construction of the barrage at Nagymaros, which was a crucial component of the whole project. However, taking this step would make them liable for an enormous amount of compensation which would be claimed by Austria, Czechoslovakia and various organizations in Hungary. Austrian banks lent the Hungarian authorities $580 million in 1986 in return for twenty years' electricity supply due to start in 1995. Despite the fact that completion of the barrage would be cheaper than cancelling it, Hungary's Parliament voted 186–7 to terminate the project. Czechoslovakia, having completed 70 per cent of its half of the project, the Gabcikovo Dam, quickly demanded $1.7 billion in compensation as the dam at Gabicikovo will lose at least a quarter of its generating capacity as a result of the Hungarian decision. Austria, seeking return of its investment and compensation for the envisaged lost electricity supply, is seeking over $2.5 billion compensation.

DDT Common name for Dichloro-diphenyl-trichloroethane, an organochlorine compound which became widely used as a contact insecticide⬦. Regarded as the most extensively applied of all pesticides⬦, it was first synthesized in the late nineteenth century but entered into widescale use only in the 1930s. DDT has a broad spectrum – it is active against mosquitoes⬦, flies, fleas, lice, cockroaches⬦ and other disease-carrying and destructive insects, specifically attacking the insects' central nervous system. It has traditionally been used as the major control agent in malaria⬦ eradication programmes. Uncontrolled worldwide use has led to many insects developing resistance to it.

DDT is a fat-soluble, markedly stable compound and is essentially non-biodegradable in the environment. Together with its breakdown products it is highly toxic to some species of fish⬦, poses an extreme hazard to birds, and is now a common contaminant of groundwater⬦. Its impact on wildlife has been disastrous mainly because of bio-accumulation. DDT builds up in the fatty tissues of all animals and is concentrated as it passes to animals higher up natural food chains, with toxic effects. Owing to its widespread use, DDT is now present in virtually all foods and living organisms. In humans, it has been found in significant and sometimes dangerous levels in mothers' milk and is particularly potent in areas where the diet is deficient in protein.

In humans fatal poisoning is rare, but symptoms from acute poisoning caused by swallowing more than 20 g include nervous irritability, muscle twitching, paralysis of the tongue, lips and face, convulsions and coma. In the 1960s, DDT was revealed to be strongly carcinogenic and use in the United States was banned in 1971. Use of DDT is now prohibited in many industrialized countries, but it continues to be widely used as a pesticide throughout the Third World⬦.

DEATH RATE The number of deaths recorded in any given year divided by the mid-year population of the country in question gives the death rate. It is used for comparative purposes and is usually multiplied by 1,000 and expressed as the Crude Death Rate (deaths per thousand population). Since 1965, global death rate figures have been falling: from 13·3 in 1965–70 to 9·9 in 1985–90. As standards of living improve, health services get better and populations age, Crude Death Rates can eventually be expected to increase. Europe has shown an increasing Crude Death Rate for the past

twenty years – up from 10·3 in 1965 to 10·8 in 1990 – but is the only region to exhibit an increase.

DEBT The global debt crisis arose when several developing countries⬦, especially those with outstandingly heavy financial obligations, found themselves driven into a position where they were unable to repay or service their debts. At least fifteen countries fell behind in debt-service⬦ payments, including Argentina, Bolivia, Brazil, Costa Rica, Dominican Republic, Ecuador, Honduras, Ivory Coast, Liberia, Nigeria, Panama, Peru, Tanzania, Zambia and Zaïre. At the start of the 1980s, 109 developing countries owed a collective debt of $579 billion. By 1986 the total had risen to $1.02 trillion, climbing to $1.39 trillion by the end of 1989 and rising at 3 per cent annually. Fortunately, favourable fluctuations in exchange rates and a weakening of the US dollar led to a $7 billion reduction in the year's-end total. Despite the continual increase in the total, the World Bank⬦ reported that the situation was improving, although it forecast a further $30 billion rise in the debt by the year 2000.

Of the total debt, $1.24 trillion – or some 87 per cent – was held by countries belonging to the Organization for Economic Co-operation and Development (OECD)⬦, either directly or through international organizations. Claims on Central and Eastern European countries amounted to around $150 billion. These countries held claims on developing nations totalling around $120 billion.

The sheer size of the indebtedness of many developing countries became a useful lever, giving them a weapon to use in negotiations with donor governments and agencies. The inability or conscious decision not to repay outstanding loans threatened the stability of the international banking system, forcing the donor governments into action to ease the economic plight of the developing nations and protect the commercial banking sector. Most of the increased debt in the latter half of the 1980s resulted from new loans to developing countries from 'official' sources, mainly governments, which offset the fall in amounts owed to private sources such as commercial banks. This reinforced the trend toward ensuring that outstanding loans were with official sources. Official loans accounted for 14 per cent of the debt of nineteen severely indebted nations designated by the World Bank as Upper Middle-Income Countries⬦ in 1982; by 1988 this figure had risen to 30 per cent.

Many developing countries have been forced into borrowing more simply to pay their debts. Other nations try to 'reschedule' their debts – this means that they are unable to meet their contractual obligations and must therefore negotiate new repayment terms. Providing further loans to pay off debts seems an unlikely way out of the crisis, countries simply fall deeper into debt and enter a downward 'debt spiral'. The best way out would be to write off some of the debt – a move proposed by many, including the Soviet Union and European creditor governments, led by France, who are urging their commercial banks to write down the value of Third World◇ debt.

After the full scale of the debt crisis became apparent, total repayments tended to be far larger than new loans and there was an annual net flow of money from the Third World to the developed world. During 1989, developing countries in Africa, Asia and Latin America paid in excess of $133 billion to northern creditors. In all, between 1983 and 1989, a surplus of approximately $165 billion flowed from 'recipients' to 'donors'. In 1990 the trend was reversed and the amount of money flowing into developing countries in the form of aid◇, new investment and loans exceeded the interest charges, debt repayments and profits flowing out to the tune of around $9 billion.

DEBT PURCHASE Third World◇ debt◇ can be bought on the open market at considerable discount, and it is this that has sparked the concept of 'debt swaps'. Creditors would rather receive some return on their loans than run the risk of getting nothing and therefore sell debt at reduced rates. Most of the Latin American debt is owed to commercial banks, whereas most of that of African countries is owed to official sources. Consequently it is Latin American debt that dominates the markets. In February 1989 Argentinian debt was quoted as 18 cents (US) to the dollar. Bolivian at 11 cents, Brazilian at 29 cents, Mexican at 35 cents, Peruvian at 5 cents and Venezuelan at 33 cents.

DEBT RELIEF May take the form of either refinancing, rescheduling or cancellation of all repayments. A loan is refinanced when the creditor country makes a new loan to enable the debtor nation to meet the debt service◇ payments on an earlier loan. A loan is rescheduled when the amortization◇ or interest payments, or both, on the outstanding portion of the loan are rearranged to make payment easier.

Most of the high-profile debt-relief initiatives, including the Baker and Brady plans, have been aimed at the so-called Middle-Income Countries (MICS)◇, such as Argentina, Brazil and Mexico, which owe most to commercial banks. Low-Income Countries◇, whose debts tend to be with the governments of donor nations, and other MICS such as Egypt, the Ivory Coast and Jamaica, which also owe the bulk of their debt to donor governments, have separate arrangements to deal with their indebtedness.

Baker Plan – The first major economic plan to try and solve the Third World◇ debt problem, proposed in 1985 by James A. Baker, then Treasury Secretary of the United States. The plan was based on the willingness of commercial banks to increase lending and the provision of new, strictly controlled loans from the International Monetary Fund (IMF)◇, World Bank◇ and the other regional development banks. New financial arrangements were suggested that would allow voluntary exchanges of debt◇ for other obligations, or for equity between banks and debtor nations. The plan failed to make much impact, primarily because commercial bank lending continued to decline and multilateral lending stagnated.

Brady Plan – The Brady Plan, proposed in 1989 and named after the incumbent US Treasury Secretary, Nicholas Brady, called for $70 billion in debt relief to reduce the amount owed by MICS. The World Bank and IMF agreed to provide about $12 billion each over the following three years to support debt reduction. Japan also pledged about $4.5 billion. The plan recommended a number of government measures to improve the volume of debt-conversion schemes. The yearly total of $29 billion proposed under the scheme would not go far, considering that thirty-nine nations were potential users of the Brady Plan.

If the money were used to buy back debt at the average market discount price of 36 cents (US) per dollar, commercial debt would be reduced by a mere $80 billion, assuming that there were no rises in market prices. The majority of the $221 billion debt owed to commercial institutions would remain untouched. Furthermore, the debtor nations would have to pay interest on the initial $29 billion and the effective end result would be a reduction in interest payments of around 14 per cent. A secondary facet of the plan called for $30–35 billion of public money to be used used to guarantee lower-value and lower-interest bonds that could be exchanged for commercial bank debt. This would cut the amount debtors have to pay each year by $6 billion. Public guarantees on the new bonds were meant to stimu-late banks to accept a reduction in the amount they were owed. By the end of 1990 the plan had resulted in savings of at least $22 billion for the MICS, and another $5 billion worth of aid◇ loans were can-celled for the poorest members of the group.

DEBT SERVICE   The repayment, composed of interest and amortization◇, due in respect of a loan. This is often expressed as a debt-service Ratio, the ratio of debt service payments to earnings from exports of goods and services in any period. In 1975, the collective debt service of the Third World◇ amounted to 9 per cent of their export earnings. By 1980 it had risen to 13 per cent and by midway through the decade it had climbed past 20 per cent.

Debt-service payments by developing countries exceeded loan disbursements between 1983 and 1989, resulting in a net transfer to the North◇ of $165 billion.

DEBT SWAP   Several ingenious methods to help alleviate the debt◇ burden of Third World◇ countries appeared during the 1980s, mainly mak-ing use of the availability of discounted debt. The United Nations Children's Fund (UNICEF)◇ pro-posed a plan of Debt Relief for Child Survival. Under the plan, banks would be encouraged to make over a portion of their high-risk Third World loans to UNICEF through beneficial tax deductions. UNICEF would then assume the claim on the devel-oping country, with repayments being made in local currency which could be used to finance UNICEF projects in the country concerned.

In a so-called debt-for-nature swap, Bolivia agreed to exchange $650,000 of its $4 billion exter-nal debt for an agreement to preserve 1.5 million hectares of forest◇ and a 122,000 ha wildlife reserve. The US-based Conservation International environmental group purchased the debt from an American bank at an 85 per cent discount and gave it to the Bolivian government. In return, the author-ities agreed to protect the rainforest◇ and set up a $250,000 trust – using local currency – to adminis-ter and maintain the reserve. With Conservation International acting in an advisory capacity, com-mercial operations will be allowed on some of the land◇, provided that any development is sustain-able.

In the world's largest 'debt-for-nature' swap the Worldwide Fund for Nature (WWF)◇ agreed to pay off $5.4 million of Ecuador's $11 billion national debt. WWF was to buy the debt cheaply from a New York bank, $3 million of the money saved would be used to finance projects in the Galapagos Islands, the rest being spent on projects on mainland Ecuador. Similar deals were negotiated with Bolivia, Costa Rica, Peru and the Philippines. The debt-for-nature swap has also been extended to Africa: the WWF has concluded deals with Madagascar and Zambia. WWF (US) has agreed to buy $2.1 million of Madagascar's national debt at a 55 per cent reduction. A similar conversion deal with the Zambian government has been agreed. An anonymous Swiss donor has pro-vided the WWF with the funds necessary to buy $2.27 million of Zambia's debt from a group of European banks at an 80 per cent discount. The local-currency equivalent will go towards protection of elephants◇ and rhinoceroses◇ in Zambia.

Debt-for-equity swaps are those in which foreign investors buy up discounted Third World debt in international financial markets and exchange the debt with the debtor country in return for shares in local commercial operations. Commercial banks are now taking advantage of this idea and establishing special Debt-for-equity Funds. Creditor banks form an investment company and become indirect inves-tors, maintaining an interest in the country to which they had originally made the loan. Three of these funds have been set up in Brazil, two in Chile, and one each in the Philippines and Venezuela, all with around $50–100 million of capital. In 1990 a debt-for-equity fund was set up in Argentina, with $1 billion in funding. Fourteen banks from Europe, North America and Japan were amongst the insti-tutions buying shares. Argentine debt is converted into shares in Argentine companies, and the fund envisages exchanging three dollars of debt for every dollar of shares acquired.

In one of the most innovative debt-swap ventures, the Dutch football club PSV Eindhoven, which is owned by the transnational◇ Philips organization, acquired discounted Brazilian debt and used it to pay for the transfer of a Brazilian international foot-baller, Romario Farias.

DECLARATION ON SOCIAL PROGRESS AND DEVELOPMENT   Adopted by the United Nations◇ General Assembly in 1969, the Declara-tion aims to promote human rights◇ and social justice throughout the world. It stresses the need to eliminate illiteracy◇, hunger◇, poverty and unem-ployment; the improvement of living standards, both material and spiritual, is the avowed goal.

Individual citizens are also urged to play an important part in the defining and achieving of each nation's development.

The Declaration states that development must be inherently people-orientated and stresses that development programmes devoid of human considerations are doomed to end in failure; in many instances the plight of the very people who were the intended beneficiaries has been worsened.

DECOMMISSIONING A term used to describe the process of dismantling and disposing of old nuclear reactors. To date, experience of this process is limited and the full costs of such operations, the dangers involved, and the methods needed to accomplish them are still being discovered. Nuclear facilities have a lifespan of around thirty years, and as Britain's nuclear industry was the world's first, it has had to take a lead in decommissioning. The process involves disposing of whole nuclear plants and all the integral parts, including disposal of the reactor vessel itself. Experience has already shown that this usually has to be undertaken by remote control and often underwater. High radioactivity◇ levels may prevent this action from being feasible for over a hundred years. Large volumes of radioactive waste◇ will be produced as a result of decommissioning and the problem of disposing of this waste has yet to be solved.

The International Atomic Energy Agency (IAEA)◇ reported that in 1990, 143 nuclear facilities in seventeen countries were at some stage of decommissioning. Moreover, 64 nuclear reactors and 256 research reactors could become in need of decommissioning by the year 2000.

Despite the need to decommission its nuclear plants and research facilities, the UK has actually fallen behind in the technology needed to dismantle the actual reactor core chambers. Japanese engineers are at the forefront, having already used robotic tools to dismantle a small experimental pressurized water reactor. In late 1990, German scientists began to dismantle the core of a nuclear power◇ plant at the Centre for Nuclear Research in Karlsruhe. Using a remote-control robotic cutting arm, the engineers hoped to develop and practise techniques which would be suitable for all kinds of reactors, although prospects of this seemed unlikely as the reactor in question had operated for only eighteen days and was not particularly radioactive.

The costs of decommissioning are astronomical, if not prohibitive – an estimated $480 million for a single pressurized-water reactor at Sizewell in

Britain alone. In 1989, experts calculated that in the UK the costs of decommissioning and dealing with spent fuel will be in the region of $960 million per nuclear plant. The Worldwatch Institute◇ in the United States estimates that decommissioning costs will work out at an average of $1 million per megawatt of generating capacity. No decommissioning costs were calculated during the building stages and planning of any nuclear reactors, and as no funds were set aside for this eventuality – despite the knowledge that reactors were likely to have a lifespan of less than fifty years – the bulk of decommissioning costs will therefore generally have to be met from future funds.

As well as the technical problems involved, the authorities are now having to face mounting economic pressures as the cost-effectiveness of decommissioning is also being called into question. In Germany, critics of the $100 million cost of dismantling the Karlsruhe plant point out that at least 1,000 tonnes of radioactive material will be removed from the installation and the bill for transport and storage will also have to be met. They claim that for the same money spent merely on dismantling, the plant could have been maintained in a mothballed state for 200 years.

DEEP WELL WASTE DISPOSAL Waste, especially hazardous waste◇, can be disposed of by burial in deep repositories. Underground disposal is an economically and environmentally acceptable alternative to landfill sites or chemical or thermal treatment plants.

By 1985 only one natural deep inactive mine site in the Federal Republic of Germany had been identified as meeting safety criteria for the disposal of hazardous waste.

Deep-well sites, in comparison, are custom made for waste disposal. The wastes are injected under high pressure into porous rocks, generally at a depth greater than 600 m. The wastes displace any liquids or gases in the rock strata and become trapped by the pressure of the overlying rock. There are over 140 deep wells operating in the United States alone. Other nations – such as Canada, the Federal Republic of Germany and Spain – also make use of deep-well disposal, although it is banned in Japan and the Netherlands. Wastes from the petrochemical industry forms the bulk of the material disposed of in deep-well sites.

Three drawbacks with the process have been identified. Waste, or water runoff contaminated by it, can enter deep aquifers, polluting water⟿ supplies. Overpressurized wells can also erupt, as happened in the United States in 1968. In areas prone to seismic activity, the risks of earthquakes is greatly enhanced when deep well injection is practised.

DEFORESTATION   In 1950 just over 100 million hectares (ha) of the world's forests⟿ were cleared, but forests still covered around one-quarter of the world's land⟿ surface. By 1975, well over 200 million ha had been destroyed to meet the needs and demands of the burgeoning human population. By the year 2000, between 600 and 700 million ha could have disappeared, and forests are expected to cover only one-sixth of the land area. The world's forests, coniferous, temperate and tropical, are all under threat, but it is the destruction of tropical forests which is currently having the greatest impact.

Tropical rainforests⟿ play a critical role in regulating the global climate. They help to maintain the balance of gases in the atmosphere, producing vast quantities of oxygen and using up vast quantities of carbon dioxide⟿ during photosynthesis. They are a storehouse of genetic diversity and provide a wide array of goods and materials for human and industrial use. Tropical rainforests cover only 6 per cent of the total land surface of the planet but contain at least half of all species of life on the earth. Since 1945 over 40 per cent of the world's rainforests have been destroyed, and as a result, an estimated fifty species or more of plants and animals become extinct every day. During the 1980s, deforestation was accelerating and 7.3 million ha of tropical forest were being cleared annually for agriculture. Another 4.4 million ha were selectively felled each year for timber. Between 1980 and 1990 at least a million sq km of forests in developing countries⟿ were destroyed.

At these rates, all tropical forests would be cleared in 170 years' time. A report from the World Resources Institute⟿ published in 1990 suggested that the annual rate of tropical deforestation was in fact between 16.4 and 20.4 million ha, considerably more than the 11.4 million ha reported by the Food and Agriculture Organization (FAO)⟿ at the beginning of the decade. Deforestation rates are accelerating in Brazil, Cameroon, Costa Rica, India, Indonesia, Myanmar, the Philippines, Thailand and Vietnam. In Africa, deforestation was held to be occurring thirty times faster than reforestation⟿.

Commercial logging is directly responsible for only 20 per cent of the deforestation in tropical rainforests. Related activities, such as road-building and damage to other trees as logs are extracted, increases the toll, particularly as they allow shifting cultivators greater access to areas of forest. The landless poor, who invade the forests, burn and clear land and cultivate a plot for a short while before moving on to repeat the process, are the main cause of forest destruction. The impact of their actions is global. Burning the forests not only reduces uptake of carbon dioxide by the trees, it releases over 2 billion tonnes of carbon dioxide into the air each year, increasing global warming⟿ in the process. By comparison, the burning of fossil fuels⟿ in industrialized countries releases 5.6 billion tonnes per year.

Forest cover in tropical nations such as Brazil, Colombia, Indonesia, Ivory Coast, Malaysia, Mexico, Nigeria, Peru and Thailand is disappearing at an average rate of about 80,000 ha per year. In Brazil alone, 35,000 sq km are being destroyed annually. Yet forests play an important role in protecting watersheds, restricting soil erosion⟿ and sedimentation⟿, and in the recycling⟿ of nutrients. Even though the soil on which they stand is poor and of little use for agricultural production, tropical forests have an added value in that they are a source of timber, of food for the local population⟿, a repository of germ plasm for medicinal and crop-breeding uses, provide a wide variety of materials for human and industrial use, and are a major tourist attraction. To solve the tropical deforestation dilemma, a balance must be found between the immediate need for food, land,⟿ energy⟿, minerals⟿ and foreign exchange and the concerns of those in distant lands who are calling for the decimation to stop. However, it must not be forgotten that many of the industrialized nations clamouring for the cessation have already destroyed their own indigenous forests during their national development process, primarily to help fuel economic growth. Broadly speaking, since pre-agricultural times the world's forests have declined by 20 per cent: from 5 billion to 4 billion ha. Temperate forests, mostly in the developed North⟿, have lost by far the highest proportion of their area – some 33 per cent. In comparison, tropical evergreen forests, which are currently under the most intense deforestation pressure and which the world's conservationists are now trying to protect, declined by only 5 per cent.

In most tropical countries, it appears more beneficial for nearly all concerned to clear forests rather than conserve them, although this is not true. In parts of Brazil and Central America, forests have been cleared for cattle-ranching to produce beef for the so-called 'Hamburger Connection'◇. The revenue from beef production covers approximately 35 per cent of the cost of the ranches, the rest is met from subsidies provided by the Brazilian government in the form of tax concessions and subsidized credit. Between 1966 and 1983, these incentives to deforest are estimated to have cost the Brazilian government in the region of $5 billion.

The major organizations charged with saving the tropical rainforests, – the International Tropical Timber Organization (ITTO)◇, the World Bank◇, the FAO and other agencies involved in the implementation of the Tropical Forestry Action Plan◇ are, according to many critics, merely offering solutions which hand money and resources to the rich and powerful, the very people who stand to gain from the short-term, unsustainable exploitation of the forests, as evidenced in the Hamburger Connection, rather than promoting the sustainable management being advocated. In economic terms, rainforests are worth more standing than felled. One hectare of forest in Peru used for harvesting fruit and latex, together with a very restricted amount of logging, was found to produce a sustainable yield of $7,000 per ha, seven times what would have been realized if it had been used for logging alone.

A proposed World Forest Convention, when adopted in 1992, will introduce measures to use, replant and protect tropical forests and encourage preferential trading and financial arrangements to assist economies in tropical countries.

DENGUE FEVER Also known as breakbone fever: a disease of the tropics and subtropics caused by a virus transmitted to man via the bite of a mosquito◇, *Aedes aegypti*. At least 30 million people suffer from dengue; upper estimates are twice as high. The disease is characterized by painful joints, fever and an irritating rash, symptoms beginning a week or so after the bite. The symptoms recur in milder form after a few days. There is no specific treatment for dengue; the disease usually runs its course over seven days. It is rarely fatal, but patients are debilitated and need lengthy convalescence. They are given painkillers and calamine lotion to soothe the rash. The global incidence of the disease is showing a marked increase, particularly in the Caribbean region.

DESALINATION The removal of salt from brine to produce fresh water.◇ Desalination is an extremely energy-intensive, expensive process which is practised only where there is no alternative supply of fresh water. It is used to produce irrigation◇ water in arid regions in which fresh water is in very limited supply but sea water is readily available, notably in the wealthy Gulf states in the Middle East. Several methods are employed, the most common being the evaporation of sea water by heat or by reduction of the pressure on it, so-called flash evaporation. The water vapour produced is then condensed to give relatively pure water, the dissolved salts being left behind in the evaporating vessel. Large quantities of heat are needed for this method – hence its use on a large scale only in oil-rich, water-poor desert countries. Freezing is another technique, pure ice forming when brine is frozen. This method theoretically requires less energy◇ than evaporation, but the process is considerably slower. Two other techniques, reverse osmosis and electrodialysis, employ membranes to separate water from the dissolved salts, but both are prohibitively costly owing to the large quantities of power needed to drive the processes. The most promising method under development is a solar-powered still which will drastically reduce the costs of the power needed to effect any of the desalination processes.

A few desalination plants are operating elsewhere in the world apart from the Middle East. Drainage water flushed through large-scale irrigation schemes to wash out salts can significantly raise the salinity of water in rivers and other watercourses. In the United States so much of the water from the Colorado River is used for irrigation and agricultural purposes that the US government was forced to build a desalination plant to reduce the salinity of the water in the river before it flows across the border into Mexico.

DESERTIFICATION The process by which arid or semi-arid land◇ becomes desert through climatic change or human action. Removal of the precious topsoil layer by artificial means or erosion by wind, water or desiccation results in a lowering of the ground's water-storage capacity and fertility, thus causing crops to fail. Over intensive farming and the destruction of natural trees and vegetation helped to create the dustbowl in the United States and have caused deserts in Africa and South-West Asia to spread by several kilometres a year. Solutions to the problem are now based on effective management of surface cover and sound farming practices. Traditional peasant agricultures do not usually cause desertification and place minimum pressure on the land, but when peasants are forced into repeated use of marginal lands, then the desertification process is unavoidable.

The 1977 United Nations⟳ Conference on Desertification (UNCOD) first drew attention to the problem, and a decade-long programme to counter desertification followed. The United Nations Environmental Programme (UNEP)⟳ claimed that desertification was causing $26 billion worth of loss in food production every year, and that damage could be prevented by an annual expenditure of $4.5 billion. By late 1989 only 10 per cent of this figure had been raised from the international community.

During the late 1980s, critics from institutions such as the World Bank⟳ and the Institute of Development Studies (IDS)⟳ in the United Kingdom began to express the belief that UNEP's original figures were inaccurate and based on unscientific observations. In 1977, UNEP claimed that the Sahara Desert was irreversibly extending southwards every year, a statement now disputed by a World Bank report. Original figures were based on a study in 1975 which concluded that the Sahara had moved south 100 km between 1958 and 1975. In 1989 two geographers from University College London questioned this observation, pointing out that the 1958 data came from extremely limited material that was available from weather stations at the time, and that the 1975 figures were taken when there was a drought⟳. Comparative satellite studies of the Sahara during the 1980s do not show an advancing desert, and several researchers on long-term projects in the Sahel report very little irreversible degradation. When the rains return productivity recovers. During the dry years 1982–4 the vegetation front did move south, but it moved north again during 1985–7 when the rains returned. Researchers in Australia report similar cyclical findings.

A 1984 UNEP review resulted in the agency's head, Mustafa Tolba, saying '35 per cent of the world's land surface is at risk . . . each year 21 million hectares is reduced to near or complete uselessness.' Critics also dispute these figures, alleging that calculations are based on inaccurate data and a flawed definition of desert. The UN definition – 'the diminution or destruction of the biological potential of land that leads ultimately to desert-like conditions' – also includes land that is waterlogged. Most of UNEP's 1984 figures came from a questionnaire sent out in 1982, and African governments were completing it at the height of a drought. The figure of 21 million hectares at risk is viewed with great scepticism, researchers claim that this total includes land that is permanently too arid to support agricultural production. Several national plans for creation of 'green belts' and massive tree-planting schemes to halt or reverse the desertification process are now being reassessed.

In 1989, the UN Food and Agriculture Organization (FAO)⟳ reported that improper land use under conditions of low and fluctuating rainfall were the basic cause of the desertification process which was affecting 350 million ha of farmland worldwide.

DESERT LOCUST CONTROL ORGANIZATION FOR EAST AFRICA (DLCOEA)
The DLCOEA was initiated by a convention signed in 1962 by the governments of Ethiopia, France (for Djibouti), Kenya, Somalia, Tanzania, and Uganda. Sudan joined the organization in 1968. During 1965 and 1966 the participating governments signed a co-operative agreement with the Food and Agriculture Organization (FAO)⟳. This reflected the depth of concern over the potential threat posed by the desert locust⟳ and the need for control in the area of the world most at risk from these insects.

Each country remains responsible for the control of locusts within its borders. The DLCOEA co-ordinates and carries out any control, particularly against flying swarms, which become beyond the capability of national teams. All governments contribute financial support, and funding is also received from non-member countries. In addition, assistance in the form of expert personnel and equipment is provided by a variety of donors.

DEVELOPING COUNTRIES
Countries that do not have sophisticated industrial bases to their national economy and have comparatively low *per capita* incomes and levels of social development. The economies of these countries are mainly agriculturally based, often dependent on only one or two commodities,⟳ and are characterized by abundant, cheap, unskilled labour and a scarcity of capital for investment. Where the nation's economy is heavily dependent on a single agricultural crop, in years of crop failure or poor world demand the impact on the economy can be catastrophic.

Around 70 per cent of the world's population live in the developing countries, almost all of which are in Africa, Asia, Latin America and Oceania. Many communities outside the major towns are poverty-stricken and hunger⟳, disease and illiteracy⟳ are still commonplace. The Development Assistance Committee of the Organization of Economic Co-operation and Development (OECD)⟳, the United

Nations⊙, the International Monetary Fund (IMF)⊙ and the UN Conference on Trade and Development (UNCTAD)⊙ all produce listings of developing countries, none of which totally coincides with any other.

DEVELOPMENT ASSISTANCE COMMITTEE (DAC) A specialized committee of the Organization for Economic Co-operation and Development (OECD)⊙, the DAC provides a forum for consultation among the OECD's seventeen main donor countries and the European Commission. It aims to increase the resources channelled to the developing countries⊙ and to improve overall effectiveness. The DAC is not an aid⊙ agency, its role is to improve, harmonize and co-ordinate the aid policies and programmes of its members. For this purpose, regular reviews are held during which the quantity and quality of each country's aid programme and policies are reported on and examined. Any grant or loan that contains an element of 'conditionality' is recorded by the DAC. Around 75 per cent of all official aid registered with the DAC is bilateral aid, much of it in the form of grants. The members are Australia, Austria, Belgium, Canada, Denmark, European Commission, Finland, France, Federal Republic of Germany, Italy, Japan, the Netherlands, New Zealand, Norway, Sweden, Switzerland, the United Kingdom and the United States.

DEVELOPMENT DECADE The United Nations⊙ defines each ten-year period as a Development Decade in order to focus attention on the developing countries⊙ and to try and accelerate the development of their 'human and natural resources' by swift and effective international action. The first Development Decade started in 1961, the Second ran from 1971 to 1980 and the Third Decade finished in 1990. A mid-term review is held to help assess progress and identify needs and policy changes or improvements that can be implemented in the next Decade.

DIABETES Over 50 million people around the world suffer from diabetes, and the disease is on the increase in both developed and developing countries⊙. Simple lack of care in following a healthy lifestyle is one reason, and a movement away from breast-feeding⊙ infants may also be a crucial factor.

Diabetes is a condition in which the human body is unable to make use of carbohydrates, such as sugar⊙ and starch, in the diet because the pancreas does not produce enough of the hormone insulin⊙. As a result, sugar accumulates in the blood and tissues, eventually causing defects in various parts of the body. The usual symptoms are the passing of excessive amounts of urine, a prodigious and almost insatiable thirst, loss of weight despite eating large quantities of food, and a general feeling of fatigue. Damage to the eye is one of the most prevalent and dangerous complications: diabetes is the major cause of blindness in the world.

Despite the extensive research that has been carried out into the disease, exactly why the body fails to produce enough insulin remains a mystery. However, it has been established that there is an inherited genetic component which plays a part in causing diabetes, although the means by which this passes from generation to generation is not understood. Diabetes is known to be 'activated' by other factors such as viral infections and obesity. In individuals who contract the disease in adulthood, its appearance is usually stimulated by obesity. A simple way of reducing the risk of developing diabetes is therefore to maintain a normal body weight, this is particularly necessary for older people who have a history of diabetes in their family. Similarly, the World Health Organization (WHO)⊙ advises that women who give birth to babies weighing more than 4.5 kg should pay special attention to their weight if they are to avoid the disease in later life.

A study carried out in Denmark has indicated that breast-feeding may be an important factor in reducing the incidence of diabetes. The predilection for using artificial milk rather than breast-feeding, which has occurred in most developed countries over the past three decades, may be the causal factor behind the increasing levels of childhood diabetes observed over the same period.

Today, around 4 per cent of the population of developed countries suffer from diabetes, but in developing countries⊙ the situation is very different, with 35 per cent of adults being affected in some areas.

With education⊙, access to insulin and maintenance of a suitable diet, the condition can be controlled. In areas where health services are good, the life-expectancy of diabetics is virtually the same as that of non-diabetics.

DIELDRIN Dieldrin, and the closely related aldrin and endrin, are chlorinated hydrocarbons⊙ that have been widely used as insecticides⊙ around the world, both in the home and outdoors. Like all other similar chemicals, dieldrin, being synthetic, persists in the environment and accumulates in food chains. Dieldrin has been found in rainwater, ground and surface water⊘, soil and food crops. All three chemicals are toxic to birds, beneficial insects, fish⊘ and mammals. Several species of small mammals and birds have declined as a result of dieldrin poisoning. Dieldrin is so highly mobile that once released into the environment, its dispersal is uncontrollable.

In humans, symptoms of poisoning include dizziness, nausea, anxiety and muscle seizures. Skin rashes can also occur from dermal exposure. As well as being acutely toxic, in 1962 dieldrin was shown to be carcinogenic. As a result, in 1976 dieldrin and the closely related aldrin were banned for all agricultural uses in the United States. In Europe the regulations over the use of dieldrin vary enormously, ranging from a comprehensive ban in France to voluntary restrictions in the United Kingdom and the Federal Republic of Germany. The World Health Organization (WHO)⊘ classifies dieldrin as 'extremely hazardous' and the World Bank⊘ recommends against its use. Chemical companies in Europe continue to manufacture dieldrin and export it for use throughout the Third World⊘.

DIESEL ENGINE

The diesel engine is the most fuel-efficient vehicle power source currently available for mass-produced vehicles. Although they do not produce the same cocktail of poisonous and destructive gases emitted by petrol engines, diesel motors produce an exhaust containing high concentrations of 'particulates', mainly soot. These particulates contain polyaromatic hydrocarbon (PAH) chemicals which are carcinogenic if inhaled regularly. Diesel engines also produce oxides of nitrogen and sulphur, which are major contributory factors to the production of acid rain⊘.

Regulations to restrict the level of emissions of nitrous oxides and particulates from all vehicles in the European Community⊘ will come into force after 1991. Exhausts will have to be fitted with filters to remove particulate matter and PAHs. Filters, using ceramic honeycombs and platinum⊘ catalysts, reduce the levels of soot and other harmful particles by 80 per cent. They also reduce emissions of hydrocarbons by 60 per cent and carbon dioxide⊘ by 50 per cent. Of necessity, the original filters clogged up after about 550 km of driving and needed to be cleaned regularly; consequently they were cumbersome as well as expensive. In 1990 a new system, capable of being 'regenerated' – cleaned regularly by channelling hot exhaust gases under pressure through the filter while the unit is in use on the vehicle – was developed which could reputedly filter out 95 per cent of particulate matter with only a negligible 1 per cent rise in fuel consumption.

Attempts are also being made to reduce the sulphur content of all diesel oil sold in Europe, most of which contains around 0.3 per cent sulphur. Plans to lower the content to 0.05 per cent will mean a rise in price, as more crude oil⊘ will have to be refined to power the desulphurization process.

DIOXINS

A family of 210 closely related chemicals, the most notorious of which is tetrachlorodibenzo-paradioxin (TCDD). They have no industrial uses whatsoever and are not deliberately manufactured. They occur as contaminants in several industrial processes, most notably during the production of the herbicide⊘ 2,4,5-T, and are formed as waste products in industrial and waste-treatment processes. TCDD is one of the most toxic chemicals⊘ ever isolated: to some mammals it is 70,000 times more poisonous than cyanide, and also carcinogenic. TCDD was amongst the chemicals released during the accident at the chemical plant in Seveso⊘ and was a contaminant in Agent Orange, a defoliant sprayed liberally during the Vietnam War⊘ which led to a deterioration in the health of all who were exposed to it. TCDD was also found to be a contaminant of drinking water⊘ in the Love Canal⊘ area in the United States, site of a massive chemical-waste dump, and a major pollutant at several other landfill sites in the country. In several industrialized nations, buildings have been contaminated by dioxins following fires in which electrical insulating equipment containing polychlorinated biphenyls (PCBs)⊘ have been burnt. Incomplete incineration of chlorinated wastes, domestic rubbish and plastics also release dioxins.

The danger posed by dioxins to human health remains a matter of controversy. In 1984 the Carcinogenic Assessment Group of the USA's Environmental Protection Agency (EPA) found dioxins to be highly carcinogenic. The industries responsible for the inadvertent production of dioxins do not dispute this basic premise but argue that the compound is not harmful to humans in the low doses in which it occurs, insisting that its worst effect is to cause a severe skin disease, chloracne, in those exposed to it. Several major studies have shown that TCDD is carcinogenic, mutagenic and teratogenic, and causes a variety of adverse health effects.

DIRTY DOZEN The 'Dirty Dozen' are, in fact, a group of eighteen chemicals singled out in 1985 by the Pesticide Action Network (PAN) as examples of the most dangerous pesticides⟿ which pose a significant human health hazard. PAN, a worldwide citizens' coalition of groups and individuals opposed to the irrational spread and misuse of pesticides, campaigned to have the use of these chemicals phased out around the world. The Dirty Dozen comprised campheclor (toxaphene), chlordane/heptachlor, chlordimeform, ethylene dibromide, DDT⟿, dibromochloropropane, the 'drins' (aldrin, dieldrin⟿ and endrin), HCH/lindane, ethyl parathion, paraquat,⟿ pentachlorophenol and 2,4,5-T. All these poison the wider environment and tend to accumulate in food chains, causing death and disease in many species of wildlife. Traces of some have been found in Antarctica⟿, far removed from any site of application. By 1990 the Dirty Dozen had been banned or severely restricted in at least sixty countries. Nevertheless, these chemicals are still being produced in bulk and exported for use in Third World⟿ countries where controls over chemical use are less stringent or non-existent.

DISARMAMENT The Geneva Protocol of 1925, which is observed by 120 nations, bans the use in war of asphyxiating, poisonous or other gases and the use of bacterial agents. It marked the first time methods or weapons of war were to be regulated along lines that were acceptable to the majority of the world's population. Later, in 1930, the League of Nations⟿, forerunner of the United Nations⟿, attempted to introduce the first-ever programme actually to reduce the weapons of war, albeit unsuccessfully. After the cessation of the Second World War and the build up of arms that followed, several attempts to introduce binding legislation to control or reduce weapons, or to promote peaceful coexistence, have been made. While the goal of the UN has been complete disarmament, in practice most measures agreed upon have been for 'arms control'⟿. Only three agreements – two multilateral, the other bilateral – have brought about actual dismantling and destruction of weapons.

Despite almost fifty years of dialogue between the Soviet Union and the United States, and even though several bilateral agreements and treaties have been agreed to limit nuclear weapons, it was not until the end of 1987 that the two sides finally reached accord on a means actually to destroy some nuclear weapons, as embodied in the INF Treaty.

Biological Weapons Convention (1972) – A multilateral agreement recognized by 99 states that prohibits the development, production, stockpiling and use of biological agents and toxins in armed conflicts. It also calls for the total destruction of any accumulated stocks.

Elimination of Intermediate-Range Nuclear Forces Treaty (INF) (1987) – The INF Treaty bans all Soviet and American ground-launched ballistic and cruise missile systems with a maximum range of between 500 and 5,500 km – a total of 2,587 missiles in all – and provided for their destruction over a period of eighteen months to three years. For the first time the Treaty included a process of verification acceptable to both sides, one of the major stumbling blocks to all previous attempts at arms reduction. The Treaty does not include sea-launched or air-launched missiles and will cut nuclear arsenals by only 4 per cent.

Conventional Forces in Europe Treaty (CFE) (1990) – A landmark multilateral agreement between the sixteen members of the North Atlantic Treaty Organization (NATO)⟿ and the six Warsaw Pact⟿ countries, each nation signing on its own behalf, covering the reduction of conventional forces in Europe. The intention is to reduce both sides to parity in tanks, guns and other offensive weapons. Both sides are pledged to destroy large numbers of tanks, combat aircraft and other non-nuclear weapons. The Treaty covers an area extending from the Atlantic coast in the west to the Ural Mountains in the east.

With regard to actual disarmament, the CFE Treaty was not as substantial as first thought, partially due to the geographic limitations set. The Soviet Union reportedly moved 60,000 tanks, guns and armoured vehicles east of the Urals to avoid having to destroy them. With only 21,000 tanks left in the CFE area, the Soviets will have to destroy or convert only around 8,000 to meet its obligations. Similarly, to reach the agreed parity limits, the Western nations will be able to deploy several hundred attack helicopters.

DIVISION FOR ECONOMIC AND SOCIAL INFORMATION (DESI) The DESI is a specialized part of the United Nations⟿ Department of Public Information which deals with information about the UN's role in development. It initiates and runs information programmes in support of the various UN development initiatives, maintaining an active liaison with development information sections or departments of member governments of the UN, the media, Non-Governmental Organizations

(NGO)⇨ and academics, as well as organizing national and international colloquia and symposia.

DRIFT NETS Formerly made of natural fibres, such as hemp, cotton⇨ or flax, fishing⇨ nets were visible and acoustically detectable to marine mammals. From the 1950s onward fishing nets began to be made of cheaper, extremely thin but extremely strong plastic filaments. Most nets are now made from a synthetic monofilament nylon which is both invisible to fish⇨ and marine mammals and non-biodegradable. The new drift nets were quickly taken up by the world's largest factory-fishing fleets and individual fishing units in the United Kingdom, the United States and elsewhere.

Drift net use focused attention on the Japanese, South Korean and Taiwanese fishing fleets, especially the Japanese. Their Red Squid fishing fleet in the Pacific was reported to be setting over 50,000 km of monofilament nets every night for seven months of the year. Around 100,000 seals, dolphins and turtles and up to 500,000 seabirds became tangled up in them each year. Many nations – especially those in the Pacific, led by Australia and New Zealand – took steps to ban the use of drift nets within their 200-nautical mile Economic Exclusion Zone (EEZ)⇨. In 1989 the Japanese agreed to a temporary halt to the use of drift nets to catch tuna and squid in the South Pacific, later endorsing a United Nations⇨ resolution which called for a ban on the use of drift nets in the Pacific by July 1990 and a worldwide ban effective from July 1992. A proposal put forward at the end of 1990 to abolish the use of drift nets in the Mediterranean was not adopted following opposition from the United Kingdom, acting in the belief that any such ban might be extended to the North Sea and Atlantic, where British fisheries would suffer. At the same time, the United States authorities decided to ban the use of drift nets within their own territorial waters as well as by US fishermen in international waters.

Used drift nets or parts of nets that are lost or discarded at sea are called 'ghost' nets, as they continue to trap and kill marine animals and seabirds for years. It is estimated that between 1,200 and 122,000 tonnes of commercial fishing nets are lost or discarded at sea each year. Annex V of the MARPOL Convention⇨, which came into force in December 1988, is supposed to cut down the number of nets lost or discarded by prohibiting 'the disposal into the sea of all plastics, including but not limited to synthetic ropes, synthetic fishing nets and plastic garbage bags'.

DROUGHT A term used to describe a prolonged absence of natural precipitation. This can either be due to natural weather conditions or can arise as a consequence of the alteration of the environment such as the removal of trees and destruction of vegetation. Lack or insufficiency of rain can cause water⇨ shortages, crop damage and depletion of groundwater⇨ and soil moisture. Drought is the most serious constraint to agricultural production in nearly every part of the world. The term does not necessarily refer to complete lack of precipitation: in the UK, fifteen days with less than 0.2 mm of rain constitutes an 'absolute drought'.

Droughts can have a devastating and long-lasting impact. In the sub-Saharan region of West Africa drought conditions have persisted, almost continually, since the late 1960s. The drought which began in the Sahel region of Africa in 1968 led to the loss of well over 100,000 lives within five years and irreparably altered life in the Sahelian zone. The Ethiopian drought of 1973–4 alone claimed the lives of 200,000 people and precipitated the overthrow of Emperor Haile Selassie.

In 1982 there were serious droughts in five continents. In addition to droughts in northern and southern Africa, Australia, Italy and Spain suffered their worst droughts for over fifty years. Nicaragua's crops were devastated by a dry spell in the middle of what is usually the rainy season. Thailand experienced a summer drought, while crop production in India was set back by a delay in the start of the annual monsoon⇨.

Four types of drought are recognized:

In hot and cold desert regions of the world, drought is more or less permanent. In these areas, plants are specially adapted to the harsh environment and crops can be produced only through continual irrigation⇨ using groundwater sources.

Regular, seasonal drought occurs in parts of the world where there is a well-defined cycle of rainy and dry seasons. Where crops are grown, planting must be carried out so that the crops develop during the rainy season.

In some instances, droughts occur where there is an unusual and unexpected rainfall failure. This can happen anywhere but is a frequent phenomenon in humid and subhumid climates. These unpredictable droughts tend to be irregular, of limited duration, and often localized.

Under certain circumstances, especially when high temperatures prevail, frequent showers may not be sufficient to replace the amount of water lost through evaporation or transpiration through plants. The result is a prolonged, gradual water deficiency that eventually reduces crop yields.

DUSTBOWL A name given to an extensive area of arable land in the Great Plains in the United States that was devoted to large-scale monocropping of cereal crops, mostly wheat⟳. Agricultural production was extended into areas where rainfall was frequently below 25–30 cm per year. As a result of a combination of the intensive monocropping⟳ and the prolonged droughts⟳ that struck the area during the 1930s, massive soil erosion⟳ became commonplace and vast quantities of precious topsoil were blown away to form frequent dust storms. Millions of hectares of land were laid to waste across five states, and a single four-day storm in 1934 blew away 300 million tonnes of soil. Thanks to a virtually continuous application of huge quantities of agricultural chemicals and water, much of the land has been rehabilitated to a point where it can be used to grow crops once again.

DUTCH ELM DISEASE A serious tree disease caused by a fungus, *Certacystus ulmi*, which is carried by two species of bark beetles. First described in the Netherlands in 1919, the disease spread rapidly. By 1929 it had crossed to the United States and Canada.

The beetles use the bark of trees killed by the disease as breeding sites. They fly from tree to tree, inadvertently carrying spores of the fungus with them. The fungus blocks the vessels in the tree that carry water to the leaves, which subsequently wilt and die. Preventive measures can be taken to avoid the disease but they are expensive and therefore not widely implemented.

The disease was originally introduced into England in 1931 when, during a six-year period, 20 per cent of the elm trees in the south of the country were killed. During the 1960s a new virulent strain of the fungus developed in the northern USA and Canada. This new strain was thought to have been introduced into Britain in 1964 by beetles imported on logs from Canada. Two years later, nine million British elms out of a total of 23 million had died, and virtually all the elm trees in southern England were wiped out.

E

E NUMBERS A system of identification for food additives introduced in Europe (hence the 'E' for European) during the late 1970s to designate additives that are regarded as safe for human consumption. All processed food produced after January 1986 must have E numbers or the actual name of the additive displayed in the list of ingredients. Taste additives, the largest group, do not have E numbers. The authority for allocating an E number lies with the European Council of Ministers, acting on the advice of the European Commission and its Scientific Committee for Food.

Previously, consumers in Europe had little or no information about which chemicals were being used in the food and drinks they were buying. The initiative to develop the E number system was taken by the food industry in response to proposals to ban all food additives which were suspected of causing ill health. Under the system, E100 numbers are generally colouring agents; E200s are mainly preservatives and acids; E300s are antioxidants and acid regulators; E400s include emulsifiers, stabilizers, thickeners and bulking agents. In addition, some additives are represented on labels by numbers with no E prefix, signifying that they are under consideration for being ascribed an E number. Other additives – including aspartame, caffeine, gelatine, iodine, iron⟳ and some B-group vitamins – have no designated E number. Introduction of the system has basically shifted the onus of responsibility on to consumers to decide which additives are safe for them to eat.

EARTHWATCH A wide-ranging programme designed to monitor the state of the world environment, assess the impact of natural and man-made changes, identify the causes, suggest preventive measures, post warnings of impending hazards, and communicate the information to governments, scientists, industrialists and other concerned agencies. Established and co-ordinated by the United Nations Environment Programme (UNEP)⬦, Earthwatch operates through three main components: the Global Environment Monitoring System (GEMS)⬦, the International Register of Potentially Toxic Chemicals (IRPTC)⬦ and the International Referral System for Sources of Environmental Information (INFOTERRA)⬦.

EAST AFRICAN DEVELOPMENT BANK (EADB) An independent, intergovernmental bank initially founded in 1967 as part of the now-defunct East African Community. Membership was originally restricted to members of the Community – notably Kenya, Tanzania and Uganda – but was later opened to other nations. The Bank aimed to provide finance for industrial development projects in East Africa in a manner designed to create an equitable balance in industrial capacity between member states. Its mandate was broadened in 1980 to include loans to promote agriculture⬦, tourism⬦, transport and communications.

EASTERN EUROPE In addition to enormous problems of trade and employment as they move towards a free-market economy, the former centrally planned, Communist-led nations of Eastern Europe are having to face up to massive and widespread environmental problems. Any attempts to build sound economies in these countries will depend heavily on the success of the attempts to repair their natural environment and conserve natural-resource bases. In Hungary, investment in environmental protection amounts to less than 1 per cent of the Gross Domestic Product (GDP), while environmental damage reduces GDP by 2–5 per cent. Poland is believed to lose the equivalent of 10 per cent of its Gross National Product (GNP)⬦ as a direct result of environmental pollution. Investment of 4–5 per cent of national income could improve the environment on a par with that seen in Western Europe, but eliminating major pollution in Poland will cost at least $20 billion.

The cost of cleaning up the badly polluted Eastern European nations will not only be high; it will have to be borne primarily by external sources, for Eastern Europe is heavily indebted already: to the tune of $150 billion in 1990. At the first meeting of East and West Environment Ministers in 1990, the Eastern delegates announced their governments' intention to instigate environmental protection and improvement programmes and confirmed that they would all participate in the activities of the European Environment Agency⬦. The European Community⬦ also announced that one third of the $580 million set aside for developing Eastern Europe would be spent on improving the environment. A code of conduct was also initiated which will prevent EC-based companies moving into Eastern Europe to take advantage of lax environmental controls.

The true extent of environmental degradation in Eastern Europe was also pointed out at the joint meeting by ministers of the countries concerned. In Czechoslovakia, life-expectancy is seven years lower than the EC average. The lignite – or brown coal⬦ – used in the country contains a high level of sulphur, with the result that the nation has the world's worst sulphur dioxide⬦ pollution⬦, exceeding pollution limits set in the West by twenty times. Over 70 per cent of the nation's trees are being damaged by acid rain⬦. In addition, departing Soviet troops have left behind a legacy of extensive environmental damage, including waste oil⬦ which has permeated into underground water⬦ supplies.

The Polish Minister described the industrial belt of Silesia in Poland as 'the most polluted part of Europe', and added that one third of Polish rivers are unfit for any use. The former East Germany was represented at the meeting and reported that 700,000 tonnes of waste were imported into the country from neighbouring West Germany. The unification of Germany now means that the former Eastern part will have to comply with the strict environmental protection legislation that prevailed in the West. In 1990, 55 per cent of forests⬦ in the former East Germany were damaged, 40 per cent of waste water was discharged untreated, and most of the former nation's nuclear facilities were closed on safety grounds. As part of the vast clean-up bill that will have to met by the West's taxpayers, huge amounts of revenue and technology will have to be devoted towards cleansing the 15,000 waste dumps, many of them illegal, into which most of East Germany's toxic wastes were poured untreated.

Economic reforms in the Soviet Union will necessitate swingeing changes in industry and environmental protection. Part of the problems of the past lay in the fact that a state monopoly on natural resources⬦ such as oil and timber led to extensive overconsumption. Consumption of natural resources per unit of national income in the

Soviet Union is estimated to be two to three times that in other countries. As well as improving efficiency of resource use, as part of bilateral agreements on environmental protection the USSR is committed to a 30 per cent reduction in sulphur emissions in the European part of the country, a 50 per cent cut in the territory adjoining Finland, and a similar reduction in pollution discharged to the Gulf of Bothnia. In addition, nature reserves in the USSR cover only 0.25 per cent of the total territory, whereas it is recognized that to conserve the nation's full range of biodiversity⌾, 6 per cent of the land area needs to be protected. A Resolution has been adopted by the Supreme Soviet that would extend protection to 3 per cent of the land⌾ by the year 2000, but at a cost of $1 billion.

The cost of improving the environment in the Soviet Union, as anywhere else, will be high. For many it will be too high, and many industries, unable to meet the USSR's stricter pollution-control levels, will be forced to close down. Foreign-currency reserves of $20 billion will be needed to facilitate the move toward a more sustainable and environmentally friendly economy, currency which the Soviet Union does not have to spare. The success of environmental protection and improvement measures throughout Eastern Europe will therefore depend heavily on contributions from the West. However, in 1991 the donor countries were still locked into a policy debate about the weight to give to market-led environmental improvement compared to that devoted to investment in public infrastructure.

ECOLOGY The study of the relationships between all living organisms, plants, animals and humans, and the environment, including the ways in which human activities affect other wildlife populations and alter natural surroundings. Modern ecology dates from the work of Charles Elton in the 1930s and is concerned with the relationship of different species both with each other and with biological, physical and chemical components of the environment, or habitat, in which they live. A community of organisms, of either the same or differing species, and the habitat in which they live is called an ecosystem⌾. The effect of human intervention on such ecosystems can be predicted to some extent, enabling effective conservation of wildlife and management of wildlife resources. An understanding of ecology has enabled scientists to develop methods for the biological control⌾ of pests and means to improve crop production.

ECONOMIC COMMUNITY OF WEST AFRICAN STATES (ECOWAS) Established in 1975, with sixteen member states, ECOWAS's aims are the production of a common Customs tariff and commercial policy; promotion of free circulation of people, labour, services and capital within the Community; harmonization of agricultural policy; and to secure joint development of transport, communications, power, industry and other infrastructure.

ECONOMIC EXCLUSION ZONES (EEZs) Agreed as part of the Law of the Sea,⌾ EEZs extend 200 nautical miles (370 km) seawards from all coastlines, the countries concerned being able to exercise control over various matters such as fishing⌾ and pollution⌾ control within their own zones. Shipping has right of passage in these waters, but rights over the mineral⌾ resources on the seabed in EEZs remain a matter of international debate. Where EEZ boundaries overlap, the EEZ is deemed to extend to a point equidistant between the nations concerned. Ocean areas claimed under EEZ regulations total about 24.5 million square nautical miles – almost equal to the earth's land⌾ surface – and cover 99 per cent of living marine resources commonly exploited by world fisheries. Most of the world's major fishing⌾ grounds lie within the EEZs; several small island states, notably those in the Pacific Ocean, have benefited substantially from the introduction of the zones. As a result of the EEZs, areas of 'open ocean' were reduced by approximately 30 per cent. The Law of the Sea has yet to be officially ratified by the requisite number of states to become international law. The United States is the only nation not to recognize exclusive fishing rights within declared EEZs.

ECONOMIC GROWTH An expansion in output of a nation's economy, traditionally measured in purely economic terms, usually determining the increase in the Gross National Product (GNP)⌾. Economic growth is regarded as desirable because many politicians and economic authorities consider it to be the best way of raising living standards. However, it tends to have drawbacks such as increased levels of pollution⌾ arising from greater industrial activity, depletion or despoliation of natural resource reserves, and environmental degradation which, at present, are not taken into account when the GNP is calculated. For comparative purposes and to allow an evaluation of improvements in living standards, economic growth is measured in *per capita* terms. In a developing country⋄ the rate of economic growth may be high but the actual rate of growth *per capita* may be much lower if the population⋄ is increasing at a relatively rapid rate.

Until 1990, improvements in the living standards of nations continued to be outwardly judged by the level of improvement in *per capita* GNP performance. The United Nations Development Programme (UNDP)⋄ then refined this process and began producing its annual Human Development Report⋄ containing a Human Development Index, which takes into account social and demographic as well as economic pointers in its compilation.

ECONOMIC AND SOCIAL COUNCIL OF THE UNITED NATIONS (ECOSOC)

One of the principal organs of the United Nations⋄, responsible to the General Assembly, ECOSOC determines the UN's economic and social activities. The fifty-four-member council seeks to improve living standards and to help solve economic and social problems throughout the world – by initiating studies, preparing reports and making recommendations in these fields, organizing international conferences and co-ordinating the activities of specialized agencies. Six functional commissions are concerned with statistics, population⋄, social development, human rights⋄, the status of women, and narcotic drugs. There are also five regional economic commissions:

Economic and Social Commission for Asia and the Pacific (ESCAP) – Originally launched as the Economic Commission for Asia and the Far East (ECAFE) in 1947, the title was changed in 1974. Its activities include the establishment of regional centres for training and research purposes, the promotion of schemes and projects for regional co-operation, the distribution of information and the provision of advisory services over a wide range of subjects.

Economic Commission for Africa (ECA) – Established in 1958 to assist in Africa's social and economic development. Its activities include collecting information, research, recommending national government and intergovernment action, and the provision of advisory services. It also promotes studies into the economic and technological development in the region.

Economic Commission for Europe (ECE) – Established in 1947 to assist in the economic reconstruction of Europe and to maintain the economic relations of European nations both among themselves and with

other countries in the world. It has become a central agency for the promotion of trade, the exchange of technical and statistical information, and research on and analysis of all aspects of economic development in the region.

Economic Commission for Latin America (ECLA) – Established in 1948 to assist in the economic development of Latin America.⋄ ECLA activities include the collection of information for research, recommending government and interregional action, and co-ordination of advisory services. It was strengthened in 1962 by the creation of the Latin American Institute of Economic and Social Planning (ILPES).

Economic Commission for Western Asia (ECWA) – Established in 1973 to initiate and participate in the process of economic reconstruction and development in western Asia. The Commission aims to raise the level of economic activity in the region, promote trade and maintain and strengthen the economic relations of the countries in the region both among themselves and with the outside world.

Economic Co-operation among Developing Countries (ECDC) – The idea of fostering economic co-operation among the developing countries⋄ of the world was first proposed at a meeting of the Group of 77⋄ in 1976. As a result, a Committee on ECDC was subsequently established by the United Nations Conference on Trade and Development (UNCTAD)⋄. The Group of 77 attach considerable importance to this concept, and in 1979 they drew up a 'programme for collective self-reliance' which covers a broad range of issues, including mutual trade, regional development, transfer of technology, special action for the least-developed⋄ and geographically disadvantaged developing countries⋄, and the creation of new institutions and transnational⋄ enterprises.

ECOSYSTEM

A scientific term used to describe a community of organisms and their environment. It can apply equally to a small, isolated patch of land⋄ or an entire planet. The term was first introduced by an Oxford ecologist, A.G. Tawnsley, in 1934 to describe the basic functional unit in ecology⋄. An ecosystem includes all closely related living organisms in a designated area, as well as the physical environment. These are deemed to interact and to be inseparable.

ECOSYSTEM CONSERVATION GROUP
(ECG) The ECG is composed of experts and official representatives from the United Nations Environment Programme (UNEP)⊙, Food and Agriculture Organization (FAO)⊙, Educational, Scientific, and Cultural Organization (UNESCO)⊙ and the World Conservation Union⊙ (IUCN), who meet periodically to discuss matters of common interest. The group attempts to develop a degree of co-ordination with regard to the conservation⊙ components of the various programmes run by each organization.

EDUCATION
The 1948 Universal Declaration of Human Rights affirmed everyone's right to a basic level of education, yet United Nations⊙ predictions suggest that by the year 2000 the illiterate population of the developing world alone will have reached one billion.

In 1985, 105 million children aged six to eleven were not attending school. Of this total, 70 per cent were in the Least-Developed nations⊙ and 60 per cent were girls. By 1990, although almost all the world's male children were receiving at least one or two years' primary education, 100 million children still had no chance of schooling whatsoever. If trends continue, it is estimated that by the end of the century 200 million children will not be receiving a proper education.

The average global expenditure on educating a child is $350, but this masks a huge discrepancy between what is spent in the industrialized world ($6,000 per child) and in the Least Developed Countries ($2 per child).

At the end of the 1980s, one in four adults – over 900 million people – were thought to be illiterate. The 1990 World Conference on Education set an ambitious target of cutting adult illiteracy⊙ to 15 per cent by the year 2000. The Conference also emphasized the cost-effectiveness of education. Primary schooling is known to improve worker productivity both in the factory and in the field, as well as providing the skills for self-employment and entrepreneurship. In the United States, at least 70 per cent of all cancer⊙ deaths are linked to behaviour that can be changed by education, and the only effective means to combat Acquired Immune Deficiency Syndrome (AIDS)⊙ currently available is via education. Similarly, education can help prevent degradation of the environment and is a useful tool in population⊙ control. Females who receive a basic education usually have fewer children, and their offspring's chances of survival increase significantly.

ELECTROMAGNETIC FIELDS
Scientists from around the world have discovered that simply exposing seeds briefly to electrical or magnetic fields before they are planted helps to improve crop yields by more than 10 per cent, although the reasons for this remain little understood.

All plants and animals are exposed to the earth's weak electromagnetic fields, which influence their growth. Researchers in Japan have found that exposing seeds to strong magnetic fields causes the plants to grow more flowers and increase yields. Kidney and soya beans exposed to magnetic fields ten times greater than the earth's germinated more quickly and were ready for harvesting in a much shorter space of time than normal, with yields rising by 33 per cent. Soviet scientists have proved that merely watering crops with magnetized water⊙ can raise yields by 15 per cent. It is thought that magnetized water raises the microbiological activity of the soil, making it easier for plants to take up nutrients such as nitrogen, potassium and phosphorus.

Exposure to electrical fields produces similar effects. Tomato and gourd seeds in India soaked in water and exposed to electrical fields of differing strengths for less than one minute show increased germination rates and yield rises of 15 per cent. Similar results have been obtained by scientists in China, who found that treated seeds showed an average increase in yield of 13.5 per cent over untreated seeds. The treatment proved effective for wheat⊙, rice⊙, cotton⊙, soya beans, corn, tobacco⊙, cucumber and tomato. It is possible that exposure to electrical fields may induce a heating effect in the seeds, helping them to germinate earlier, thus increasing the growing period and subsequently raising yields. Similarly, as seeds with a high moisture content exhibit the best results, the growth promotion may be due to a magnetization of water within the seed.

ELECTROMAGNETIC RADIATION
The world now boasts a wide range of sources emitting low-level electromagnetic radiation, including high-voltage power lines, electric blankets, microwave ovens, radar dishes and radios. Minute electrical currents are also generated by all living organisms, and in animals they are important regulators of the central nervous system. Magnetic fields are developed whenever an electric current is passed and the brain and all other organs produce very low-level magnetic fields detectable outside the body as a

result. Deep in the brain, the pineal gland is sensitive to all magnetic fields, produced internally or externally, and alters its output of neurohormones in synchrony with any changes. These hormones govern brain activity and the function of other glands that are responsible for controlling all normal bodily growth and functions.

The 'electromagnetic smog' created by modern-day electrical devices exposes individuals to significantly higher radiation levels than normal: the average United States citizen receives a daily electromagnetic radiation dose 200 times as high as that from natural sources. In particular, the extremely low frequency (ELF) radiation (or microwaves) produced by high-voltage power lines is of growing concern. In the mid-1970s evidence emerged to indicate that ELF fields one million times lower than those commonly found under high-voltage power lines caused significant biological effects. A study of employees at the US Embassy in Moscow, which was receiving a regular bombardment of microwaves as part of Soviet surveillance operations, revealed a 40 per cent incidence of blood cell disorders amongst employees.

Low-frequency electromagnetic radiation was assumed to be harmless, as it is non-ionizing; the only danger was thought to lie in its ability to heat tissues, as happens in a microwave oven. During the 1980s, however, the effect of exposure to this low-level radiation began to be more fully appreciated. A 1981 World Health Organization (WHO)⌀ report on the health effects of microwaves concluded: 'exposure led to the appearance of autonomic and central nervous system disturbances', producing a variety of deleterious effects. Exposure to abnormal magnetic fields disupts biological cycles and causes effects similar to chronic stress syndrome. This can ultimately result in weight loss, weakened immune systems, decline in fertility, low birth weights, diminished resistance to infections, and serious psychological disorders. Normal cell division is also affected, increasing the risks of cancer⌀.

In 1982, three years after a high-voltage overhead electricity power line was constructed a few hundred metres from a dairy farm in New York, cows on the farm had produced dead or defective calves and all the chickens produced defective, unsaleable eggs. Three members of the farming family developed auto-immune diseases, including Hashimoto's disease, an irreversible disorder of the thyroid gland. A US study later found that cancer rates are five times higher among children living near high-voltage electricity pylons.

The strength of magnetic fields produced by power lines depends on a variety of factors including the voltage, local topography, weather conditions and local geological structure. Voltages in overhead power lines vary from country to country: the Soviet Union and USA use up to 765 kV, while the United Kingdom uses 400kV.

Despite its use of microwaves to 'bug' foreign embassies, the Soviet Union regards all forms of ELF radiation as harmful and sets exposure limits 10,000 times stricter than those in the United States. It is also thought to have introduced legislation to specify a one-kilometre safety corridor for its power lines.

Building on the knowledge that pulsed microwaves can be highly psychoactive and can cause neurological changes, the United States has embarked on a major programme to develop electromagnetic weapons and hopes to produce low-level microwave beams capable of immobilizing and disorientating opposing soldiers.

EL NINO A poorly understood recurrent climatic phenomenon that affects the Pacific coast of South America but appears to have a dramatic influence on weather patterns much farther afield as well. A surge of warm ocean waters recurs every seven to fourteen years in the eastern Pacific Ocean. The effects are generally first felt in December (the name is Spanish for 'the Christ Child').

The phenomenon is thought to begin with a reduction in the trade winds in the tropical South Pacific. This reduces the pushing effect on surface waters in the ocean, allowing warm surface water in the eastern Pacific to accumulate. Where warm water accumulates, the upwelling of colder water is prevented. Starved of the nutrients that this upwelling normally produces, the surface waters are impoverished. Many seabirds starve and thousands of fish⌀ die, dramatically affecting commercial fisheries as a result. Research has also indicated that salinity and the exchange of heat from the ocean's water to the air in a region to the north of Papua New Guinea, where El Niño is thought to originate, could be an important factor. Salt water has a greater density than fresh water, and an incoming block of salt-dense water could deflect the warm current off-track and so trigger the El Niño effect, causing severe climate disruption elsewhere in the world.

A wide variety of disasters have been blamed on the El Niño effect, including a famine in Indonesia in 1983, bush fires in Australia arising from droughts◇ caused by El Niño, rainstorms in California and the destruction of the anchovy fishery off the coast of Peru. Climatic changes in Central and North America and even in parts of Africa, including the droughts in the Sahel, have been said to have been heavily influenced by the El Niño effect.

ELEPHANTS The African elephant has long been hunted for its tusks, which are the major source of ivory◇. Over the past fifty years or more, the natural habitat of the elephant has also disappeared as a result of human population◇ expansion and agricultural activities. In 1990, figures produced by experts from the Ivory Trade Review Group (ITRG) and the Worldwide Fund for Nature (WWF)◇ indicated that there were only 625,000 African elephants left in the wild. If hunting◇ trends continue, they estimate that wild populations will be extinct by the year 2005. The Asian elephant, which is exploited more for work than for its tusks, is also in decline. There are only an estimated 30–55,000 animals left in the wild, less than 10 per cent on the population of African elephants.

For the African elephant to survive, many consider it essential that the ivory trade be stopped, but several nations and some officials from the Convention On Trade in Endangered Species of Wildlife (CITES)◇ are in favour of controlled trade continuing. Half of the CITES Ivory Unit's $400,000 annual budget was derived from the ivory trade. In 1989, 60 tonnes of ivory were reportedly traded, worth an estimated $8.6 million. Following a fierce debate in 1990, the African elephant was moved from Appendix II of CITES (which allows controlled trade) to Appendix I (which allows no trade whatsoever). However, despite these efforts to minimize hunting, illegal hunting continues and other nations – including Botswana and Zimbabwe, where elephant numbers are increasing to the extent that animals can be culled annually – have taken a 'reservation' allowing them to trade in ivory legally.

Problems of population dynamics may well see the fears of extinction◇ realized. Male elephant tusks provide the most ivory, so males are selectively killed. In 1979, the mean weight of tusks on the market was 10 kg. By 1988 it had dropped to 4.5 kg, meaning that twice as many animals were needed to produce one tonne of ivory, and younger and younger male elephants were being killed. As a result, in some regions there are as many as 99 females to 1 male. This does not help in maintaining the breeding patterns of an animal with an extremely slow reproductive cycle. Females do not begin to reproduce until they are twenty-five to thirty years old.

Exploitation of elephants has been unsustainable since the 1950s, but the ITRG maintains that Africa can still sustainably produce 50 tonnes of ivory annually. In 1987, however, around 350 tonnes left the continent. The demand for ivory is highest in Japan, where ivory stamps are traditionally used instead of handwritten signatures. In 1988, Japan imported 30 tonnes of worked ivory and 110 tonnes of raw ivory. No economic pressure can be brought to bear on Japan, and it is highly unlikely that scientific reason will overcome hundreds of years of Japanese tradition, despite the fact that alternative 'ivory' from vegetable and artificial sources is now available.

EMPLOYMENT Many people consider that the right to work is a basic human right◇, but employment opportunities in most developing countries◇ have traditionally lagged well behind population◇ growth. Globally, the labour force is increasing at around 2 per cent per year. The International Labour Office (ILO)◇ suggest that between 20 and 25 per cent of urban adults in the Third World◇ are without regular work. Furthermore, the total labour force in the developing world is expected to rise dramatically: from 1.2 billion in 1980 to 1.9 billion by the year 2000. If this is true, it means that to attain an acceptable level of employment in the Third World, more jobs will have to be created than currently exist within the entire industrialized world.

One of the traditional solutions to this dilemma has been the migration of labour, and the economies of many developing countries have benefited substantially in the process. For all the developing countries, the percentage of export earnings contributed from remittances from nationals working overseas doubled between 1970 and 1982. It now stands at just under 2 per cent, running at around $22 billion. This avenue has suddenly begun to close and there has been a mass exodus of foreign workers from northern Europe and the Gulf states following economic depression and the downturn in oil◇ prices. Between 1977 and 1982, 90,000 migrant foreign workers opted for repatriation and left France. Some 200,000 left the Federal Republic of Germany in 1982 alone, and the same year Iran and Iraq also saw 200,000 foreign workers repatriated.

In the developing world relatively few people have paid employment, although this does not mean that they are idle. The so-called 'informal sector' or 'black economy' predominates in many parts of the Third World. Estimates for Africa and Asia alone show that only 10 per cent of adults work officially for regular wages, the majority of the rest rely on subsistence agriculture⚬, payment in kind, or some form of bartering of goods or services, in order to survive. Additionally, the role of women is often belittled or overlooked entirely. An ILO study found that women's employment between 1945 and 1989 declined by 67 per cent in Algeria and Cape Verde and by 50 per cent in Bangladesh, Bolivia and the Maldives. The reason for this was traced to the fact that women are often viewed as unpaid, unofficial workers on family farms and businesses. This effectively rendered millions of working women invisible in 62 of the 83 nations studied, despite the fact that in many of these countries women do the bulk of the agricultural work.

No matter which country, developed or developing, many people are forced to work under appalling conditions and for pitifully low wages. Discrepancies in pay and conditions are likely to be maintained over the next decade or so, despite the best efforts of those campaigning for equality. However, unless extremely careful and well-considered planning is carried out by the global community, the next century may see a dramatic change in working practices and a possible further polarization, both within individual countries and between North⚬ and South⚬. In the developed countries during the next century, many women, all those people over forty-five, the disabled and the young – hundreds of millions of people – will probably be unemployed as industries become more and more automated, with only jobs in the service industries likely to increase significantly. Automation creates leisure time for the wealthy but does away with work for the poor, thereby increasing the gap between rich and poor. However, as the developed nations come to terms with a shrinking workforce, increasingly aging populations and growing dependence on machines, a complete change in the work ethic may be necessary. Meanwhile, in the South countries will probably be facing an ever-increasing number of young people clamouring for work. Newly Industrializing Countries⚬ such as Brazil and Mexico may well have potentially more jobs and markets than other countries, but they may have to concentrate on labour-intensive, low-technology industries to provide an acceptable lifestyle for the majority of the population.

ENDANGERED SPECIES   There is no universally accepted set of definitions to determine the status of wild plants and animals. The meaning of terms like 'threatened species' varies according to the organization reporting. Groupings differ; it is therefore difficult to assimilate data from different sources with any degree of confidence.

The following definitions are those employed by the Species Survival Commission of the World Conservation Union⚬ (IUCN) and are accepted for use by international bodies such as the Convention on Trade in Endangered Species of Flora and Fauna (CITES)⚬.

Extinct – Species not definitely located in the wild during the past fifty years.
Endangered – Species in danger of extinction, whose survival is unlikely if the causal factors continue operating.
Vulnerable – Species believed likely to move into the 'endangered' category in the near future if the causal factors continue operating.
Rare – Species with small world populations that are not at present 'endangered' or 'vulnerable'.
Indeterminate – Species known to be 'endangered', 'vulnerable' or 'rare', but not enough information is available to determine which category is appropriate.
Insufficiently known – Species that are suspected of belonging to one of the above categories but are not definitely known to be due to lack of information.
Threatened – A general term used to denote species which are in any of the above categories.

ENERGY   The first direct human use of energy came with the discovery and adaptation of fire. Combustion is a process for turning chemical energy into heat energy. The first fuel exploited was wood⚬, and well over 2 billion people still rely on the non-commercial use of fuelwood⚬ for their main source of energy. Fossil fuels⚬ such as peat⚬, coal⚬, oil⚬ and natural gas⚬ have been extensively used for about 8,000 years. They are all burned to produce commercial energy to power industrial processes and provide people with light and heat. The Industrial Revolution, the development of motorized transport, mechanization, and the spread of technology during the twentieth century have resulted in a huge increase in demand for fossil fuels, some of which may soon be exhausted. If 1986 production rates are maintained, coal reserves will last for 220 years, natural gas for 59 years and oil for slightly over 30 years.

Initially, coal was the main source of commercial energy, accounting for 80 per cent of global use in the 1920s. This fell dramatically as reserves of oil, and later natural gas, were discovered and exploited. The change to reliance on oil led to coal providing only 30 per cent of commercial energy needs in 1984, while use of natural gas, already providing about 20 per cent of global energy, was showing an annual 4 per cent increase.

The global consumption of commercial energy increased threefold after 1950 but slowed down during the early 1980s following the oil crisis, reaching around 310,000 petajoules by the middle of the decade. Consumption is heavily concentrated in the industrialized countries of the West and East. These regions, housing 30 per cent of the world's population⬦, use 80 per cent of global commercial energy production. By 1990, the world was using a total of 320,000 petajoules of energy. Projected demands for the year 2000 vary widely, ranging between 400,000 and 500,000 petajoules. Proven commercial energy resources amounted to 30 million petajoules at the end of 1986.

Pollution⬦ problems associated with the use of fossil fuels and dwindling reserves have focused attention on alternative sources of energy and the question of how future demands are to be met. Nuclear energy⬦, first produced on a commercial scale in 1954, already produces 17 per cent of the electricity generated around the world (5 per cent of global energy production), and alternative sources of renewable energy⬦ are being evaluated and developed – by 1986 hydroelectric power⬦ already supplied 21 per cent of global electricity (7 per cent of global energy). More efficient use of the energy that is already being produced would ease matters. In the industrialized world, over half the energy used for transport, industry, agriculture⬦ and households is wasted. Between 1973 and 1985, energy-conservation measures and improved industrial efficiency in developed countries within the Organization for Economic Co-operation and Development (OECD)⬦ reduced energy intensity (defined as the amount of energy used to produce a unit of Gross Domestic Product [GDP]) by 20 per cent.

Simple measures to improve inadequate technology or curb profligate human usage can have a significant impact. Under a 1990 US law governing the use of energy-efficient refrigerators, 90 per cent of inefficient pre-1987 models will be outlawed. This is forecast to save $28 billion by the year 2000 and avoid the need to build twenty-five large power stations. Studies in industrialized countries suggest that if all currently developed energy-conservation measures were implemented, energy efficiency would be 30 per cent higher by 2000, resulting in a 25 per cent drop in energy consumption without any fall in industrial productivity or standards of living.

ENVIRONMENTAL DAMAGE TREATY An initiative proposed by the European Committee on Legal Co-operation of the Council of Europe⬦. The international agreement aims to improve the levels of compensation to be paid for environmental damage caused across international borders in Europe. Accidents which result in large-scale environmental damage often affect several nations. The Treaty will complement existing special arrangements dealing with environmental damage caused by nuclear accidents, hazardous waste⬦, and marine pollution⬦.

ENVIRONMENT LIAISON CENTRE (ELC) An international Non-Governmental Organization (NGO)⬦, founded in 1975 and based in Kenya, which aims to strengthen the capabilities of NGOs working in all fields of the environment and sustainable development⬦, particularly NGOs in the Third World⬦. The ELC provides information, advises on project design and implementation, administration, fund-raising and media relations. It is able to provide limited financial assistance to project initiatives designed to help build NGO collaboration and encourage or widen networks, or to support international conferences. In addition to helping forge links and co-ordinate activities between NGOs from around the world, the ELC plays a key role in facilitating NGO input into and support for the United Nations Environment Programme⬦, the UN Centre for Human Settlements and other specialized UN agencies. The ELC has a membership of about 250 environment and development NGOs from seventy countries and maintains permanent contact with many others.

ESPERANTO An artificial language invented by a Polish philologist, L.L. Zamenhof, in 1887. It was intended to be a universal medium of communication and is recognized as the most successful of all artificial languages, being spoken by over 100,000 people. Its grammar is regular and pronunciation is in line with spelling.

ESSENTIAL DRUGS   In 1977, the World Health Organization (WHO)⟡ launched a policy on essential drugs, emphasizing that most serious health problems could be effectively overcome or treated using relatively few drugs, all of which had proven therapeutic values and were safe to use. To demonstrate the problem they were trying to overcome: during the late 1970s over 1,000 brands of medicine were available for prescription in Norway. In India and other developing countries⟡ the figure was in excess of 15,000. From the tens of thousands of different drugs produced around the world, the WHO compiled a list of around 250 'essential' drugs, those which were considered prerequisite to the needs of the sickest and poorest people in the world.

Implementation of the WHO policy would lead to the elimination of undesirable drugs, curb over-medication and promote the use of cheaper 'generic' drugs. Brand-name drugs usually cost considerably more than the basic, generic preparation. Doctors in Britain, for example, commonly prescribe a brand of the drug frusemide that costs up to nine times more than the generic version, causing the National Health system to lose almost $10 million a year. However, the WHO has been unable actively to promote the essential drug policy, as it is merely an agent of its member governments, many with vested interests. Major transnational⟡ pharmaceutical companies also oppose the scheme, as it would severely restrict their operations. The world's leading drug-producing nations – France, the Federal Republic of Germany, Italy, Japan, the United Kingdom and the United States – control about 75 per cent of the global trade in drugs. They also contribute over half of the WHO's budget. Under $5 million is allocated to the essential drugs programme each year, less than is normally spent on commercial promotion of any new drug.

ETHANOL   Ethanol, or ethyl alcohol, produced through fermentation processes, is the active ingredient of alcoholic drinks. It has widespread industrial uses and is also increasingly being used as a fuel. Many agricultural crops such as corn⟡ or rice⟡, and virtually all vegetative residues, can be converted into ethanol. In the United States, ethanol is produced from corn for use as a vehicle fuel. Around six tonnes of corn produced on a hectare of land⟡ will yield about 2,200 litres of alcohol. The ethanol can be used directly as a fuel, or mixed with ordinary petroleum (or gasoline) to produce 'gasohol'. Car engines need no modification if the gasohol mixture contains less than 20 per cent ethanol.

Added to petrol, alcohol increases its octane rating and eliminates the need for lead⟡ additives. Engines running on alcohol are 18 per cent more powerful than those running on petrol, and emissions of hydrocarbons and carbon monoxide are significantly reduced. Gasohol is widely used in the USA and in several developing countries⟡, notably Brazil, Kenya and the Philippines. Brazil has the largest programme for making alcohol from biomass⟡, using its vast supplies of sugar cane. The country's National Alcohol Programme (PROALCOOL) intends to produce in the region of 11 million cubic metres of ethanol annually. More than a million vehicles in the country run on pure alcohol, and virtually all passenger cars run on gasohol. Fermenting surplus sugar⟡ could produce enough biofuel to satisfy 2 per cent of the current global market for petrol.

EUCALYPTUS   A genus of tropical and subtropical evergreen trees, containing almost 600 species, native to Australia but now widely planted elsewhere. Also known as gum trees, they are fast-growing, drought-resistant and commercially important sources of timber and other products. The wood can be used for fuel, building and fencing, the bark for paper.⟡ All parts of the tree contain essential oils; the oil of eucalyptus commonly used in proprietary medicines as an inhalant comes from the leaves of the blue gum tree.

In their natural habitat, eucalypt forests prosper on poor soil, are adapted to frequent fires and tolerate numerous environmental and ecological conditions, supporting a wide range of arboreal fauna. They are being logged extensively; the bulk of Australian eucalyptus timber is converted to wood chips to make paper pulp. The wood chips are exported to Japan: approximately 5 million tonnes annually.

Commercial forests⟡ are long-term investments, harvesting occurring every forty to eighty years. They are also particularly damaging. Soil nutrients are severely depleted and many of the unique Australian fauna are unable to utilize young trees. Forests will not be suitable for them until they reach the age of 150, by which time the animals may well have become extinct.

Eucalyptus trees take up a great deal of water⟡ when growing and are used to reclaim marshy land⟡. They are also used extensively in reforestation⟡ projects, and 40,000 sq km around the world is now under eucalyptus. The fast-growing trees are used to protect areas threatened by desertification⟷, even though eucalyptus is not an ideal candidate for this purpose. Its thirst for water inhibits the growth of other plants, depriving farmers of fodder and preventing the replenishment of groundwater⟷, causing streams to run dry. Eucalyptus trees are also relatively vulnerable to a variety of pests, they produce toxins which, in areas of low rainfall, are not flushed from the soil; and they commonly deplete soil of nutrients, returning far less in leaf litter than they absorb through their roots.

EUREKA PROJECT   A European advanced-technology co-operation programme. Originally proposed by the French government as an alternative to the collaborative Strategic Defence Initative (SDI)⟷ programme adopted by the United States, the Eureka project was launched in 1985, with seventeen European nations participating, and aims to find applications for peaceful means of joint research initiatives. It was designed to promote international co-operation between governments and industry. By 1990, $8 billion had been spent, including $272 million on 74 environmental-improvement projects. The Japanese government's Human Frontiers Programme⟷ is a similar venture.

EUROPA NOSTRA   An international federation of approximately 2,000 independent Non-Governmental Organizations (NGOS)⟷ in twenty-three European countries devoted to the preservation of Europe's natural and cultural heritage. Established in 1983, with an office in London, it acts in an advisory capacity to the Council of Europe⟷. It also organizes international conferences, exhibitions, expert studies and annual European Conservation Awards. The name originates from the Italian, meaning 'our Europe'.

EUROPEAN ATOMIC ENERGY COMMUNITY (EURATOM)   An organization formed by six members of the Western European Union (WEU) – Belgium, France, the Federal Republic of Germany, Italy, Luxembourg and the Netherlands – to promote co-operative research and development into the peaceful uses of atomic energy. Its governing bodies, the European Coal and Steel Community (ECSC)⟷ and the European Economic Community (EEC), were amalgamated into the European Commission in 1967.

EUROPEAN BANK FOR RECONSTRUCTION AND DEVELOPMENT (EBRD)   The EBRD was established in 1990, following the significant political changes that occurred in Eastern Europe, with the aim of helping to promote economic reforms and the movement towards democratization in former Communist-dominated countries. The twelve member states of the European Community⟷, in conjunction with the European Investment Bank⟷ and the European Commission, own 51 per cent of the new bank, which had an initially proposed capital of $12 billion. A further twenty-nine countries are also members, the United States (10 per cent), Japan (8.5 per cent) and the Soviet Union being the major participating nations. The idea of the bank was originally proposed by France, and its structure, aims and envisaged function are similar to those of the International Bank for Reconstruction and Development (IBRD)⟷, better known as the World Bank, which was set up to help rebuild Western European economies shattered by the Second World War.

The Bank, with headquarters in London, began operating in 1991, but according to the EBRD's French head, Jacques Attali, it is not expected to make any profits for at least two years.

EUROPEAN COAL AND STEEL COMMUNITY (ECSC)   Established in 1951 by six of the seven members of the Western European Union (WEU): Belgium, France, the Federal Republic of Germany, Italy, Luxembourg and the Netherlands. Of those nations that signed the 1948 Brussels Treaty which formed the WEU, only Britain refused to join. The aim of the ECSC – based in Luxembourg, and operational since 1952 – was to increase efficiency in the coal⟷ and steel industries by removing trading restrictions. The original six signatory nations also formed the European Atomic Energy Community (EURATOM)⟷ and were signatories to the 1957 Treaty of Rome⟷ which formed the basis of the European Economic Community (EEC). The ECSC, EURATOM and the EEC were gradually merged into a single body, and in 1967 the three executive bodies were amalgamated into a single European Commission based in Brussels.

EUROPEAN COMMUNITY (EC)   The EC is basically an organization of Western European states designed to foster economic co-operation and common development, with the eventual aim of economic and monetary union, together with a measure of political unity. Agreements have been reached on the removal of Customs tariffs, the setting of common tariffs for non-member states, and the abolition of barriers to free movement of labour, services, and capital.

The organization incorporates the European Coal and Steel Community (ECSC)◇, the European Atomic Energy Community (EURATOM)◇ and the European Economic Community (EEC), the governing agencies of which were merged to form the European Commission in 1967. The six originating members – Belgium, France, the Federal Republic of Germany, Italy, Luxembourg and the Netherlands – adopted a common agricultural policy (CAP)◇ in 1962, which has since undergone several revisions. Four permanent institutions were also created to improve integration and apply the treaties establishing the ECSC, EEC and EURATOM. The European Parliament and European Court of Justice, both sited in Brussels, interpret the Rome Treaties which formed the basis of the EEC, and a Council of Ministers drawn from the parliaments of member states decides on the Commission's proposals.

A Common Fisheries Policy was adopted in 1983, and the European Monetary Union aims to work towards acceptance of a single European currency, although attempts to develop a common foreign policy have met with only limited success. The EC has a highly developed set of rules governing trade and economic agreements with non-member countries. Aid◇ to the signatories of the Lomé Convention◇ and to the dependencies of EC member states is channelled through the European Development Fund◇. The European Investment Bank◇ was also established to make loans to the ACP countries◇ as well as to EC member states themselves, and to some developing countries◇ loosely associated with the Community. Since 1976, the allocation of financial and technical aid to Latin American◇ and Asian countries has been growing steadily. Criticism has been levelled about the adverse effects on world food markets of exports of heavily subsidized food under the EC's controversial CAP and the continued protectionist measures under the same policy which work against imports of some foods from the developing countries. The EC has also operated worldwide food aid programmes in cereals, dairy and other products, using its surplus production.

In 1973 the original EC members were joined by Denmark, Ireland and the United Kingdom, followed by Greece (1981), Portugal and Spain (1986). Greenland left in 1985 having achieved independence from Denmark. Austria applied for membership in 1989 and several newly democratic Eastern European nations have indicated their desire to join.

EUROPEAN COMMUNITY: INSTITUTIONAL ORGANIZATION

European Commission – The Commission, sited in Brussels, is made up of seventeen members appointed by individual governments (two from France, Germany, Italy, Spain and the UK) and is supposedly independent of any national interest. Members are elected for a four-year term. The Commission acts as an advisory body and is responsible for implementing policies decided by the Council of Ministers, to which it puts proposals for consideration. The current twelve national EC governments can exert influence on the decision-making process through the Council of Ministers, via the Commission, or through the Committee of Permanent Representatives of the Twelve, which is also based in Brussels. The Commission, which is answerable to the European Parliament, meets weekly; decisions are taken by a simple majority.

Parliament – At its foundation the Parliament was merely consultative, but it has since assumed increasing power. Members became directly elected by the voting public in member countries in 1979. It is still not a true legislative body but has the authority to dismiss the Commission and reject the Community budget in its entirety. The Parliament conducts its business in a number of centres. The full Parliament meets in Strasbourg, most specialized committees meet in Brussels, and the Secretariat is in Luxembourg. There are 518 parliamentary seats; elected members serve five-year terms sitting as political groups. Socialists form the largest single political grouping, but they are outnumbered by the aggregate strength of members from the centre and right parties. The Parliament consults the Council of Ministers and exercises some control of this body.

Council of Ministers – The Council of Ministers, which meets in Brussels and Luxembourg, is the Community's principal decision-making body. Composed of one Minister from each of the member states who decide on Commission proposals, it passes its decisions on to the Commission. Decisions may be taken on a qualified majority basis, whereby members' votes are weighted according to population◇. Heads of States meet twice a year in the European Council or Summit.

Judiciary – The European Court of Justice, based in Luxembourg, adjudicates on disputes according to the Treaties of Rome⌀ and all further EC legislation. Thirteen judges rule on questions of Community law and the validity of actions taken by the Commission, Council of Ministers and individual governments. Judgement is by majority vote and is binding on all parties. A Court of Auditors is also located in Luxembourg to review matters of a financial nature.

EUROPEAN COMMUNITY: OPERATION

Very few texts can be adopted without the opinion of the European Parliament having first been sought, but both the Commission and the Council of Ministers can issue:

Regulations – which automatically become Community law.
Directives – which are binding on member states (as to the ends but not the means).
Decisions – which are binding on those to who they are addressed.
Recommendations/Opinions – neither of which are binding.

EUROPEAN DEVELOPMENT FUND (EDF)

The main instrument of financial and technical co-operation between the European Community (EC)⌀ and the developing African, Caribbean and Pacific (ACP) countries⌀ which are signatories to the Lomé Convention⌀. The terms of the Convention are regularly renegotiated, and each new agreement has a separate EDF which lasts for the duration of the Convention. EDF VI, which came into force in 1986, provided $8.5 billion for the period up to March 1990. All Community members contribute to the Fund, which is administered by the European Commission. The aid⌀ provisions of Lomé build upon and improve those laid out in the Yaoundé Conventions⌀. More flexible in nature, they allow the forms of aid to be tailored to the requirements of specific projects and individual recipient countries. The EDF is not part of the Community budget as such but is made up of separate financial contributions from the overseas aid budgets of the member states. Under the Treaty of Rome⌀, EC members are bound to contribute to each EDF the sums agreed each time by the Council of Ministers.

EUROPEAN ENVIRONMENT AGENCY

The European Environment Agency, established in 1990 will be called upon to monitor and verify all the environmental data being produced by individual European Community (EC)⌀ members, as well as those nations in Eastern Europe⌀ that have agreed to participate. With a budget of $5.75 million, the Agency will be expected to collate and assess the data and ensure that uniform criteria are used in assessment procedures. It is envisaged that the Agency, which is expected to be working in all nine official Community languages, will be called upon to orchestrate the directive of the EC Environment Ministers that gives the public freedom of access to environmental information. Any information on the environment held by public authorities, or their contractors, is to be made available to anyone who requests it. A reasonable charge may be made for any database searching that has to be carried out.

Data on soil, air, water⌀, natural flora and fauna, and any information on matters affecting natural resources⌀, may be covered by the directive, which also lists data which may not be revealed. These are classified information such as military secrets, commercial secrets, matters of public security, information that is *sub judice*, personal data, and information that could lead to environmental degradation. Details concerning the release of genetically modified organisms will be included in the data that must be given out. The EC has agreed that the description of organisms, purpose and place of release, methods of monitoring the release, and ways of dealing with emergencies may not be considered confidential.

EUROPEAN ENVIRONMENTAL BUREAU (EEB)

An international, independent Non-Governmental Organization (NGO)⌀ based in Brussels, which aims to co-ordinate NGO's activities and strengthen their effect and impact on European Community⌀ environmental policy and projects. It achieves its objectives through a mixture of lobbying, generalized and specific publications, and a series of seminars and regular conferences.

EUROPEAN FREE TRADE ASSOCIATION (EFTA)

An association originally founded in 1960 and based in Geneva, to promote economic development, foster free trade of industrial goods between members, and uphold liberal, non-discriminatory practices in world trade. Denmark and the United Kingdom left in 1972 to join the European Community (EC)⌀, followed later by Portugal. EFTA membership now encompasses only six states –

Austria, Finland (an associate member in 1981, becoming a full member in 1986), Iceland, Norway, Sweden and Switzerland – although the agreement also extends to Lichtenstein. Tariffs on non-farm trade between members was abolished in 1967. By 1973, free-trade agreements had been negotiated between all EFTA members and the EC, and by 1984 tariff-free trade in industrial goods was achieved between all eighteen members of the two groups. Individual countries are free to negotiate independent trade agreements with non-member states. EFTA's main goal is the formation of a European Economic Space (EES) through negotiation with the European Community, the aim being to have the EES in place by 1993.

EUROPEAN INVESTMENT BANK (EIB)
An independent public institution, established under the Treaty of Rome⊃ within the European Community⊃ chiefly to ensure the balanced and harmonious development of member nations through co-ordinated investment in their less-developed regions and modernization of industry or introduction of new enterprises. The bank may also invest in some states associated with the Community, including Mediterranean countries and the African, Caribbean and Pacific (ACP) countries⊃ that are signatories to the Lomé Convention⊃.

EUROPEAN MONETARY SYSTEM (EMS)
The EMS – or Snake, as it is sometimes called – was established in 1979 and is a voluntary system of semi-fixed exchange rates between the member countries of the European Community⊃ to stabilize their currencies. The devaluation or revaluation of the currency of any country requires the agreement of all the other member nations. The system is based on the European Currency Unit (ECU), which has now entered global markets in its own right. The ECU is a weighted average of all the different currencies of the member states. In 1990, most member countries of the EC were members of the EMS, which was viewed by many as the first step towards full monetary union within the EC and the eventual adoption of a single European currency.

EUROPEAN SPACE AGENCY (ESA)
An organization established in 1975 following the merger of the European Space Research Organization (ESRO) and the European Launcher Development Organization (ELDO). The aim of the ESA is to encourage weather, telecommunications and other space programmes. This includes the launching of geostationary satellites. Before the first successful launch of the ESA launcher *Ariane* in 1979, all ESA satellites were launched by the North American Space Administration (NASA).

EUTROPHICATION
Eutrophication is a natural process through which bodies of water⊃ become enriched with nutrients which then stimulate plant and algal growth. The process normally takes thousands of years, but increased nutrient loading as a result of human activities can cause eutrophication to occur much more rapidly. The over-enrichment of areas of water – rivers, streams or lakes – is primarily due to a cocktail of nitrogen-containing fertilizers⊃ being washed from soil by rain and phosphates from detergents being discharged in municipal sewage. Phosphorus is primarily responsible for eutrophication in most lakes; nitrogen causes the process to occur in shallower bodies of water and in coastal sites. The increased presence of both these chemicals encourages the rapid growth of algae and plants, to the extent of eliminating oxygen from the water. Where algal blooms – a common sign of eutrophication – occur, they make the water uninhabitable for fish⊃ and other forms of aquatic animal life. Pollution⊃ from sewage, nutrients contained in animal wastes, runoff from land⊃ and industrial wastes (inorganic nitrates and phosphates) and phosphates from water-softeners all promote the eutrophication process. Nutrient overload caused through human neglect is particularly damaging, as much of the excess nitrate and phosphate settles to the bottom of watercourses and lakes to stimulate growth at a later date.

Eutrophication caused by human activities and neglect has transformed the ecosystem⊃ in thousands of lakes throughout Western Europe, Japan, and industrialized parts of North America.

EXCHANGE RATES
The value of one nation's currency in terms of another's. International trade and tourism⊃ make it essential for currencies to be convertible at a more or less stable rate of exchange. Until the outbreak of the First World War and for a period between 1925 and 1931, currencies were backed by, and convertible into, gold⊃. In 1947 the International Monetary Fund (IMF)⊃ was founded: members' currencies were fixed in terms of gold and could not be changed without full IMF agreement. Special Drawing Rights (SDRS)⊃ were introduced in 1970 in an attempt to increase world liquidity, but they proved unable to maintain the stability of

the monetary system. In 1971 the decision to suspend the convertibility of the US dollar into gold led to an agreement to allow currencies to float in value. Floating rates avoid the problem of large, destabilizing devaluation changes.

The fluctuations in currency exchange rates since the early 1970s, when most countries began to float in value against each other, has caused significant difficulties for almost all nations, notably developing countries◇. The exchange rates of many developing countries are 'pegged' to a major currency, or to a basket of currencies. The floating of the major currencies of international commerce produces uncertainty about the real value of exports and imports.

The aim of systems such as the European Monetary System◇ is to fix currencies in terms of each other. This has several advantages, but it also has the drawback that weak countries face balance-of-payments problems when their exports start rising in price compared with those of other countries.

EXPANDED PROGRAMME ON IMMUNIZATION (EPI)

A programme, co-ordinated by the World Health Organization (WHO)◇, which attempted to vaccinate 80 per cent of the world's children against the six major childhood diseases by the end of 1990. Tetanus kills 800,000 newborn children every year, and whooping cough (or pertussis) proves fatal for 600,000 of the 5 million who contract the disease annually. Somewhere in the world, usually in the developing countries◇, a child dies from measles◇ every fifteen seconds. Tuberculosis◇ claims up to 10 million victims, and there are 275,000 cases of poliomyelitis◇ in the world each year; the disease is the leading cause of disablement in the Third World◇. All these six diseases – which, even when they are not fatal, exacerbate malnutrition◇ and adversely affect normal growth in children – can be prevented by a series of vaccinations.

The EPI began in 1974, when immunization coverage for children in the developing world was below 5 per cent. By 1989, around two-thirds of the world's children were being reached and over 2 million lives were being saved each year as a result of the immunization programme. A third dose of the polio vaccine was reaching 67 per cent of the world's children during the first year of life. A third dose of the vaccine against diphtheria, pertussis and tetanus reached 66 per cent, the tuberculosis vaccine (BCG) found 71 per cent, and 61 per cent of the world's

infants received shots of measles vaccines. As the EPI progressed, further goals were set, including those of reducing measles cases by 90 per cent and eliminating neonatal tetanus by 1995. In addition, the programme now aims to provide immunization with EPI vaccines for 90 per cent of the world's children by the year 2000.

The EPI initiative is one of the world's best-supported international co-operation programmes, with global institutions such as the United Nations Childrens Fund (UNICEF)◇, the World Bank◇, the United Nations Development Programme (UNDP)◇, the Rockerfeller Foundation, Rotary International and the Save The Children Fund◇ playing major roles.

In 1990, around 46 million infants every year were still not being fully immunized against the six childhood killer diseases. Some 2.8 million children were dying every year and another 3 million were disabled, yet it cost only $10 dollars per child to vaccinate them against all six diseases.

EXPORT CREDITS GUARANTEE DEPARTMENT (ECGD)

Exports of capital goods are frequently dependent on long-term credit being given to customers, and this is becoming increasingly significant for the developing countries◇. The ECGD is an agency of the British government, established in 1919 and run on commercial lines. It provides insurance for exporters against the main risks involved in selling overseas, and gives guarantees to banks providing export finance. Export credits are extended to developing countries as part of its normal operations. When the length of credit exceeds one year, these are included in British net private aid◇ flows, which count towards the 1 per cent of Gross National Product (GNP)◇ aid target agreed by the United Nations◇, but the business and repayment terms the ECGD is prepared to insure are governed by strict commercial criteria in order to meet the requirement to run on a solvent basis.

The ECGD also operates an investment guarantee scheme, which came into being as a result of the Overseas Investment and Guarantees Act in 1972. Insurance can be given for new investment overseas against the risks of war, expropriation and restrictions on the remission of profits. Cover cannot be offered against commercial risks and the scheme is meant to apply solely to developing countries. Other donor governments have established similar agencies, including COFACE (France), HERMES (Germany) and the EXPORT-IMPORT BANK (United States).

EXTINCTION A term referring to the complete disappearance of a species of plant or animal from the planet. It is a natural biological phenomenon and is part of the process of evolution. Over 90 per cent of all species that have ever lived have disappeared, and eventually every species on earth, including *Homo sapiens*, may become extinct. Mass extinctions have happened in the past, but the rising pace of extinctions is due solely to human activities such as increasing levels of pollution⬦, hunting⬦ or habitat destruction.⬦ Historically, extinctions, even 'mass extinctions', were lengthy processes – the dinosaurs disappeared over a period of 2 million years. Today, species are being lost in a matter of centuries, and some over a much shorter time span.

As wildlife habitats disappear, so do the indigenous species. Tropical forests⬦ are increasingly threatened with destruction, yet it is here that animals and plants reach their greatest diversity. New species are described every year, many with very restricted ranges. Tropical forest loss could account for the extinction of a third of the world's species of wildlife. Threats to other regions of biodiversity⬦, such as coral⬦ reefs and wetlands⬦, exacerbate the situation. Wildlife species are currently being lost at an alarming rate – 400 times faster than at any other time in history.

In 1988, figures from the World Conservation Union⬦ (IUCN) indicated that a total of 795 species of mammals, reptiles and amphibians were threatened with extinction. In addition, 1,073 species of birds, 596 species of fish⬦ and a conservative estimate of 2,125 species of invertebrates were also in danger of disappearing for ever.

Projections of a 15–33 per cent loss of global wildlife species by the year 2000 have been put forward, with as many as 130 species becoming extinct each day, although these have been labelled as exaggerated by many scientists. No one really knows what the exact extinction rate level is or will become – except that it will be high.

No authoritative and fully comprehensive list of threatened animals is available, and it is virtually impossible to say with any degree of certainty which species are extinct or nearly so. The concept of 'Red Data' books to monitor the status of wildlife was proposed only in 1964, and the first – 'Mammals' – appeared in 1966; since then the IUCN has regularly produced the most up-to-date listings of threatened species. However, what data are available have never really been fully collated, and the Species Survival Commission of the World Conservation Union, although in the process of trying to do just that, report that it will be at least the mid-1990s before any comprehensive analytical results are published.

More is known about the status of mammals than any other group because more money is available to fund studies of the larger mammals and primates than for other smaller, less attractive animals or insects. Nevertheless, a quarter of the world's mammals are either threatened or so little studied that there is no information on which a calculation of their status can be based.

EXXON VALDEZ An illegally piloted supertanker, fully laden with 1.2 million barrels of crude oil⬦, which ran aground in Prince William Sound in Alaska in 1989. The ship had just left Valdez, the southern terminal of the Alaska Oil Pipeline. It spilt 267,000 barrels of oil. Handling of the spill, the resources available to combat it and the attitude of the oil industry roused a great deal of controversy. It was ten hours before the first containment booms and oil-removing equipment reached the scene. Seven oil-skimming vessels should have been available, but only two were deployed. The oil industry also failed to produce the full 7,000-metre-long containment boom which it had said would be available should a spill occur.

The oil eventually covered 25,000 sq km of coastal and offshore waters. Fish⬦, kelp, and marine mammals all died in their thousands. By mid-1990, 35,000 dead seabirds from 89 species had been found, but these were believed to represent only 10–30 per cent of the total killed. Some 10,000 sea otters, 16 whales and 147 bald eagles were amongst the larger animals to perish as a result of the spill, with salmon, black cod and valuable herring-spawning grounds also being decimated. The massive clean-up⬦ operation that took place was believed to have cost the Exxon Oil company, which owned the stricken vessel, well over $2 billion.

Hydrocarbons toxic to marine organisms appear in the sediment of sub-Arctic waters at least one year after an oil spill. Oil compounds survive longer in colder waters, harming most fish species and impairing reproduction in marine crustaceans. Wildlife in the region will continue to suffer the effects of the spill well into the 1990s, and the public concern aroused by the accident has cast doubts over the further exploitation of the fuel resources which lie under the Arctic National Wildlife Refuge in Alaska.

F

FARAKKA BARRAGE In the 1950s India decided to construct a barrage at Farakka on the River Ganges near the border with Bangladesh. The barrage was planned to divert water⬦ from the Ganges down the Hooghly Channel to flush away silt which, during the dry season, threatened to clog up the port of Calcutta. Work went ahead during the 1960s, despite protests from Bangladesh, and the barrage opened in 1975. Arising from the international disagreement over the barrage, an Indo–Bangladesh Joint Rivers Commission was established in 1971 to decide how best to use their common waters. However, in 1976 India unilaterally used the Farakka Barrage to divert part of the Ganges flow. Bangladesh complained that this caused wells in the south-western part of the country to dry up and drinking water to be contaminated by salt water encroaching further up the low-lying delta, the rise in salinity threatening to destroy the Sundarbans, Bangladesh's largest natural forest⬦. In addition, fisheries were being depleted. Eventually Bangladesh appealed to the United Nations⬦, a move which ultimately led to new negotiations with India. In 1977 a 'water-sharing' agreement was reached under which Bangladesh would receive 60 per cent of the Ganges waters during the dry season. The agreement, which was originally scheduled to last for five years has since been extended twice, although the effectiveness of both the agreement and the barrage itself remain matters of great controversy.

FARMING SYSTEMS If the food needs of the present and future population of the earth are to be met, which, if any, of the world's farming systems will be able to produce the levels of food required? Which, if any, are truly viable in the long term?

In Africa, where constraints are worst, agricultural production has reached crisis point. In tropical Africa agricultural and livestock⬦ production is curtailed by a variety of natural factors. Most African soils are low in clay and organic matter and are particularly susceptible to erosion. According to the United Nations Environment Programme (UNEP)⬦, 742 million hectares (ha) – more than a quarter of the continent – is undergoing moderate or serious desertification⬦ and may soon be totally unproductive. Overall, soil erosion⬦ rates in Africa have increased twentyfold in the last thirty years. Furthermore, diseases such as onchocerciasis⬦ and trypanosomiasis⬦ effectively remove vast amounts of land from agricultural production or prevent the rearing of productive livestock.

In Asia, increased production has relied on the 'Green Revolution'⬦ based on the use of scientifically engineered High-yield Varieties (HYV)⬦ of crops and a massive increase in cultivated land through irrigation.⬦ Asia can now support half the global population⬦, primarily because it has two-thirds of the world's irrigated land⬦. However, irrigation is not a universal answer to the quest for sustainable agriculture⬦. Irrigated land in the world tripled between 1950 and 1985, the 177 million ha of newly irrigated land drawing heavily on both surface and underground water⬦ resources. Globally, 1.3 billion cubic metres (cu m) of water are used for irrigation annually, but 3 billion cu m are withdrawn, so over half the water is lost during storage or transport. Moreover, poorly managed irrigation systems can prove counter productive because of waterlogging, build-up of salts and eventual loss of soil fertility. Major problems are now appearing throughout Asia. In Africa, a *Club du Sahel* study of irrigation in the Sahel concluded that 'the development of new irrigation areas has barely surpassed the surface area of older ones which had to be abandoned', a situation paralleled in many newly-irrigated areas around the world.

In North America, as in Europe, intensive monoculture⬦ farming has been a major factor in agricultural development. Yet every decade, 7 per cent of the world's soil is lost through the impact of these large-scale monocrop farming systems. Production can be maintained only through the addition of vast quantities of agricultural chemicals. Expanding food production by mining the soil and depleting underground water supplies can produce results, but only for a limited period. Farming systems in the United

States have exhausted vast tracts of land, and an area twice the size of California has been rendered unproductive. Nearly 20 per cent of all groundwater⬦ withdrawals for agricultural purposes in the USA are in excess of aquifer recharge. In Europe farmers have turned to mixed farming as a means of intensifying production. This maximizes land use but poses several problems, particularly of weed and pest control.

Almost all the steps which have led to the increase in agricultural productivity in the developed world – such as mechanization and use of chemicals – have had high energy⬦ requirements and have therefore been extremely costly. The Food and Agriculture Organization (FAO)⬦ estimates that, to accommodate the necessary improvement in agricultural performance in the developing countries⬦ needed to keep up with the demand for food, energy requirements will increase 50 per cent by the year 2000. Fertilizers⬦ will account for 60 per cent of this increase, mechanization, irrigation and pesticide⬦ use will account for the rest.

FAST BREEDER REACTOR   A nuclear reactor in which natural uranium⬦ enriched with U-235 or plutonium-239 is used without a moderator, the chain reaction being sustained by fast neutrons. The reactor core is surrounded by a blanket of natural uranium into which neutrons escape. These collide with U-238 nuclei to form U-239, then decay to form Pu-239. Because more Pu-239 can be produced in the blanket than is required to enrich the fuel in the core, the reactors are termed 'breeder' reactors and they are more economical in terms of uranium use than thermal reactors. Their main disadvantage is the temperature at which they operate, which is so high that a liquid metal, usually sodium, is needed as a coolant. Leakage of the metal would be disastrous and the plutonium⬦ produced is toxic, difficult to dispose of and a potential source of material for the production of nuclear weapons.

FAUNA AND FLORA PRESERVATION SOCIETY (FFPS)   An independent, voluntary organization based in London which aims to ensure the international conservation⬦ of wildlife, especially those species of plants and animals deemed to be threatened with extinction⬦. The organization started out in 1903, originally as the Society for the Preservation of the Wild Fauna of the Empire (SPFE). The FFPS attempts to raise public awareness of the plight of endangered wildlife, through a variety of publications, meetings and exhibitions and other informative activities, as well as promoting both the establishment and proper management of wildlife reserves and the enforcement of laws governing the protection and conservation of wild animals and plants.

FERTILITY RATE   A crude calculation used for demographic purposes: an estimate of the number of children an average woman would bear during her reproductive lifetime, assuming that age-specific fertility rates prevailing at the time of the calculation remained constant. Globally, the fertility rate has fallen from 4.86 in 1965 to around 3.28 in 1990. By the end of the 1980s regional differences remained marked – fertility rates ranging from 6.22 in Africa to 1.83 in Europe. The latter indicates that women in Europe were tending to have fewer than two children, which represents the population replacement level. If this trend continues, national populations⬦ will begin to decline.

During the 1980s several authorities in Europe actively encouraged couples to have more children to prevent this decline, using various incentives and covert measures to achieve their ends. In France, the figure of 1.8 children per woman remained almost constant between 1975 and 1990. As in other countries, this figure is used as a basis for local and national planning policies and for the population-increase policies of successive governments. National insurance pension funds also use the figure to help determine pension-financing arrangements. In 1990, the head of France's National Institute of Demographic Studies resigned amid a political row. He claimed that the Institute was using a method of calculation that led to the 1.8 figure even though it was known that the total number of children born to women stood at 2.1.

In 1990 Rwanda became the nation with the highest fertility registering a rate of 8.3. Meanwhile, the fertility rate in the most populous countries had begun to stabilize, at around 4.2 in India and 2.3 in China, the latter in part due to the government's wide-ranging birth control⬦ policies. Kenya, held by many to have the world's highest fertility during the 1980s, exhibited a sharp decline: down to 6.7 average lifetime children per woman. Moreover, countries with traditionally high fertility rates were also showing marked declines. Bangladesh had not only reduced its fertility rate to 4.9 children (down from 6.7), but also the desired family size in the country had dropped: down to 2.9 children.

FERTILIZER Any substance containing some of the twenty or so chemical elements essential for plant growth that can be added to soil to maintain or improve soil fertility. Both nutrients and humus are needed to maintain the physical structure and overall productivity of the soil. Fertilizers are essential components of most modern intensive farming systems⟜. They may be organic – manure, compost, bonemeal, blood, or fishmeal – or inorganic – usually compounds of nitrogen, phosphorus and potassium. Nitrogen is generally added as industrially manufactured ammonium nitrate, phosphates are derived from naturally occurring rocks, and potassium is mined from potash deposits.

Globally, the average consumption of fertilizers rose from 62 kg per hectare (ha) in 1975 to 86 kg/ha in 1985. Application levels vary widely. By 1986 India was applying 57 kg/ha and Japan 427 kg/ha, but overall rates had risen markedly with the widespread introduction and use of High-Yield Variety (HYV)⟜ seeds. In Africa an average of 32 kg/ha were being applied compared with 386 kg/ha throughout Europe. The Food and Agriculture Organization (FAO)⟜ estimates that the future rate of growth will be about 8.5 per cent annually, with agricultural production doubling between 1980 and 2000. Plants take up most nutrients in solution through the roots and for this reason fertilizers are often applied in aqueous solution. But only half of all fertilizer added to land⟜ is taken up by crops – the rest is lost, generally being washed away and causing pollution⟜ problems in watercourses, mainly eutrophication⟜ of surface waters and nitrate contamination of drinking water⟜.

FILARIASIS A collective term used to describe at least eight different infections caused by various species of filarian roundworms, all of which are prevalent in the tropics and subtropics. The two most significant of these diseases are river blindness (onchocerciasis⟜) and lymphatic filariasis (which causes elephantiasis). These diseases threaten the lives of 1 billion people, mainly in Africa, South-East Asia and, to a lesser extent, Latin America.⟜

Lymphatic filariasis is caused by *Wucheria bancrofti* worms transmitted to humans via the bite of various types of mosquito⟜, including *Aedes*⟜, *Culex* and *Anopheles*.⟜ The parasite larvae enter the bloodstream and develop into adult worms which collect in the human lymphatic system, eventually blocking lymph vessels and causing surrounding tissue to swell; this leads to the condition known as elephantiasis. Some 90 million people in 76 countries suffer from lymphatic filariasis and 900 million are at risk in Africa, the eastern Mediterranean and Latin America. Lymphatic filariasis is treated with two drugs, diethylcarbamazine (DEC) and suramin, although they are not easy to administer and have unpleasant side-effects. Mass drug-administration programmes, coupled with a vector-control programme, have led to a marked reduction in infection rates in areas where the disease is endemic. The incidence of the disease has been vastly reduced in the Solomon Islands and eradicated in Sri Lanka.

FIRST WORLD A vague, rarely used term which usually refers to the industrialized countries of North America and the democratic states of Western Europe, together with the developed nations in Oceania – Australia, Japan and New Zealand. Between 1940 and 1960 the world's nations became roughly divided into three camps, often labelled 'East', 'South'⟜ and the 'West', which is synonymous with 'North'⟜. The latter group comprise the First World, although several of the traditionally neutral countries – Austria, Finland, Ireland, Sweden and Switzerland – are often taken to be included in the category.

FISH Fish have existed on earth for 450 million years and are one of the oldest forms of life. They have been extensively hunted or cultivated by humans for nearly 5,000 years, and fishing⟜ is one of the oldest productive occupations. Of some 20,600 known freshwater and marine species, about 9,000 are currently used for a variety of purposes, nutritional and commercial. Of these, only 22 marine species are routinely harvested in quantities exceeding 100,000 tonnes per year, with five groups – herrings, cods, jacks, redfishes and mackerels – accounting for half the annual catch.

Fishing and related activities provide employment for over 25 million men and women around the world. In economic terms, global fish exports earn well in excess of $17 billion annually. As well as being a major source of food, fisheries make an essential contribution to the production of many industrial goods, such as lubricants, paints and glues. Oils and fats from fish are also used in the manufacture of margarine and soaps.

Of the global commercial fish catch, 20 per cent is eaten as fresh fish, 22 per cent is frozen for human consumption, 14 per cent is cured and 13 per cent is canned. It is commonly reported that around 30 per cent of the world's commercial fish catch is directly reduced to oil, fertilizer⊕ or fishmeal for use as a high-protein foodstuff for livestock⊖. Although 69 per cent is said to be for human use in one form or another, in reality the viscera and other wastes from fish are also used to produce animal feed. So, in total, roughly half of the world's fish catch ends up as animal feed.

In nutritional terms, fish is the primary source of protein for many people throughout the world, including half the population of Asia. Fish and fish products have a protein content of between 15 and 20 per cent, and fish contains the highest protein per gram and per calorie of all commonly consumed foods. In production terms fish is also very cost effective, yielding 3–30 times as much protein as animal husbandry for the same energy⊕ expenditure. As well as containing easily digested protein, fish has a high content of the amino acid lysine, which makes it an extremely beneficial supplement to the low-protein, high-carbohydrate diets commonly found in developing countries⊕. Fish also contains a variety of other nutrients, including vitamins A, B, D, calcium, phosphorus, iron⊕, iodine and fluorine.⊕ Fish oil has been shown to be an important factor in helping to prevent heart attacks. Yet although seas and oceans cover around 70 per cent of the world's surface, fish and marine animal products provide only 23 per cent of all animal protein consumed (6 per cent of the world's total protein supply, both plant and animal) and less than 1 per cent of the world's calories.

The place of fish in the diet varies widely around the world, and the reasons for this are not just the proximity of good fishing grounds. *Per capita* fish consumption in the Soviet Union is twice that in the United States. This situation has arisen because the Soviets, faced with the difficulty of expanding their livestock output in the 1950s and 1960s to meet the demand for proteinaceous meat, decided that it would be easier to provide protein by investing in distant-water fisheries. Consequently they were at the forefront of developing factory ships, exploited all ocean stocks and caused national fish consumption to rise accordingly.

In Japan, fish traditionally ranks high in the diet, although the traditional annual consumption of around 90 kg *per capita* is falling markedly and is now around 40 kg as the country moves towards more western eating habits and away from the traditional fish and rice⊕ diet. Due to the low land: people ratio in Japan, a decision was taken to devote scarce land resources to the production of rice, meaning that animal protein had to be obtained

from the sea as far as possible. Taste preferences are also an important factor in determining what role fish plays in a nation's diet. Americans for example eat over 35 per cent of the world's tuna catch, whereas the Japanese consume around 50 per cent of the global squid catch.

Although, in general, people in the Third World⊕ eat less fish *per capita* than those in developed countries, it represents a much larger proportion of their total animal protein supply and is a much sought after commodity. In the developing countries, poorer people tend to spend proportionately more of their household budget on fish than on all other kinds of meat. Owing mainly to overfishing, fish consumption in the Western world has been falling since the early 1970s but according to the Food and Agriculture Organization (FAO)⊕ the global demand for fish will rise from 92 million tonnes in 1990 to 113 million tonnes by the year 2000, with over half this demand coming from developing countries.

FISH FARMING   Fish farming, the breeding of fish⊕ in a controlled environment, is the major form of aquaculture⊕ in the world. In the 1980s, 10 per cent of the world's fish requirements were provided by farms, notably varieties of carp, trout, salmon and shellfish. Cultivation of fish is a long-established tradition among rural populations in the Far East, especially in China, where around half of the fish eaten, some 4 million tonnes, is produced from fish farms. These occupy 10 million hectares of land⊕, often as an integral part of a complex farming system⊕ on the country's 40 million hectares of irrigated land. Fish are raised with pigs or poultry; the animal waste is used to fertilize fish ponds, or in flooded fields to grow rice⊕. The fish eat harmful insects, destroy weeds and improve soil quality, improving grain yields by up to 10 per cent in the process. Following active promotion by international development agencies, the Food and Agriculture Organization (FAO)⊕ has decided to dedicate half its fisheries budget to aquaculture. Fish farming is increasing worldwide by about 5 per cent a year.

FISHING   Since 1950 global yields of fish⊕, both marine and freshwater, have risen almost fivefold: from 20 million tonnes in 1950 to around 100 million tonnes in 1990, exceeding global beef production by a wide margin. Increases over the last two decades have been slight compared with those observed in the 1950s and 1960s – in 1969 production was 69 million tonnes, rising to only 72 million tonnes by 1980. Throughout the 1970s the spectacular growth rate of catches slowed dramatically through a combination of overfishing, water◇, soil and air pollution◇, high fuel costs, inadequate equipment and facilities, and a lack of integrated production and marketing policies. The productivity of many traditional marine fishing grounds plummeted, with eleven oceanic fishery stocks becoming severely depleted. Although the overall size of catch remained more or less constant as new deep-water ocean stocks began to be exploited, annual increases slowed to 1–2 per cent, below the rate of population increase.

Saltwater sources provide over 80 per cent of the world's total fish catch. Developing countries◇ with extensive coastlines accounted for only 27 per cent of global marine fish catches in the 1950s, but today they account for around half of all catches, thanks mainly to the effect of the 200-nautical-mile Economic Exclusion Zones (EEZ)◇ which protect their fishing grounds.

There are three main facets of fisheries;

Deep Sea Fishing – Largely monopolized by a handful of industrial nations up to the early 1970s. Since then, and following the introduction of the EEZs, developing countries have been building up fishing fleets and onshore installations, and training personnel to exploit their fisheries potential to the full. Licensing systems have been introduced allowing industrialized nations to fish in EEZs bringing much-needed foreign exchange into Third World countries. Resource-poor nations are now faced with the problem of how to stop poaching. Mauritania has an EEZ of around 130,000 sq km containing what is regarded as one of the richest fishing grounds in the world. In 1981, 120,000 tonnes of fish were officially declared as being caught in the area, but the actual total catch was estimated to be nearer 1.25 million tonnes.

In total, marine fish catches rose from 18 million tonnes in 1950 to around 84 million tonnes in the late 1980s.

Artisanal fishing – Mostly carried out in inland freshwater and offshore coastal marine fishing grounds. Usually catches are landed by single fishing units owned or hired by a fisherman or fishing family. In 1982, approximately 25 per cent of the total world fish catch was contributed by artisanal fisherfolk. Small-scale artisanal fishing is of prime importance in the developing countries, supplying at least three-quarters of domestic demand. In the 1980s, artisanal fish catches represented two-thirds of the total catch in Asia and 85 per cent of the African catch.

Catches of freshwater fish rose from 2.2 million tonnes in 1950 to over 13 million tonnes by 1989. *Aquaculture* – Fish husbandry has been growing in importance over the past decade or so but still produces only around 10 per cent of the total global fish catch. Some 7 to 9 million tonnes of fish are produced annually, mostly in China, India and other Asian countries. Even in these countries, however, the land◇ used for aquaculture◇ is estimated to be only 10 per cent of the potential area. In most developing countries the water temperature is ideal for aquaculture, and the Food and Agriculture Organization (FAO)◇ estimates that yields from fish farming◇ could be raised to around 40 million tonnes per year.

The FAO concludes that a global investment of $1.5 billion annually (roughly what is spent per day on armaments) until the year 2000 would see the annual fish harvest rise to 130 million tonnes. Early reports suggested that the ultimate potential global catch was as high as 400 million tonnes, but revised estimates suggest that the Maximum Sustainable Yield◇ from the oceans of 'conventional' fish – those for which a large market already exists and excluding squid and octopus – is about 100 million tonnes.

As with all forms of food production, the problem of waste is one which needs to be overcome. A significant portion of all fish is inedible for humans and around 40 per cent of fish catch is in effect waste. Commercial fishing is also notoriously wasteful. Trawlers, especially shrimp boats, throw overboard an estimated 10 million tonnes of finfish, many of which do not survive. The FAO calculates that half of these wasted fish could be marketed. Post-harvest losses are also enormous. At least 10 per cent of catches are lost, mainly due to spoilage caused by a lack of facilities for immediate chilling, combined with poor storage and marketing. For cured fish, losses of 10–40 per cent are common as a result of insect infestation and physical deterioration during the curing process.

FLOODS Floods, a naturally occurring phenomenon, affect millions of people, although localized flooding is occurring with greater frequency as a direct result of human activities.

The most dramatic floods seen globally are those created by tidal waves or tsunami. These are produced by upheavals on the deep ocean floor, travel at speeds of 900 km per hour and can reach 30 m above normal sea level. Whole settlements and coastal communities may be totally wiped out if no warnings are given. The effects of tsunami are mostly retricted to coastal areas in the Pacific Ocean.

On land, floods are caused when high-water stages cause watercourses to overflow their natural or artificial banks on to land⊖ that is normally dry. In many instances, this occurs as a natural process, such as when a river inundates its floodplain. The Nile River in Egypt and the Yellow River in China are examples of rivers which traditionally flooded a major plain, until dams⊖ and embankments were constructed to control the natural flooding cycles. Floodplains have extremely fertile soil and have thus proved irresistible sites for human settlements and farming operations. When flood-prevention measures fail in these areas, the results can be catastrophic.

Floods may also be caused by excessively high rainfall over short time spans, or ice jams in rivers during spring melts.

Where drought⊖ occurs, particularly in areas where human activities such as deforestation⊖ have been a contributing factor, the ground becomes unable to absorb the heavy rainfall that sometimes follows the breaking of the drought. This results in localized floods, with such areas often experiencing a drought–flood cycle. The flooding may occur at some distance from the site of deforestation, and over 20 million hectares of land are now flooded annually due in part to deforestation in neighbouring countries.

Between 1980 and 1985 the number of floods reported on a regional basis was: 82 in Asia; 54 in the Americas; 11 each in Africa and Europe and 4 in Oceania. Floods cause crops to be destroyed and sewage systems to overflow. Hunger⊖ and disease are thus added to the loss of life and destruction of homes and property that arise directly from inundation and water damage. The 162 flooding incidents reported between 1980 and 1985 were estimated to have caused the direct deaths of 33,000 people, made at least 20 million homeless, and caused in excess of $20 billion worth of damage.

FLUORIDE   Fluorides originate from naturally occurring minerals⊖ in the earth's crust. Generally, they are insoluble in water⊖, but under certain geological circumstances – which are common in China and India – the fluorides dissolve in water and enter underground supplies. In small amounts, fluoride has been shown to help prevent tooth decay, and many industrialized countries now deliberately add it to water supplies for this reason. In higher concentrations fluoride is a health hazard, causing blue skin and fluorosis, a disease which can cause human bones to become soft and crumbly and new bone formations to be deformed. Mottled teeth are a common sign of excess fluoride intake. The World Health Organization (WHO)⊖ recommends concentrations of 0.6–1.7 mg of fluoride per litre of water for the prevention of tooth decay. Chronic fluorosis is caused when long-term regular daily intakes of fluorides rise above 15 mg per litre.

Due to the geological make-up of soils in many parts of China and India, the water supplies of 25 million Indians and 40 million Chinese contain dangerously high levels of fluoride. The dangers of high concentrations of fluorides have been known for over seventy years, and although several methods of removing them have been developed, they have mostly been complicated and expensive. In India in 1989, the Nalgonda Technique was developed: a simple, low-cost ($0.20 per cubic metre of water) defluoridation technique that can be used either on a small scale or for larger community-sized systems.

FOOD AID CONVENTION (FAC)   Together with the Wheat Trade Convention, the FAC forms the International Wheat Agreement. Under the FAC, founded in 1967, the twenty-one contracting parties committed themselves to give a specific minimum amount of domestic grain (or cash for purchase of grain from another convention member state) to developing countries⊖ every year over the Convention's lifetime. The Convention obligations were originally in the region of 7.5 million tonnes per year, and small quantities of other cereals such as rice⊖ were included. The International Wheat Council⊖, based in London, was charged with administering the Convention and providing the secretariat.

FOOD AND AGRICULTURE ORGANIZATION (FAO)   Founded in 1945, the FAO is the largest of the specialized agencies of the United Nations (UN)⊖, with a membership of 158 countries. It concentrates its activities on finding ways to eliminate hunger⊖ and poverty throughout the world, aiming to increase the efficiency of the production and distribution of all food and agricultural products, to improve the condition of rural populations and to raise levels of nutrition in all the developing countries⊖. The FAO collects, analyses and disseminates information, advises governments on policy and planning, and provides opportunities for governments to meet and discuss food and agriculture problems. It carries out a major programme of technical advice and assistance for the global agricultural community, predominantly in the developing world, giving direct, practical help on behalf of governments and development-funding agencies. By 1989, well in excess of $33 billion of foreign and domestic capital had been channelled to the agricultural sectors in more than a hundred developing countries.

The FAO's governing body is a Conference, composed of all member nations, which meets every two years. The Conference elects a Council of forty-nine member nations – which serve three-year rotating terms – to oversee the day-to-day running of the organization, working together with the Director-General, who is head of the Secretariat based in Rome. The Council is advised by five specialized Committees – on commodity◇ problems, fisheries◇, forestry◇, agriculture◇, and world food security – which are open to all member states. The constitution allows all member countries to have an equal say in determining the FAO's programme and activities. Funding comes from three main sources: contributions by member nations, the Trust Funds set up by member countries, and the United Nations Development Programme (UNDP)◇. The contributions of member countries to the Regular Programme, or internal budget, are set according to a scale agreed by the FAO Conference. Three-quarters of the field programme is dedicated to agriculture and, in the main, to increasing crop production.

After forty-six years, the Soviet Union decided to join the FAO's 158 other members in 1990. The Soviet government will have to pay $30 million as its share of the $574 million FAO budget over the 1990–92 period.

FOOD IRRADIATION When radiation from any source strikes any material, it transfers its energy◇. Radiation above a certain level, known as ionizing radiation, transfers sufficient energy to create positive and negatively charged particles (ions) which alter the chemical structure of biological material. Ionizing radiation, carefully regulated, can be used to preserve food as a result of these chemical changes. The chemical alterations produced in irradiated foods, which are usually exposed in packaged form, allow the food, especially fresh fruit and vegetables and other perishable produce, to

be kept longer – beyond levels achieved through most preservation methods commonly in use.

Only lower-energy ionizing radiation is used in the irradiation of food destined for human consumption, and provided the irradiation is properly controlled, food should not become radioactive. However, it is possible for some trace metal compounds in the food to be made radioactive.

Two radioactive materials have been identified as having energy levels low enough to be safe for the irradiation process: Cobalt-60 and Caesium-137. These materials are not commonly available and are therefore expensive.

The amount of energy deposited on or in the food via the irradiation process is measured in kiloGray (kGy). In 1983 a joint Expert Committee of scientists from the World Health Organization (WHO)◇, the International Atomic Energy Agency (IAEA)◇ and the Food and Agriculture Organization (FAO)◇ set a General Standard which applies to all food treated up to an overall average dosage of 10 kGy, as food is thought to be safe up to a point when 10 kGy is actually absorbed. (This level is equal to 100 million times the amount of radiation received from a chest X-ray, or 10 times the annual dose from natural background radiation.)

The chemical changes and biological effects which occur as a result of the irradiation process are not yet fully understood, but it is recognized that there are three levels of activity depending on the degree of exposure:

Low dosage (less than 1 kGy)
• Vegetable sprouting is stopped.
• Fruit ripening is delayed.
• Insect pests are killed.
Medium dosage (1–10 kGy)
• Some micro-organisms are killed and their numbers reduced.
High dosage (above 10 kGy)
• Food is sterilized (virtually all bacteria and viruses can be killed or inactivated).

Bacteria are much less sensitive to radiation than people, and the dosage needed to kill them is much higher than that needed to cause death in humans; there is therefore a risk to operators in irradiation factories. Less than 10 kGy can be fatal to humans but 580 kGy are needed to achieve a 90 per cent kill rate of the food-poisoning◇ bacterium *Staphyloccocus aureus* in a packet of beef. Another more toxic food-poisoning organism, *Clostridium botulinum*, is even more resilient and needs 3,600 kGy to kill 90 per cent in a cooked joint of meat.

Some micro-organisms are extremely resistant to attack by radiation and one bacterium, *Deinococcus radiodurans*, needs over 5,000 kGy to kill 90 per cent.

Most significantly, irradiation allows food that has been contaminated or spoiled to be either sterilized or at least made edible. It is the fact that spoiled food can be altered to make it fit for human consumption, and the opportunities for exploitation that this offers, that upset many opponents of irradiation. They also claim that the appearance, feel, texture, taste and smell of many foods is altered as a result of the process. Milk and fatty foods are particularly susceptible to change. Radiation also damages at least six vitamins, especially A, C, D, E and K. Toxins and hazardous novel chemicals may also be produced in the food as a result of irradiation. The production of aflatoxins⟳, carcinogenic chemicals produced by moulds, was found to be stimulated by irradiation at approved dose levels.

FOOD POISONING A range of acute illnesses caused by harmful bacteria or poisonous food, more correctly known as foodborne diseases. These are usually infectious or toxic, and enter the body through ingestion of food. Most foodborne diseases arise from microbiologically contaminated food and the World Health Organization (WHO)⟳ receives reports of hundreds of thousands of cases each year from all around the world. In the developed world, underreporting means that there are an estimated ten real cases for each one officially reported; in the developing world the ratio is 100:1. The lives of millions of people are at risk, and in the Third World⟳ as much as 70 per cent of diarrhoeal diseases can be attributed to contaminated food. Foodborne diseases lead to impaired digestion and malabsorption of nutrients, lowering resistance to other infections and diseases. Of the 2.6 billion people who travel each year for either business or pleasure, the WHO estimates that between 20 and 50 per cent suffer from diarrhoea related to consumption of contaminated food. Contamination is frequently caused by incorrect handling, storage or preparation. Often, it cannot be avoided. Despite all that modern science has to offer, it is still impossible to guarantee pathogen-free poultry. Contaminated poultry is a major source of food poisoning, causing 10 per cent of all foodborne illnesses in the United States.

Raw foods are a particular risk. The three main foodborne pathogens are *Salmonella*, *Campylobacter* and *Listeria*. The most frequent cause of foodborne

disease is contamination with one of the many species of *Salmonella*. Various strains are found in cattle, pigs and poultry. Humans infected with these strains often have to let the sickness take its course, for many pathogens have developed resistance to antibiotics, as antibiotics are not just used to treat animals but are also fed to livestock⟳ regularly as growth-promoters to produce more meat more quickly. This allows the microbes to develop resistance. Consequently, food poisoning is on the increase and is becoming more lethal: incidence has dramatically risen over the last decade in both the developed and developing world. During the 1980s the incidence of *Salmonella*-related poisoning doubled and cases caused by *Campylobacter* tripled.

FOOD SHORTAGES According to some leading research organizations, the 1990s may see growing problems of food shortages. Food yields grew at 3 per cent annually from 1950 to 1984, but since then the rate of increase has been falling steadily. By the end of the 1988–9 crop year, cereal stocks had tumbled to 306 million tonnes, following the largest recorded year-to-year drawdown. The total volume held represented 17 per cent of estimated world consumption, equal to the minimum level which the UN Food and Agriculture Organization (FAO)⟳ considers necessary to maintain global food security. Although food production rose by 3.2 per cent in 1989, the global harvest was only 1 per cent greater than that in 1988 and global cereal stocks declined for the third consecutive year.

Researchers from several international institutions conclude that food production will not keep pace with population⟳ increases up to the year 2000. Population projections see the number of people increasing by 1.7 per cent per year during the 1990s, representing an increase of 940 million people as compared to an increase of 850 million during the 1980s. Scientists are beginning to realize that some of the world's ecosystems⟳ that have traditionally been exploited to produce excesses of food may be approaching their inherent ecological and technological production limits. Given the constraints of soil and water⟳, some ecosystems are already transforming the maximum possible amount of solar energy into food.

Water and fertile soil are the basic factors which govern food production. Many of the world's most productive areas for human agriculture⟳ have been worked in an unsustainable fashion. Of the world's total consumption of fresh water, 70 per cent is already devoted to farming at levels that cannot be maintained in the long term. The water table under the North China Plain is falling by 1–2 m each year. In the United States, over a quarter of irrigated land is using more water than is replenished causing water tables to fall by up to 1 m annually. In the Soviet Union, the Aral Sea⟷ is drying out owing to the use of its feedwaters for irrigation, although much of the land⟷ is used to grow cotton⟷, not food crops. In many parts of the world, irrigation⟷ projects will have to be stopped or curtailed because irrigation water is being diverted for drinking purposes, although, overall, the use of irrigation is likely to continue to grow.

Soil is also a diminishing resource owing to over-exploitation. In the United States, 16 million hectares of farmland which has been severely eroded and depleted through overfarming is being converted to grassland or woodland. This total represents 11 per cent of the country's total cropland. The Soviet Union was also forced to stop farming 13 per cent of its cropland during the 1980s because of erosion or the need to rehabilitate the land. China is similarly losing vast quantities of land through a combination of erosion and human settlement.

Food surpluses from North America and the 'mountains' of stored food in Europe seen in the early 1980s had all but disappeared by the end of the decade, and by 1990 world grain stocks had shrunk to approximately sixty days' supply.

FORESTS Forests and woodlands cover nearly one third of the earth's total land⟷ surface. The wood⟷, fuel, food and income they provide are basic to the well-being and survival of hundreds of millions of people and various forms of wildlife. Forests can be divided into three major categories according to climate and tree type.

In the colder northern climes or highlands are the boreal forests. These contain softwood, coniferous trees such as pine, spruce, aspen and larch. These trees are often planted commercially as a source of wood for paper⟷ pulp. These forests harbour relatively few species of plants and wildlife, generally being dark and dank with dense, infertile floors made up of a carpet of needle-like leaves.

In warmer zones, temperate forests dominate. These are more diverse, made up of a mixture of conifers and hardwood, deciduous trees, with closed forest and open woodland. Broadleaf trees, such as oaks, maples, and hickories, predominate. Temperate forests are lighter than boreal forests and consequently support a wider variety of plants and animals. Organic matter, from rotting vegetation

and animal remains, produces a range of nutrients which tend to stay in the soil, making the land excellent for cultivation.

In the much warmer and humid tropics, tropical forests are the richest, densest and most diverse of all. Containing hardwood trees such as teak and mahogany, they are luxuriant and host a great diversity of wildlife species, sheltering up to a hundred per hectare. Due to the richness and diversity of life and the rapid means in which nutrients are recycled and used, tropical forest soil is fragile and poor. Virtually all the nutrients in the system are locked up in the wide variety of vegetation, so tropical forest land is not good for farming purposes.

Forests ameliorate harsh climatic effects, control soil erosion⟷, protect watersheds, maintain soil fertility, and help control pests. They yield food, oils, spices, gums, resins, tannins, dyeing materials, animal forage, beverages, pesticides⟷, green manure, rubber⟷, animal products and medicines. Yet every year around 245,000 sq km of land is cleared of trees. If this trend continues, an estimated 2–5 million of the earth's plant and animal species will disappear as result over the next fifty years.

Virtually everywhere, natural forests have been decimated during the development process. Since the beginning of the agricultural age, the world's forests have declined by 20 per cent. The greatest loss has been seen in earth's temperate forests: over 35 per cent have been cut down. Over the last 200 years, 97 per cent of the virgin forests of the continental United States have been destroyed. The last primeval forests are in the Pacific Northwest but over 90 per cent of these have already been logged. In Europe, the only tract of truly indigenous forest that still exists is a small patch on the border between Poland and the Soviet Union. What is left of forests in Western Europe is also under threat from pollution⟷. The European Commission reported that in 1987 14.7 per cent of trees in Europe were damaged. This total fell to 12.8 per cent in 1988, only because this was a year of exceptional growth conditions. Serious damage to trees in Germany, Greece, Italy, the Netherlands, Scotland and Spain was commonplace. In an annual assessment of its own forests in 1989, authorities in the Federal Republic of Germany found that 53 per cent of the trees showed signs of damage.

FOSSIL FUELS The mineral fuels – commonly coal⟷, oil⟷ and natural gas⟷ – that occur in rock formations. Other potential fossil fuels, such as oil shales, tar sands and peat⟷, have not been as widely or as intensively exploited. Fossil fuels form as the result of the deposition millions of years ago of the remains of vegetation (coal) and living organisms (oil and gas) which were buried under subsequent soils and rock deposition and later subjected to intense heat and pressure.

Fossil fuels play an important role in supplying current energy⌀ needs, but reserves are finite. Gas and oil are rapidly being exhausted. Due partly to the oil price crisis during the 1970s, oil use has been falling. Since 1979, consumption has fallen by 14 per cent and is forecast to fall a further 15 per cent by the end of the 1990s.

Known oil reserves are unevenly distributed, with the Middle East commanding 55 per cent of the total. Oilfields in Antarctica⌀ do not belong to anyone and have not been exploited, remaining a matter of international debate. Oil reserves at the end of 1988 stood at 128 billion tonnes and if prevailing rates of extraction were maintained they would be exhausted by the middle of the next century.

In contrast, known reserves of coal are substantially higher and, at present rates of consumption, should last for another 200 years or more. Coal use is actually increasing by about 3 per cent a year, despite fears about the environmental consequences. The burning of coal without proper pollution-control measures is environmentally destructive, creating acid rain⌀ and exacerbating the greenhouse effect⌀.

Natural gas which is usually found with oil, accounts for 20 per cent of the world's energy needs. This share will probably increase as gas is used to replace oil, although storage and transport problems will have to be overcome if such use is widespread. Global production has been rising steadily: up to 6 per cent annually during the 1980s. Natural gas is relatively abundant, but reserves of this fossil fuel are showing the greatest rate of depletion: known reserves are likely to be exhausted within 50–60 years.

FREE TRADE   International trade, free of all tariffs or quotas, was first proposed in Britain in 1846. According to traditional economic theory, free trade allows nations to specialize in those commodities which can be produced most efficiently, so world production is maximized. The idea was revived in 1947 with the General Agreement on Tariff and Trade (GATT)⌀, but unrestricted trade in any sense has never been achieved. In the 1980s the continuing recession in world trade caused protectionist measures which discourage foreign imports

by imposing heavy duties, making imports uncompetitive with home-produced goods.

FRIENDS OF THE EARTH INTERNATIONAL (FOEI)   An international voluntary organization established in the Netherlands, with a network of thirty-three autonomous national branches in many countries throughout the world. The FOEI Secretariat, based in London, together with the associated national organizations, aims to promote policies and action to protect the natural environment, where necessary persuading governments, commercial enterprises or international agencies to change their programmes or operations. FOEI's work is carried out through direct lobbying, education⌀, and a co-ordinated range of public activities.

FUELWOOD   Virtually half the world's population, mainly those in poor rural communities, rely on wood⌀ for their primary source of energy⌀. Fuelwood is the commonest non-commercial biological fuel and is gathered from all available sites in most developing countries⌀. In parts of Africa, Asia, and Latin America⌀, the annual average consumption of wood is 1 tonne per person. Over 1.7 billion cubic metres (cu m) of fuelwood are burnt each year in the developing countries; the largest user is India (222 million cu m), followed by China (171 million cu m) and Brazil (168 million cu m). This total represents a 28 per cent increase over the amount burnt in 1975, and experts estimate that the demand will rise to over 2.1 billion cu m by the end of this century – although only 1.5 billion are likely to be available. Despite its widespread use, fuelwood is a major health hazard. Between 300 and 400 million people suffer from respiratory disease caused by inhalation of the smoke from wood fires.

Worldwide, it is estimated that of the 2 billion or more people who rely on wood for their fuel for cooking or heating, 1.5 billion currently lack adequate supplies – a figure that is expected to rise to at least 2.3 billion by the turn of the century. The roots of the crisis are increasing population⌀ pressure, degradation of woodlands by commercial forestry operations, and the clearing of forests⌀ for plantations and cattle-ranching. The situation is most acute in sub-Saharan Africa where fuelwood accounts for 80 per cent of total energy consumed but where, due to overexploitation and the effects of recurrent drought⌦, wood is in short and rapidly diminishing supply. As usual, when any resource is in short supply, the costs, in both time and money, are high. In rural communities, women may now have to spend a whole day collecting wood, venturing further and further afield to satisfy their needs for only two or three days. In towns and cities, people are paying up to a third of their income to obtain fuelwood.

FUNGICIDES Chemicals used in agriculture⌦ and industry to control fungal growths; fungicides can be particularly harmful to human health. Many of the original compounds used contained copper⌦, for example the copper sulphate used to destroy the potato⌦ blight fungus. Toxic heavy-metal compounds, such as those of arsenic and mercury⌦, have also been widely used in industry and as seed dressings. Arsenic is acutely poisonous and carcinogenic, and persists in the environment. The consumption of bread prepared from wheat⌦ and other cereals treated with akylmercury fungicides has resulted in a number of poisoning outbreaks in several countries. The largest reported occurred in 1972, when consumption of food produced from seeds contaminated with fungicide led to the deaths of 500 people in Iraq and a further 6,000 being admitted to hospital.

The safety of many commonly used fungicides began to be questioned during the 1980s, and three which enjoyed widespread use in arable farming were found to pose a major risk of cancer⌦. These three – maneb, mancozeb and zineb – are sprayed on to a wide variety of crops including wheat⌦, barley, hops, potatoes, lettuce, spinach, onions, apples, pears and blackcurrants. Investigators in the United States suggested that 125,000 additional cases of cancer would arise if use of the chemicals continued. The US-based Natural Resources Defense Council claimed that of all the chemical residues on food, exposure to fungicides poses the worst cancer risk to young children. When exposed to heat, the three fungicides produce a breakdown product which is also considered to be carcinogenic and poses a significant problem in processed foods. In Britain, the government decided to take no action to restrict the use of these long-established fungicides, all of which were approved before modern safety standards and analytic testing were introduced.

FUR TRADE Furs from a variety of wild animals have been used for wearing apparel for centuries. Furs from animals such as otters, martens and beavers are still widely obtainable from wild-trapped specimens, especially in Canada and the Soviet Union. The Soviet Union also remains the largest single producer of wild mink, although specially created mink farms in several nations now produce many of the pelts used in the global fur trade. Until the controversy surrounding the killing of harp and hooded seals off the east coast of Canada, sealskins also played a significant part in world fur trade.

There are about 36 species of large wild cats in the world, around half of which have been exploited for their fur. Large-scale trade in cat fur started at the end of the nineteenth century, but only fifty years later several species of large cats were coming under severe threat of extinction⌦ including the cheetah, clouded leopard, jaguar, snow leopard and tiger. In 1950, there were an estimated 40,000 tigers in India. By 1972, when Project Tiger⌦ was launched to try and save the species from extinction⌦, there were only 1,800 left. The snow leopard, which has a particularly thick and much sought-after fur, has suffered a similar fate, and there are probably only around 500 animals left roaming their native Himalayas.

By 1960, many of the big cats had become so rare that the fur trade was forced to turn its attention to smaller species that had previously not been exploited, such as the ocelot, margay, tiger cat, Andean cat, leopard cat, serval lynx and bobcat.

In 1988, 180,000 skins from cats considered to be endangered species⌦ were officially exported worldwide. The decline of furs as articles of clothing – or rather their elevation to the status of luxury garments – is symptomatic of the overexploitation of wild populations. A change in consumer attitudes, stimulated by the actions of pressure groups from around the world, has seen a remarkable downturn in the fur trade. The Hudson Bay House, Britain's major furmonger, has closed down and moved to Scandinavia. The trade magazine *Fur Review* was also forced to cease publication. In Finland nearly 1,000 fur factories have closed, and in the Netherlands there are only 32 registered furriers where there were once over 400. In Austria the 1989 call by the Minister of the Environment for a fur boycott reduced sales by 25 per cent and in the United States the annual Fur Expo in New York was cancelled for lack of interest.

G

G7   A label used to describe the seven governments of the most industrialized – and wealthy – nations that attend the annual Western Economic Summit. The Summit was set up in 1975 as an emergency response to the decision by the Organization of Oil Exporting Countries (OPEC)⋄ to increase oil prices. It discusses all forms of economic problems. The nations attending are Canada, France, the Federal Republic of Germany, Italy, Japan, the United Kingdom and the United States, with the European Community⋄ also being represented. Other similar groupings are the Group of 77⋄ and the G15⋄ assemblies, which both devote themselves to discussion of global economic development.

G15   A term used to describe a group of fifteen nations, all from the South⋄, that gather together to discuss development and Third World⋄ issues. The group was formed at the Non-Aligned Summit⋄ meeting in 1990 and is the third such grouping along with the Group of 77⋄ and G7⋄. Member states will devote themselves to promoting South–South co-operation and developing a common strategy towards the North⋄. Membership comprises Algeria, Argentina, Brazil, Egypt, India, Indonesia, Jamaica, Malaysia, Mexico, Nigeria, Peru, Senegal, Venezuela, Yugoslavia and Zimbabwe. A three-nation steering committee, composed of Malaysia, Senegal and Venezuela (one nation from each of the three continents in the South), guides and stimulates the group's activities.

GAIA   In ancient Greek mythology, Gaia was the name given to the mother goddess, who sprang up from primordial chaos. Gaia was also recognized as personifying the Earth being the giver of dreams and the nourisher of plants and children. The Gaia Principle is the name given to a theory – proposed by a British scientist, James Lovelock – that the biosphere⋄ is a 'super-organism' and that it, like all living organisms and systems, is capable of effecting regulatory mechanisms as and when necessary. Essentially, the Gaia Principle regards the biological and geological aspects of the earth and its atmosphere as a single, interacting system. This system has an inbuilt ability to implement corrective processes to create the environment that most favours its own stability, and to maintain it in the face of pollution⋄ and all other forms of environmental change or perturbation.

GANDHI, INDIRA (1917–84)   Indian stateswoman. A daughter of Jawaharlal Nehru, independent India's first Prime Minister, she married Feroze Gandhi (no relation of the renowned Indian leader, M.K. Gandhi), in 1942. She entered politics and rose to become leader of the Congress Party and serve two terms as the nation's Prime Minister, 1966–77 and 1980–84. She had two sons, Sanjay and Rajiv, who both became active in Indian politics.

In 1975 she was accused of electoral malpractices and threatened with the loss of her seat in Parliament. She responded by creating a state of emergency in the country and imposed strict authoritarian government. During this time, the social and economic programme followed by the government, and devised by Sanjay Gandhi, included a ruthless family-planning policy, which contributed significantly to her electoral defeat in 1977. Although Sanjay played a role in masterminding her return to power in 1980, he was killed in a plane crash the same year.

Indira Gandhi was assassinated in 1984 by members of her Sikh bodyguard following her use of government troops to clear malcontents from the holy temple at Amritsar. She was succeeded as Prime Minister by her other son, Rajiv.

GANGES   The River Ganges rises in the Himalayas in Nepal and northern India and flows 2,700 km into the Bay of Bengal through the world's largest delta in Bangladesh. Boasting a highly seasonal water⋄ flow, the river is used to irrigate the Ganges plain, which contains several of India's major cities and forms the world's second-largest agricultural area, after China's Yangtze valley. Altogether, the Ganges supports agricultural production that sustains well over 300 million people.

The river is also a major conduit for sewage and industrial discharge from the many human centres of habitation along its length. At least 114 major towns and cities discharge raw sewage into it. The Ganges is the sacred river of 600 million Hindus who seek to bathe in its waters and, upon death, wish to be cremated on its banks, their ashes being scattered on its waters. Today, many cadavers are thrown directly into the river as the scarcity and rising cost of firewood takes cremation beyond the means of most. As well as wood⋄ shortages, deforestation⋄ in the Himalayas and along the river's banks has also caused topsoil loss and siltation levels in the Ganges to rise markedly, causing severe flooding in the low-lying areas in the delta in Bangladesh. India's control of the river's flow, most notably the construction of the Farakka Barrage⋄, has caused bitter controversy with neighbouring Bangladesh.

GELDOF, BOB (1954–)

Irish rock musician with the Boomtown Rats group. He was the inspiration and driving force behind Band Aid⋄ and Live Aid (1984–6) and an abrasive and vigorous campaigner for improvements in the way aid⋄ was given to developing countries⋄. Band Aid produced a record and Live Aid was two concerts, one in London, one in New York, performed on the same day and broadcast one after the other to the world's largest-ever television audience. Both were charity events raising huge sums of money for famine prevention and relief, especially in Ethiopia. Several other similar fund-raising events, arranged by musicians, sports people and professionals from a variety of fields, were initiated and repeated on a regular basis following the lead given by Band Aid. Although Geldof is an Irish citizen, he was appointed a Knight Commander of the British Empire in 1988 for his efforts to improve living standards in the developing world.

GENERAL AGREEMENT ON TARIFFS AND TRADE (GATT)

An international, negotiated series of contracts embodying a code of practice and rules to orchestrate fair trading in global commerce which first came into force in 1948. GATT had its origins in the first tariff-negotiating conference held in Geneva in 1947 and was established following an initiative from the United States, originally as a temporary measure and primarily as a forerunner to the International Organization for Commerce, first discussed at the Bretton Woods⋄ meeting, which was planned for inauguration in 1948 but never came to fruition.

The GATT agreement is centred on four principles of 'good conduct': commercial non-discrimination, reciprocity, the prohibition of quantitative restrictions on imports, and the prompt and equitable settlement of commercial disputes and conflicts. Negotiations take the form of 'rounds'; the first round took place in 1964.

The original GATT membership totalled 23, today there are 100 contracting parties with 8 more nations negotiating membership in various working parties under special arrangements. At the final meeting of the Uruguay Round talks in 1990, 107 nations took part in the discussions. All in all, the member countries account for well over 80 per cent of world trade.

GATT has traditionally been dominated by the developed countries and focuses heavily on the protection of tariffs and trading advantages. This led the developing countries⋄ to seek another forum for negotiations; eventually, in 1964, the United Nations Conference On Trade and Development (UNCTAD)⋄ was established.

GATT's Committee on Trade and Development was established in 1965 to keep under review the application of the provisions of the Round IV agreement, which contained special measures to help promote the trade and development of the less-developed countries and to carry out other work on the trade problems of developing countries. The principle of a single tariff for all industrial manufactured goods was also first agreed upon during the 1964–7 discussions.

The major matters of concern at the Uruguay Round were developed countries' subsidies for both agricultural exports and farm support, the opening up of markets in the industrialized world to textiles from the Third World⋄, rules governing foreign investment, and global trade in services. In addition, delegates attempted to clarify the position and rules governing intellectual property. Discussions were also given over to the possibility of strengthening GATT and developing it into an International Trade Organization.

GENERALIZED SYSTEM OF PREFERENCES (GSP)

A system agreed under the auspices of the United Nations Conference on Trade and Development (UNCTAD)⋄ and later accepted by the General Agreement on Tariffs and Trade (GATT)⋄, whereby developing countries⋄ were to be given preferential access to certain markets for some of their products. Developed countries importing manufactured and processed goods from developing countries were to give the latter certain tariff advantages in their markets (either duty-free or reduced-duty access). The system is concerned purely with merchandise produced entirely, or having undergone processing, at source. The GSP encompasses around 90 per cent of industrial products subject to customs duty and taxes in industrialized nations, but covers just over half of the world's agricultural-trade products. The original aim of the GSP was to encourage industrialization in the developing world and to reduce the Third World's⟳ dependence on the export of raw, unprocessed materials. At present there are more than a dozen such schemes in operation, including those of Canada, the European Community⟳, Japan and the United States; the first scheme began in 1971.

GENE THERAPY   The name given to a technique involving the manipulation of human genes through which hereditary and genetic disorders may eventually be corrected. The first such experiments involving human patients were carried out in the United States in 1989. There are 3,000 human disorders that are caused by a single gene, including cystic fibrosis, a chronic respiratory disease, and muscular dystrophy, a degenerative muscular disorder. The techniques involved and the possibility of gene therapy becoming a viable form of treatment are matters of conjecture and concern, and the use of gene therapy poses a number of risks for any recipient. The generally favoured method of inserting a gene into human cells is via the use of a retrovirus – a naturally occurring agent capable of introducing material from itself into other organisms. Scientists can successfully implant foreign genetic material into these viral vectors. However, there is, as yet, no way of knowing where in the patient's body the modified genes will end up. There is also no method of assessing the safety of the vector viruses, and no means to get the newly inserted genes to express themselves correctly in the right cells at the right time.

GENEVA CONVENTION   An international agreement, reached in 1864, which established rules to govern the treatment of those wounded in war. The first conference was attended by sixteen countries which agreed a set of rules to cover the care and protection of non-combatants and the wounded. It was later extended to cover prisoners of war and the protection of the sick and civilians during times of conflict. The Convention has undergone several reviews – it was updated in 1909, 1929, and 1949, the last revision was in 1977. It now covers assistance for forces at sea, prisoners of war and the protection of civilians and the sick or injured during any times of conflict. Under the Convention, prisoners of war are entitled to food on the same scale as the captor country's rear-line troops. They also have the right to adequate clothing, footwear, and medical attention. They may also send and receive letters and receive food parcels and reading matter. Those permanently incapacitated may be repatriated.

Most countries accept the Convention as morally binding although it cannot be legally enforced. Neutral nations – and, more importantly, the International Red Cross – play a supervisory role in seeing that the terms of the Convention are respected, and virtually all nations agree to let internment camps be inspected by the Red Cross.

GEOTHERMAL ENERGY   The use of heat generated in the earth's interior to produce power or heat. Commonly, either subterranean hot water is pumped to the surface, converted to steam or run through a heat exchanger, or naturally occurring steam is used to drive turbines to produce electricity. Natural volcanoes, geysers and hot springs are all used, as is man-made steam. This is produced by drilling into hot rocks, pumping in cold water, which percolates into and fills underground fractures and cavities, turns to steam, becomes pressurized, and finally emerges from other boreholes.

The geothermal energy in the upper 5 km of the earth's crust is equivalent to 40 million times that contained in the world's crude oil and natural gas⟳ supplies. Although only a fraction of the total is economically exploitable, use is growing. Between 1970 and 1980 production rose by 14.5 per cent, increasing a further 17.5 per cent by 1986. In Iceland, geothermal energy is the most important source of power, and Italy, New Zealand, the Philippines and the United States all make use of it. Geothermal power currently accounts for about 0.1 per cent of the world's energy⟳ requirements and is used by nineteen countries, but there are vast resources still untapped in countries such as Indonesia and the Philippines. Other nations, such as Djibouti and St Lucia, are currently investigating how best to harness the heat energy and steam produced by local volcanoes. In Hawaii, the construction and operation of a geothermal plant on the

side of an active volcano at Pahoa is causing controversy, despite the fact that the plant will reduce the
island's dependence on oil⟠ imports. Opponents
claim that the plant will lead to the destruction of
vast tracts of lowland tropical forests⟠, although
supporters argue that only 2 per cent of the Wao
Kele forests in the surrounding area are likely to
disappear.

Like all other forms, geothermal energy has to be
managed properly, as steam can be removed faster
than it is replaced. At the world's largest geothermal
power field in Northern California, natural steam
pressure has dropped 20 per cent through overexploitation.

GHOST ACRES   Wealthy industrialized countries satisfy the food needs of their populations⟠
and derive a substantial amount of their prosperity
through their ability to import cheap foodstuffs and
agricultural products. They effectively depend on
vast tracts of land⟠, primarily located in the developing countries⟠ of the world, for the production
of their food. This land, situated in nations where
hunger⟠ and malnutrition⟠ are often rife, is
referred to as 'ghost acres', a term first coined by a
United States agronomist, George Borgstrom. Well
over 1 million sq km of valuable arable land in the
Third World⟠ are used in this way. Without
recourse to ghost acreage, countries such as the
Federal Republic of Germany, Japan and the United
Kingdom would have great difficulty in meeting the
food requirements of their population and would
have to undergo a radical change in national diet and
consumption patterns.

GLOBAL 2000   In 1977, US President Jimmy
Carter⟠ commissioned a study of global environment trends to determine what would be the major
features of life up to the year 2000. The findings
were published in *The Global 2000 Report to the
President* in 1980. The report basically concluded
that by the year 2000 the world would be more
crowded, more polluted, less stable ecologically, and
more vulnerable to disruption than it was in 1980.
The authors opined: 'for hundreds of millions of the
desperately poor, the outlook for food and other
necessities will be no better. For many it will be
worse.'

Although the report was well received and
initiated a great deal of support, President Carter
did not remain in power long enough to implement
any of its recommendations, and the succeeding
administration of President Reagan⟠ ignored it

entirely. Following his departure from office, Carter
formed the Global 2000 Project, an independent
centre which works towards implementing some of
the recommendations contained in the report.

GLOBAL ENVIRONMENT FUNDS   Protect-
ing the global environment for the benefit of all will
be an expensive business. The United Nations
Environment Programme (UNEP)⟠, for example,
has determined that it will cost $400 million
annually during the 1990s merely for Third
World⟠ nations to comply with the Montreal
Protocol⟠ to protect the ozone layer. A Global
Environment Fund, first officially proposed by Gro
Harlem Brundtland⟠, would allow a worldwide
programme of pollution⟠ reduction and environmental improvement to be established and administered. Norway has already indicated that it would be
willing to contribute 0.1 per cent of its Gross
National Product (GNP)⟠ annually – $100 million
– to such a fund.

During the late 1980s several suggestions and
ideas for the creation of specialized global funds for
environmental-improvement and pollution control
programmes were suggested. The Soviet Union
went so far as to propose that a global programme
on environment and development should be drawn
up and embodied in a Code agreed and finalized by
the United Nations Special Conference on Environment and Development, scheduled for 1992.

The problem with all these proposals was how to
produce the money to finance the operation of the
programmes. One popular suggestion was the introduction of a worldwide 'carbon tax', a levy on
emissions of carbon dioxide⟠, which contributes
markedly to global warming. The International
Energy Agency (IEA)⟠ calculated that a 20 per cent
rise in fossil fuel⟠ prices arising from such a tax
would reduce the increase in carbon dioxide in
Organization for Economic Co-operation and
Development (OECD)⟠ countries from 27 per cent
to 13 per cent. Others reported that a tax of 10 cents
(US) per tonne of carbon released through the burning of fossil fuels would produce an estimated $700
million each year. High taxes on nuclear energy⟠,
similar to the effect of a carbon tax, would provide
money whilst increasing the economic viability of
renewable sources of energy⟠ and reducing consumption of both coal⟠ and uranium⟠.

Alternative ideas included the call for an economic
price to be put on the fruits of nature, whereby
polluters would incur punitive charges for actions
leading to the loss of 'free' air, soil and water⟠. The

establishment of a Trust Fund, financed by taxes levied on use of the so-called global commons – the oceans, Antarctica⌀ and outer space – was a popular suggestion. A small tax of 0.1 per cent levied on the world's annual fish⌀ catch would produce $50 million, while a similar tax on offshore oil⌀ and gas would bring in a further $75 million. A charge of $1 per tonne for the millions of tonnes of sewage, chemicals and other materials dumped in the world's seas each year would also boost a trust fund by another $200 million. A tax of 10 cents (US) per tonne on emissions in to the atmosphere of sulphur and nitrogen would yield $30 million, in addition to any income generated by a tax on carbon emissions. Although many experts considered the sums involved to be insufficient properly to tackle global environment problems, many industrialized governments – including those of Japan, the United Kingdom and the United States – were originally opposed to any such fund.

In 1990 twenty-five industrialized and developing-country Bank members agreed to establish a fund of between 1 and 5 billion dollars to help the Third World⌀ pay for investment, new technology, technical assistance and research and training needed to help protect the environment. The new fund is to be made up of donations from eighteen developed nations only. France, which first broached the idea, promised to contribute $180 million of the new money by 1993. This followed an announcement from the heads of the UN Environment Programme (UNEP)⌀, UN Development Programme (UNDP)⌀ and the World Bank⌀, who signed a joint statement calling for the setting up of a $1 billion global environmental facility to advance their agencies' joint efforts to ensure that 'development programmes are undertaken in a manner which protects the global environment'.

This facility, scheduled to operate from 1991 to 1993, will provide new grants and provisional loans for investment. The facility will allow middle-income countries (MICS)⌀ such as Brazil and Mexico to benefit from the $400 million which will be set aside annually for environmental-improvement projects. Projects to be funded must show a clear global benefit and those initially proposed included a wetlands⌀ scheme in Turkey, a forestry conservation⌀ scheme in Cameroon and a project to cap flaring from natural gas⌀ in Nigeria. Critics claimed that the fund, although welcome, would be manipulated by the donor countries. The $200 million donated by the United States is to come out of the budget of the US Agency for International Development (USAID); the money will be administered separately by USAID in parallel with the Bank. Similarly, the largest donor, Japan, will link its donation of $240 million closely to its national aid⌀ programmes.

GLOBAL ENVIRONMENT MONITORING SYSTEM (GEMS)

The role of GEMS since its inception in 1975 has been to co-ordinate international environmental monitoring activities, particularly within the United Nations⌀ system. Co-ordinated by the United Nations Environment Programme (UNEP)⌀, GEMS aims to standardize the collection, analysis and dissemination of environmental data produced by national and international organizations from ground observation stations, ships and satellites. GEMS activities are concentrated on five main areas of environmental concern: climate, transboundary pollution⌀, terrestrial renewable natural resources⌀, oceans, and the health consequences of pollution⌀. GEMS data are used to produce periodic regional and global assessments of environmental conditions, and in 1985 the Global Resource Information Database (GRID)⌀ system was launched, making the information available in computerized, cartographic form. There are over thirty monitoring networks working within GEMS, and 142 nations and 30,000 scientists from around the world have participated in the system.

GLOBAL INFORMATION AND EARLY WARNING SYSTEM FOR FOOD AND AGRICULTURE (GIEWS)

The Global Information and Early Warning System for Food and Agriculture (GIEWS) is based in Rome and run by the Food and Agriculture Organization (FAO)⌀. Making use of a combination of field data, agro-meteorological forecasting and remote-sensing information, the system is designed to monitor the world's supply of food and provide an indication of any threats to global supplies, the FAO alerting governments to emerging food shortages or potentially adverse conditions for crop production.

GLOBAL RESOURCE INFORMATION DATABASE (GRID)

Since the mid-1980s, the United Nations Environment Programme (UNEP)⌀ has been operating a programme to collect information on the environment. Known as the Global Environment Monitoring System (GEMS)⌀, it has been collecting statistics and data from a variety of individual sources, including satellites and monitoring stations on the ground. All the data have been put into GRID, a database capable of storing 60 to 70 gigabytes of information, the equivalent of 60,000 average-length novels. GRID is in the form of a network with three regional centres: in Bangkok, Geneva and Nairobi. Eventually, all countries will be able to feed data directly into the network and have access to the information therein. Each piece of data in GRID is classified according to a geographical reference. This enables simple graphical representations of hundreds of pieces of information to be produced, usually in the form of overlaid maps. It will thus be possible to provide a reliable early warning system on the basic environmental conditions of the planet. GRID has already been used successfully to tackle important environmental and conservation◇ problems, including an analysis of elephant◇ habitats in Africa and patterns of acid rain◇ in Western Europe.

GLOBAL WARMING Global warming, mainly occurring as the result of the build-up of carbon dioxide◇ and other 'greenhouse' gases◇, has been identified by many scientists as a major environmental threat to the future of the world. Some, however, doubt that the phenomenon will have the forecast degree of impact. Initial computer models, which had predicted dire consequences as a result of the rise in sea level that would accompany a warming of the earth's atmosphere, are being steadily refined, and the dire forecasts have been mitigated somewhat. Models adjusted to compensate for the effects of cloud cover suggest that the warming due to the greenhouse effect◇ may be only half as intense as popularly predicted.

Original estimates – still held to be true – predicted that the level of atmospheric carbon dioxide would double through the continued burning of fossil fuels◇. It was initially forecast that this could lead to a temperature rise of between 1.5 and 4.5°C. Estimates now predict that the globe will warm up by, on average, 2.7°C, once the level of carbon dioxide in the air has doubled, causing a predicted sea-level rise of around 30 cm. These calculations originated from the same model at Britain's Meteorological Office which previously predicted a warming of 5.2°C. The simple addition into the model of the effect of liquid water clouds (which last longer than the ice clouds considered in the original model and will therefore block out incoming solar radiation longer) accounted for the amended prediction. The wisdom of reacting sharply to possible disasters predicted by computer models is further questioned by the same model which, using reliable

figures for the build-up of atmospheric carbon dioxide since the nineteenth century, predicted that the earth should have warmed up by 0.6°C. This is 20 per cent above the rise in temperature that actually occurred.

In the 10,000 years since the Ice Age the ambient temperature around the earth rose by 4°C, but the rate at which it is rising has increased dramatically since the beginning of the Industrial Revolution and continues to gather momentum. During the 1980s, data collected from ground-based observation stations showed that the earth warmed up by almost 0.04°C. Measurements made from space since 1979, however, suggest that, globally, there was in fact no sign of permanent warming during the 1980s. Satellites measure the temperature of thermal radiation, which is thought to provide a more reliable guide to global temperatures. The data showed that although two strong 'El Niño'◇ events produced warming in 1983 and 1987, there was a cold run of temperatures between 1984 and 1986. In 1990 the Intergovernmental Panel for Climate Change (IPCC), a group of experts from around the world, assembled to investigate the problem of global warming, concluded that although glaciers around the world were melting at increasing rates and snow cover had decreased significantly since 1980, at least thirty years of observation would be needed to confirm that global warming had begun.

In 1990 the IPCC agreed that if no steps were taken by the global community to reduce emissions of greenhouse gases, the average ambient temperature would rise 3°C by the year 3000. A 1.8°C rise by 2020 would cause sea levels to rise by 20–65 cm by 2100. Even if greenhouse gases were reduced by 80 per cent immediately, temperatures during the 1990s would still rise by 0.1°C per decade – twice as fast as they rose during the rest of the century. Scientists are in basic agreement that a natural ecosystem◇ can cope with an overall 1°C rise in temperature without being seriously disrupted, provided that the rate of change does not exceed 0.1°C per decade.

GOITRE A swelling of the neck due to enlargement of the thyroid gland. The condition usually occurs as a result of the lack of dietary iodine, which is necessary for the production of thyroid hormone. The thyroid gland enlarges substantially in an effort to increase the output of the hormone. This is the cause of endemic goitre, which is still common in parts of the world where iodine is lacking in the diet. Around 190 million people are thought to have the condition, and iodine deficiency is a major threat to the physical and mental growth and development of at least 1 billion people who live in iodine-deficient environments, primarily hilly areas in parts of Africa and Asia. Iodine deficiency, and the subsequent development of goitre, can be avoided by adding iodine to the diet. In deficient areas, programmes based on the distribution and use of iodized salt or injection of iodized oil has cured the problem. In view of the success of these programmes, the World Health Organization (WHO)✧ is aiming to eliminate iodine-deficiency disorders as a major public health problem in all countries by the year 2000 in a programme set to cost $20 million.

GOLD   Gold is an extremely rare metal, with an average distribution through the earth's crust of 0.0035 g per tonne. It frequently occurs – although in extremely low concentrations – in native metallic form. Commercially, it is worked wherever veins are found. Gold is a soft, dense metal, the most malleable and ductile known. It is a good conductor of heat and electricity and is chemically unreactive. Gold has been the traditional monetary standard and is used for coins and jewellery, teeth fillings and other dental products, and some medicinal purposes. The jewellery trade takes around two-thirds of annual production, and dentistry uses about 7 per cent of world supplies. Gold is increasingly being used for other purposes such as in coatings for electronic components and as an infra-red filter. Global production is in the order of 1,850 tonnes per year; South Africa (621 tonnes), the Soviet Union (300 tonnes) and the United States (205 tonnes) are the major producers. It is estimated that there are 35,000 tonnes of exploitable reserves in the world, and these may well begin to be exhausted over the next twenty years or so.

GOLD STANDARD   A system under which gold✧ coins were the monetary unit. This system collapsed after the First World War, and attempts to revive it were thwarted by the Depression in the 1930s. After the Second World War, the values of all the currencies of member countries of the International Monetary Fund (IMF)✧ were fixed in terms of gold and the US dollar, although in 1976 floating exchange rates became officially recognized. Although the price of gold fluctuates on the world market, holdings are still retained by every country because gold is the internationally recognized commodity✧ money, which cannot be legislated upon or easily manipulated by interested parties.

GORBACHEV, MIKHAIL SERGEIEVICH (1931–)   Soviet statesman and political leader. The son of Russian peasants, he became a member of the Communist Party in 1952. He graduated with a degree in Law in 1955 and later took a degree in Agriculture (1967). A Politburo member from 1980, he was groomed for power during Andropov's reign (1982–4), becoming Chairman of the Foreign Affairs Committee in 1984. When succeeding Chernenko as Communist Party leader in 1985, he became yet more powerful, becoming a member of the Secretariat which dealt with day-to-day Party business as well as maintaining his place as the youngest member of the policy-making Politburo. He was responsible for the introduction of far-reaching and radical reforms into the Soviet Union and did much to bring about the end of the Cold War, eventually assuming the post of President, Gorbachev's policies saw the movement towards a free-market economy, but created widespread unrest within the Soviet Union and a call for independence from many of the states. He was awarded the Nobel Peace Prize in 1990.

GRAMEEN BANK   A bank started in Bangladesh to assist the rural poor. Loans are given, mostly to women, without collateral and without the need for guarantors. Most of the Bank's business is conducted in the villages of the poor. Emphasis is placed on Bank representatives travelling to villages furthest from major offices, and on catering for the needs of the poorest families. Loans have been provided to over 500,000 people. Despite the fact that money is lent without collateral, the Grameen bank reports a 98 per cent repayment rate – far in excess of that observed in most conventional banks. The founder of the bank, Professor Muhammad Yunus, formerly an Economist at Chittagong University, started the project in 1977, with funds borrowed as a personal loan, in an attempt to counteract the abject poverty in parts of the country. He perceived that conventional banking was 'anti-poor' and that rural, resource-poor families knew how to solve their own problems if they were given the resources to do so. The Bank has now spread throughout Bangladesh and has a staff of 8,000. The borrowers also own the bank, having purchased its shares in 1986, using accumulated savings. Following the success of the Grameen Bank, similar operations, such as the Amanah Ikhtiar Malaysia (AIM), are being set up in other countries.

GRANDE CARAJAS According to environmental groups, environmental and social catastrophe are likely to follow Brazil's Grande Carajas project if it goes ahead. The $62 billion scheme will cover an area of approximately 900,000 sq km, over 10 per cent of the country's land◇ area, opening up the north-east of Brazil to industry and commercialized agriculture◇. A vast complex of mines, dams◇, railways, cattle ranches and plantations are scheduled to be built, and one sixth of Brazilian Amazonia◇ will be affected. Centrepiece of the project is the Serra dos Crajas open-cast iron ore mine, backed by a $600 million loan from the European Community (EC)◇. The World Bank◇ is also contributing substantial backing for infrastructural components of the project. Both the World Bank and EC are major funders of the scheme despite the fact that it infringes the human rights◇ and environmental-protection policies of both organizations. Japan is also contributing significant backing.

Proponents claim that the scheme will generate $10.6 billion in exports by the mid-1990s. Opponents point out that over 250,000 sq km of rainforest◇ have already been destroyed by the project and another 600,000 sq km or more are under imminent threat of destruction. The region's 13,000 aboriginal people are also threatened by a combination of disease, land invasion and loss of cultural heritage. Companies are being offered significant tax breaks to encourage them to partake in the project – this incentive, together with lax environmental legislation, is likely to fulfil critics' predictions that the project will lead to severe water◇ and air pollution◇ problems.

GREAT LAKES Five large lakes in the eastern central region of North America, mostly lying along the US–Canadian border. The lakes – Superior, Huron, Erie, Ontario, and Michigan – form the world's largest freshwater surface, containing 18 per cent of the world's liquid fresh water◇. Extensive agriculture◇, fishing◇, forestry, mining, hydroelectric power◇ generation, manufacturing, commerce and tourism◇ have developed in surrounding regions. As a result, the lakes and shore areas have become increasingly polluted. Water pollution◇ has become of particular concern, as the lakes supply drinking water to 21 million North Americans. Pollution from industries around the lakes turned them, notably Lake Erie, into 'dead lakes', where few fish◇ and other aquatic wildlife could survive.

Strict pollution controls and concerted effort have helped to clean the lakes up. In 1985, the eight US states and two Canadian provinces bordering them signed the Great Lakes Charter, followed a year later by the Toxic Substances Control Agreement, both designed to reduce pollution in the lakes and govern water use. But environmental improvement programmes have been hampered by the fact that the governments of two countries are involved, each with differing and conflicting problems and priorities. How best to curb pollution, who is most responsible, and who should pay for corrective measures are problems which have continually hindered the progress of clean-up initiatives. The complex and emotive problem of acid rain◇, which is caused mainly by industry in the United States but predominantly falls and damages forests◇ in Canada, has been a major stumbling block casting a shadow over all environmental improvement projects concerning the lakes.

GREEN BELT MOVEMENT An agro-forestry◇ project founded and run by women in Kenya. The movement, started and organized by Professor Wangari Maathi, a lecturer in veterinary medicine, aims to plant trees for a variety of reasons: to stabilize soil, for use as fuel, for landscape improvement and as a source of income for the women taking part in the scheme. Tree nurseries have been established which issue seedlings that are used to produce green belts in both rural and urban areas of the country. These seedlings are provided to women's groups which then oversee the planting and raising of the trees. For every tree that survives more than three months outside the nursery, the woman who plants and cares for it is paid 25 cents (US). Paying female tree-tenders a premium for every seedling that survives not only provides an income for women who would otherwise have none, but has also allowed the Green Belt project to attain a transplant survival rate of around 80 per cent. By the end of the 1980s the movement had seen the creation of well over 600 tree nurseries and grown 10 million trees, thanks to the involvement of 50,000 women and children in 3,000 schools.

GREENHOUSE EFFECT The atmosphere, like the glass of a greenhouse, lets much of the sun's visible and near-ultraviolet radiation pass through it to warm the planet's surface. The warm earth reradiates electromagnetic radiation⊖, but of a far lower frequency (infra-red) to which the atmosphere, like glass, is not as transparent. The infra-red radiation cannot escape and, as in a greenhouse, the heat energy is trapped, raising the ambient temperature. Human activities – such as burning fossil fuels⊖ – which increase the concentration of carbon dioxide⊖ and other similar gases in the atmosphere tend to promote the heating effect – as do the conversion and burning of forests⊖, which contribute 25 per cent of all the so-called greenhouse gases⊖.

Carbon dioxide⊖ has been increasing in the atmosphere at 0.4 per cent annually, and if this continues, atmospheric concentrations will have reached double what they were in pre-industrial times within eighty years. The concentration of carbon dioxide alone could raise temperatures by 1.5–4.5°C over the next fifty to a hundred years. The Earth has warmed up by almost 0.5°C over the last century, yet changes of only fractions of a degree can cause glaciers to advance or decline. Sea levels could rise around the world as a combination of ice melts and thermal expansion of water in the oceans. If current rises in atmospheric carbon dioxide concentrations are maintained, the polar icecaps and glaciers will begin to melt faster producing a forecast rise in sea levels of between 46 and 77 cm.

Early prognoses suggested that the greenhouse effect would cause sea-level rises of between 20 and 150 cm by the year 2030. In worst case scenarios, coastal areas of the world would be devasted. Low-lying countries such as the Maldives and Tuvalu would disappear, 15 million people in Bangladesh alone would lose their homes and the nation's agriculturally productive area would be virtually submerged, bringing the spectre of starvation to the rest of the population⊖. As more information came to light estimates of possible rise in sea level were revised toward the lower, less catastrophic end of the range.

Limiting the use of fossil fuels⊖ to reduce levels of atmospheric carbon dioxide and so limit the greenhouse effect will cost an estimated $350 billion over the next fifteen years, although some argue that fossil-fuel reductions could actually lead to savings in the long run. Irrespective of the costs, the global community simply cannot agree on the specific measures needed or how to finance the reduction or the switching to less polluting forms of energy generation. Most focus on cutting carbon dioxide emissions. Proposals for a freeze on such emissions by the year 2000 and a 20 per cent reduction in emissions by 2005 were vetoed by Japan, the Soviet Union, the United Kingdom and the United States which, between them, emit half the global emissions of carbon dioxide.

Extensive forest plantations, acting as a 'carbon sinks', have been proposed as one solution to absorbing the approximate 5 billion tonnes of carbon generated each year. Other proposals include growing algae, or issuing 'carbon permits'. Algae, grown in either fresh or saline water⊘, are ten times more efficient at carbon dioxide uptake than trees in terms of space needed, and even coastal waters, where nutrient levels and climatic conditions are favourable, can be exploited. 'Carbon permits' would be allocated to individual countries based on their population size. Countries wishing to exceed their permitted level of carbon emissions would have to acquire permits from other countries to cover the excess production. Suggestions to set up a 'global fund'⊖ to help developing countries⊖ minimize their reliance on fossil fuels and wood⊘ originally elicited little support from donor governments, although most reversed their position in 1990 and generally backed the fund.

GREENHOUSE GASES   Gases, generally produced as a result of human activities, which contribute to the greenhouse effect⊖ and so cause the ambient temperature of the world to rise. The worst of these gases is carbon dioxide⊖ which, although it is a relatively weak contributor to the greenhouse effect, contributes half the effect of all greenhouse gases because of the vast amounts vented into the atmosphere.

Molecule for molecule, one of the chlorofluorocarbons (CFC)⊖, CFC-12, is 5,400 times more effective as a greenhouse gas than carbon dioxide, but it is produced in minimal quantities compared with carbon dioxide emissions. Of the other major greenhouse gases, methane⊘ is 21 times more potent than carbon dioxide and nitrous oxide is 290 times as destructive.

Methane, produced mainly by ruminants and rice⊘ paddies, contributes 18 per cent of the greenhouse effect, with the CFC from aerosols and refrigerants, despite their potency, accounting for only 14 per cent of the total greenhouse effect arising from human sources. Lower-atmosphere ozone⊘ (12 per cent) and nitrous oxides (6 per cent), both produced in motor-vehicle exhausts, are responsible for the remainder. Several of these gases also occur and are vented into the atmosphere through natural processes.

In the late 1980s, the international community reached a degree of consensus on stabilization of carbon dioxide emissions and the phasing out of emissions of other gases such as the CFCs. Even if emissions of greenhouse gases were to be stabilized their concentration in the atmosphere would continue to increase, as they are being vented faster than they are being destroyed. Despite global action to stabilize or reduce emissions, carbon dioxide is currently increasing in the atmosphere at 0.4 per cent annually, as are methane (1 per cent annually), nitrous oxide (0.3 per cent) and CFCs (4.5 per cent). Water vapour is also a major contributor to the greenhouse effect, and its atmospheric concentration is strongly influenced by human activities.

GREEN PARTY   A name applied to environmentally friendly political parties and believed to have arisen from the German *die Grünen* environmentalist party, which first entered the Federal Republic of Germany's Parliament in 1983. The Party won 5 per cent of the national vote which, through proportional representation, gave it twenty-seven seats in the Bundestag. In Belgium the Greens gained national representation two years earlier. The Values Party, launched in New Zealand in 1972, can lay a strong claim to be the first ever 'green' party, although it never gained any seats in national elections.

Several similar parties now exist, primarily in countries throughout the European Community⌀ and elsewhere in the industrialized world. The Ecology Party in the United Kingdom changed its name to the Green Party in 1985. In the European Community elections in 1989, the UK Green Party won 2.25 million votes, 15 per cent of the total cast. This may, however, have represented the peak of achievement for the Green parties in Europe. Their success in bringing the state of the environment on to the political agenda and showing that environmental awareness and protection were vote-winning topics, caused all the mainstream political parties to pay close attention to environmental problems and introduce specific environmental protection policies into their manifestos. As a result, votes for the Green parties began to evaporate. In the first all-German elections following unification in 1990, *die Grünen* failed to capture 5 per cent of the votes and thus, under the proportional representation system, lost their right to seats in Parliament.

GREENPEACE   The modern Greenpeace movement began in earnest in 1971, when a group of

people chartered a ship and sailed to the Aleutian Islands to protest against nuclear testing.

Greenpeace International is a voluntary organization with fifteen offices around the world and a number of affiliated national branches, supplemented by autonomous national Greenpeace organizations. Members aim to monitor activities that adversely affect the environment and campaign on specific issues, such as the discharge of radioactive effluents, commercial whaling⌀ and sealing, and the disposal of toxic wastes. Co-ordinated campaigns on selected issues of environmental abuse incorporate a mixture of lobbying, scientific reports, public education⌀ and direct action. The use of direct action has sometimes led to confrontations, and many members have been arrested for their actions. In 1985, saboteurs from the French armed forces sank a Greenpeace vessel in harbour in New Zealand, killing one member of the crew. The ship was *en route* to protest against French nuclear testing in the South Pacific.

GREEN REVOLUTION   The development and widespread adoption of High-Yield Variety (HYV)⌀ strains of wheat⌀, corn⌀ and rice⌀ in the South⌀ during the 1960s and early 1970s. In 1963 the Food and Agriculture Organization (FAO)⌀ held a World Food Congress which witnessed the launch of the Indicative World Plan for Agricultural Development, which became known as the Green Revolution. It involved the introduction of HYV cereals, primarily those developed at the International Rice Research Institute (IRRI) based in the Philippines and the International Maize and Wheat Improvement Centre (CIMMYT) in Mexico. The HYV plants bred were short-stemmed to avoid wind damage and quick-maturing, making it possible to raise three crops per year on one piece of land if conditions were favourable.

For the Green Revolution, seeds had to be used with high inputs of fertilizers⌀, pesticides⌀ and water⌀, usually involving irrigation⌀. In developing countries⌀, cereal yields as a whole rose by 2 per cent annually between 1961 and 1980. Wheat yields increased by 2.7 per cent although rice yields rose only by an average of 1.6 per cent. In countries such as Indonesia and the Philippines, where irrigation was available and conditions favoured the new HYV, yields rose by over 3 per cent a year. By 1980 some 27 per cent of all seed used in the developing world was derived from HYV, although there was a massive geographic imbalance. Only 9 per cent of seeds in Africa were from improved varieties, compared to 44 per cent in Latin America.

Despite its success in raising yields, a variety of negative aspects have been associated with the Green Revolution. It is claimed that only cereals which could be stored easily and were traded on the world market were the focus of research. The food quality of hybrids is lower than that of traditional varieties, and storage of the harvest from HYVS is difficult because their high moisture content makes them much more vulnerable to moulds. HYVS do not perform well on marginal soils. They are also extremely vulnerable to pests and disease, having little or no natural resistance. Worse, they produce high yields only when used with large inputs of expensive agricultural chemicals. During the 1970s, world consumption of fertilizers almost doubled, pesticide use increased fivefold, and irrigated land⋄ rose from around 160 million hectares in 1970 to over 200 million hectares by 1981, mostly as a result of HYV use.

Increased mechanization accompanying the Green Revolution has also had negative environmental effects, mostly as a result of soil compaction in ecologically fragile areas. Almost every step associated with the use of HYV seeds is energy-intensive, and the use of mechanization to replace draught animals has high capital and operational costs, as well as leading to the loss of animal droppings, a valuable low-cost organic fertilizer⋄. Costs of irrigation and chemicals tend to mean that the produce is sent for export to recoup the cash outlay. Many farmers were prompted to forgo the cultivation of indigenous crops to plant the HYV for economic gain, causing a marked increase in prices and changes in diet preferences, tastes switching to crops that do not grow well under prevailing conditions without massive inputs of costly agricultural chemicals. Similarly, while those farmers who could afford the inputs needed to take advantage of HYV prospered, poor farmers were unable to make use of the new varieties and their plight worsened.

GROSS NATIONAL PRODUCT (GNP) The most commonly used measurement of the wealth of a country, usually taken to refer to the total monetary value of final products and services produced by a nation in any given year. The GNP comprises the total value of goods and services produced within a country (the Gross Domestic Product or GDP), together with income received from other countries (notably interest and dividend payments minus similar payments made to other countries). For many

developing countries⋄, interest and dividend payments to foreigners are normally more than similar receipts. In these circumstances, the national product is less than the domestic product. The concepts may be expressed either in gross or in net terms. For the latter, a deduction is made for the capital assets used up in producing the goods and services sold. Net national product is often referred to as national income. For comparative purposes, GNP is frequently expressed as *per capita*: the total GNP divided by the population⋄.

Based on the *per capita* figures for 1988, the four richest nations were Switzerland ($27,500), Japan ($21,020), Norway ($19,990) and the United States ($19,840). The poorest were Mozambique ($100), Ethiopia ($120), Chad ($160) and Tanzania ($160).

GROUNDNUT The seeds of a shrub, *Arachis hypogea*, also commonly known as the peanut. Groundnuts are low-growing plants native to tropical South America which push the developing seed pods underground to avoid desiccation. Consequently, groundnuts grow best on sandy, light soils which can be penetrated easily. Each pod contains up to five seeds or nuts. Now cultivated from the tropics to warm temperate regions, groundnuts require around 400 mm of rain during growth and temperatures of 23°C during their four-to-five-month growing season. They are best grown in rotation with other crops such as cotton⋄ or maize⋄, or intercropped with millet⋄, sorghum or cowpea. Harvesting should be carried out in dry conditions, for if the nuts are wet, aflatoxin⋄-producing fungal moulds may later develop on them, making them dangerous to eat. Average yields of 600 kg per hectare can be doubled with good crop husbandry.

Groundnuts are highly nutritious. They are used in cooking, eaten raw or roasted, canned and made into peanut butter. They also produce an oil which is used in cooking and for making margarine. Two thirds of world production is processed to obtain edible oils and for conversion into a protein-rich animal feed. The world's leading producers are India (7.3 million tonnes), China (5.8 million tonnes) and the United States (1.8 million tonnes); around 22.8 million tonnes are produced globally each year.

GROUNDWATER Water⋄ that has percolated into the soil or rock and become trapped within pores, cracks and fissures. Groundwater is important in the weathering process of rocks through its chemical effects and physical action when it freezes. With depth, the water table is reached; below this level, all the pore spaces are filled. Groundwater is extensively used in many parts of the world. In the United States it supplies more than half the nation's drinking water and 80 per cent of rural domestic and livestock⇨ needs.

Global supplies are being both depleted and polluted at alarming rates. Excessive deforestation⇨ causes increased water runoff, hastening the reduction of underground supplies. As a result, aquifers compact and land⇨ subsides. In coastal areas, over-exploitation of groundwater leads to intrusion of salt water from the sea. Ill-considered applications of massive amounts of agricultural chemicals and un-controlled landfill dumping are contributing to the pollution⇨ of groundwater supplies. Water moves relatively slowly through the ground, so many of the chemicals sprayed or dumped more than twenty years or more ago have yet to filter into underground water supplies. However, in Denmark for example, the concentration of nitrates in groundwater has trebled within the last thirty years owing to the effects of fertilizers⇨ and animal manure. Water becomes unsuitable for very young children when nitrate concentrations reach 45 mg per litre, a level easily attained in shallow groundwaters. Groundwater, once polluted, is notoriously difficult to purify. It contains no cleansing micro-organisms and flows too slowly for dilution or dispersal of contaminants.

GROUP OF 77 Originated as the 'Caucus of 75', a grouping of developing countries⇨ assembling for the first discussions of the United Nations Conference on Trade and Development (UNCTAD)⇨ in 1964. Formed as a conduit to express concerted Third World⇨ views on economic and development affairs, it eventually issued a joint declaration supported by 77 developing countries. It now consists of 129 countries, membership overlapping considerably with the Non-Aligned Movement⇨. The Group, which has developed a number of subgroups, argues its case at UNCTAD meetings and all other UN fora and also attempts to put pressure on the developed world to provide further aid⇨ and monetary assistance to the Third World. Occasionally referred to as G77, the Group of 77 was the forerunner of two similar groupings, G7⇨ (the industrialized countries) and G15⇨ (mainly semi-industrialized nations), all of which meet to

discuss global economic issues and promote the interests of their member states.

GUINEA WORM A parasitic nematode worm, *Dracunculus medinensis*, a serious parasite of humans in Africa, India and the Middle East. The larvae are carried by water fleas (*Cyclops*), which are often present in drinking water⇨. When swallowed by humans, they burrow into body tissues and grow to maturity. White, threadlike females reach a length of 60–120 cm and live in the connective tissue under the skin, most commonly in the feet and lower legs. They cause large blisters or ulcers into which they release larvae. When the limbs are immersed in water the larvae escape and are subsequently eaten by water fleas, in which their development continues.

Dracontiasis (or dracunculiasis) is the name of the tropical disease caused by the guinea worm. Initial symptoms begin a year after infection, resulting from the worm's migration to the skin surface. They include itching, giddiness, vomiting, breathing difficulties, and diarrhoea. The disease is common in India and west Africa but also occurs in Arabia, east Africa and Iran. At least 20 million people are afflicted, mostly in Africa, and 750,000 new cases are reported each year, mainly as a result of new reporting, detection and surveillance techniques. An estimated 60–100 million are at risk. The disease can prove fatal if the worm invades the spinal cord, brain or heart, although this is rare, with only 1,000 deaths reported annually. Dracontiasis is extremely painful and causes significant losses in work days – around a hundred days per case. There is no widely available curative treatment and infected persons do not develop immunity.

Treatment includes killing the adult worm with injections of phenolthiazine but reinfection is common. Disease transmission can be totally interrupted, and prevention of guinea worm infection was declared to be a major objective of the International Drinking Water Supply and Sanitation Decade⇨. Transmission is entirely preventable through the filtering or boiling of drinking water and avoiding contamination, through the provision of a safe, piped water supply. Contaminated bodies of water can successfully be treated with temephos. Through health education⇨ and provision of safe water the disease was eliminated from the southern part of the Soviet Union over sixty years ago. In 1989 the World Health Organization (WHO)⇨ announced a new goal of eliminating guinea worm as a public health problem by the year 2000. The same year a $76 million 'Target 1995: Global Eradication of the Guinea Worm' programme was launched by the Global 2000⌀ Project in conjunction with several United Nations⌀ and us-based international funding organizations.

GYPSUM   A mineral, hydrated calcium sulphate, used to make plaster or as a dressing for cropland. It is commercially mined, and its main industrial use is in the manufacture of plasterboard for the construction industry. Gypsum is also one of the major by-products of the cleaning of flue-gas emissions from coal-fired power stations, and gypsum production will increase dramatically as emission-control programmes come into operation. In Britain, around 3.5 million tonnes of gypsum have been mined each year in the past. At the nation's Drax power station, the largest coal-fired plant in Europe, newly installed emission-control systems are already producing 1 million tonnes of gypsum annually.

H

HABITAT LOSS   Until 1950, extensive and exploitative hunting⌀ by humans was the most important cause of species depletion. The American bison, quagga, passenger pigeon⌀, dodo and great auk were amongst those animals paying the greatest price, and overexploitation is still responsible for the slide towards extinction⌀ of the great whales and fur seals as well as numerous other land-based wildlife species.

The loss of natural habitat has now become the greatest cause of the reduction in numbers of wildlife species. Man-made destruction has been seen on a huge scale. Tropical forests⌀ are the largest storehouse of biodiversity⌀. Tropical closed forests cover 6 per cent of the earth's land⌀ surface but contain at least 40 per cent (estimates range from 40 to 90 per cent) of all the planet's species. Humankind has destroyed at least 28 per cent of the world's 'moist' tropical forests and other habitats. Countless varieties of wildlife have suffered a similar fate. In the United States alone, 98 per cent of tall-grass prairie has gone and, on a global scale, over a third of the planet's swamplands have been drained. Virtually all remaining natural habitats have been degraded from their original state.

There are many different causes of habitat loss. Whole habitats can be replaced by human settlements, harbours or roads. Conversion of land to cropland, grazing pasture or tree plantations also destroys natural ecosystems⌀. Specialized habitats, especially island sites, are also irreversibly altered by mining and quarrying. Habitat destruction as a consequence of water-resource development has occurred in many regions in both the developed and developing worlds. This can take many forms ranging from inundantion to problems caused by the overextraction of water⌀ from river systems or aquifers. Man-made pollution⌀ from a variety of sources has wrought havoc in several ecosystems around the globe, especially in aquatic sites, and the removal of basic materials such as vegetation, sand, or gravel has also quickly led to deterioration of localized habitats.

Virtually all forms of human activity have caused environmental perturbations ranging from overgrazing⌀ of livestock⌀, soil erosion⌀ and siltation arising as a consequence of deforestation⌀, to the detrimental effects of introducing non-endemic wildife species and the catastrophic environmental consequences of modern warfare.

There is no global tally of habitat loss and no authoritative figures exist, although remote sensing⌀ is beginning to make this task distinctly easier. According to a 1987 report by a us environmental group, the Sierra Club, over one-third of the earth's total landmass remains in a wild state, devoid of human settlements, roads, or other signs of permanent presence. But of this total, 42 per cent is in Antarctica⌀ or the Arctic tundra, 20 per cent is mainly desert and the rest is primarily mountain ranges. Tropical forests, which support the greatest diversity of wildlife, make up only a small and rapidly declining percentage of the total. Two regional studies by the World Conservation Union⊙ (IUCN) on species richness in Indomalaysia (South-East Asia) and tropical regions of sub-Saharan Africa during the late 1980s found a loss of original habitat of 68 per cent and 65 per cent respectively.

HAMBURGER CONNECTION Popular name given to describe the clearing of land⊙ in Central and South America to rear beef cattle, the meat from which is exported to satisfy the demands for so-called 'fast food' in the United States. The average American consumes 120 kg of meat annually, with nearly half this amount being purchased in fast food outlets, mainly in the form of hamburgers. In search of cheap meat to satisfy the demand at home, US meat-buying firms looked overseas. As a result, woodlands in Africa and, more significantly, the forests⊙ in South and Central America have suffered.

Over the last thirty years, thousands of square kilometres of tropical forest in Central America have been cleared to make way for cattle ranching. At least 65 per cent of lowland and lower montane rainforest⊙ has been cleared since 1950, at a continuing rate of around 4,000 sq km a year. Over 5 million trees have already been lost in a well defined process. First, logging companies harvest the valuable hardwoods, such as mahogany, damaging many other trees in the process. Landless peasants, making use of logging roads, then move in to clear areas and plant subsistence crops. The fragile tropical-forest land soon becomes unproductive and the shifting cultivators move on. But large tracts of poor-quality land are bought up or occupied by farmers who can afford the inputs needed to grow grass and rear cattle. Two-thirds of Central America's arable land is believed to be used to produce beef, and beef production during the 1970s more than doubled. Over the same period, beef consumption in the region fell by a third.

In Brazil, successive burning and reseeding of cleared areas with robust grasses not only keeps the encroaching forests at bay, it also allows beef to be raised at an unsustainable density of one animal per hectare. This method of farming is thought to totally destroy the productivity of the soil after fifteen years. In the past, the USA has not officially purchased uncooked meat from Brazil because foot-and-mouth, a disease of cattle which can be dangerous to humans, is endemic there. Consequently, only Brazilian meat routed through unofficial channels ends up in US hamburgers.

In Latin America as a whole, cattle ranching is accounting for the disappearance of 20,000 sq km of forest each year. In Costa Rica, where 75 per cent of the best agricultural land is now devoted to cattle pastures, new pastures are still being created despite the fact that deforestation⊙ is reducing forest cover by 60 sq km annually. If current trends continue, most of Central America's remaining forests – nearly 200,000 sq km – will have been destroyed by the year 2000. In the Amazon, about 100,000 sq km in Brazil, 15,000 sq km in Colombia and 5,000 sq km in Peru have given way to pasture over the past thirty years. Around 85 per cent of the beef raised is exported to North America. It is too lean for tastes there except for use in hamburgers, and the cheap supplies have both lowered the production price of US hamburgers and raised profits considerably.

HAMMARSKJOLD, DAG HJALMAR AGNE CARL (1905–61) Swedish politician, international civil servant. The son of a Swedish Prime Minister, he served as deputy Foreign Minister from 1951 to 1953 and headed the Swedish delegation to the United Nations⊙. In 1953 he succeeded Trygve Lie⊙ as UN Secretary-General. During his term of office he dealt with the Suez Crisis (1956) and the civil war arising from the granting of independence to the Congo (1960). He was killed in a plane crash while trying to solve the crisis in the Congo and was posthumously awarded the Nobel Peace Prize in 1961.

HARD LOAN Loans made by the Multilateral Development Banks (MDB) to developing countries⊙ at commercial rates of interest are called 'hard' loans. Every MDB has a lending programme that makes loans at rates fixed to the bank's costs of borrowing; this is known as the 'hard loan window'. Similarly, every MDB has a lending programme that offers funds to developing countries at minimal rates of interest, or interest-free, known as the 'soft loan window'. The bulk of MDB lending is of the hard loan variety.

HARRIMAN, WILLIAM AVERELL (1891–) United States diplomat. During the Second World War he was administrator of the Lease-Lend scheme and served as ambassador to the Soviet Union (1943–6), and later to the United Kingdom in 1946. He was elected Governor of New York, serving in the post from 1955 to 1958. He was instrumental in negiotiating the Nuclear Test Ban

Treaty⋄ (1963) and was US representative at the Paris Peace Talks on Vietnam (1968–9).

HAZARDOUS WASTE

HAZARDOUS WASTE Worldwide, well over 400 million tonnes of hazardous waste are generated each year, mostly by industrialized countries. Hazardous waste is any waste that possesses chemical, physical or biological characteristics that have the potential to inflict damage on either human health or the natural environment. This includes toxic waste, waste containing substances that have already proved hazardous by poisoning humans or wildlife. The bulk of hazardous wastes is composed of acidic resins, arsenic residues, compounds of lead⋄ and mercury⋄, organic solvents, pesticides⋄, and radioactive materials. American industry generates around 250 million tonnes of toxic waste each year and the European Community⋄ produces about 150 million tonnes, 30 million tonnes of which are considered to be toxic.

Several ways of disposing of hazardous waste, each with varying degrees of safety and expense, have been developed by the industrialized countries. Landfill, heat treatment, incineration⋄, underground injection, solidification and the detoxification of wastes by bioengineered organisms are the most common. Newly developed plasma arc furnaces can destroy toxic wastes, including polychlorinated biphenyls (PCBS)⋄ and dioxins⋄, almost instantly by producing temperatures four times as hot as the surface of the sun.

Traditionally, the preferred method of disposal has been to dump wastes where it is cheapest, either at sea or in Third World⋄ nations. In Europe, disposing of hazardous waste can cost up to $500 per tonne; in Africa it can be as cheap as $3 per tonne. Since 1986 approximately 3 million tonnes of toxic waste have been shipped from Western Europe and North America to other countries. Europe officially sends about 120,000 tonnes of toxic waste to the Third World each year.

Many developing countries⋄ have attempted to make capital out of disposing of wastes from industrial countries. Angola, Benin, Congo, Guinea, and Guinea Bissau are among the African nations that have concluded individual deals worth up to $2 billion to accept hazardous wastes from sources in industrialized nations. The governing authorities of the Marshall Islands⋄ have suggested to the United States that atolls in the chain which have already received fallout from US nuclear tests would make excellent dumping grounds for the 15,000 tonnes of spent uranium⋄ fuel currently stored in temporary facilities in the USA, at a cost of a mere $100 million. The suggestion has stimulated an outcry from many islanders and international environment protection groups. Many other developing countries have taken active steps to prevent the dumping – either legal or illegal – of hazardous wastes in their countries. Members of the Organization for African Unity (OAU)⋄ have set up a 'Dumpwatch' body for this specific purpose.

In Europe, Britain remains the major sink for hazardous wastes. In 1990 consignments of highly toxic chemical waste from Italy – some 200 barrels per week – were transported by road. This was the same waste that formed the cargo of the *Zanoobia* 'leper' ship which was denied permission to offload the waste at ports all over the world during 1988. A 1990 study also found that the illicit trade in toxic waste in the UK was three times as great as the legal trade, and that once in the country, less than 25 per cent of the waste was incinerated or disposed of safely. The majority was simply dumped in landfill sites.

Unless disposed of carefully, hazardous wastes pose a significant and long-term danger. Landfill sites are a particular problem as persistent chemicals can cause pollution⋄ of surface and groundwaters⋄, contamination of land⋄, and mass exposure of whole communities to the dangerous effects of highly toxic chemicals⋄. The Love Canal⋄ site in the USA is a good example, although there are now numerous similar sites in Europe and North America where clean-up⋄ bills are calculated to be in the order of billions of dollars, the Federal Republic of Germany alone estimating landfill clean-up costs to be in the order of $10 billion.

If properly treated, most wastes can be rendered harmless. High-temperature incineration is the most successful method but it has drawbacks, especially when proper operational temperatures are not maintained. The long-term solution lies in reducing amounts of waste rather than safe disposal.

HEALTH FOR ALL

HEALTH FOR ALL Health For All by the year 2000 is a target set by the World Health Organization (WHO)⋄ to try and raise the overall standard of health for the world's population⋄. During the last forty years, since the inception of the WHO, world health has indeed shown a marked improvement. Even in the developing countries⋄ life-expectancy⋄ has risen from 41.0 to 59.7 years because of health improvements and medical advances.

Irrespective of the progress achieved on all fronts, the task remains far from complete. At present, 50 million people die annually from all causes, including diseases. If current trends are maintained, 200 million people will die prematurely during the 1990s from preventable causes, mostly in the developing world.

In 1990, at least 20 per cent of the world's population – around 1 billion people – were diseased, in poor health or malnourished, according to the WHO. Of this total by far the greatest proportion are found in developing regions of the world. In sub-Saharan Africa 160 million people were infected with a variety of diseases, in South and East Asia another 500 million were suffering from disease and malnutrition◇. In Latin America a further 90 million people were ill. In developing countries, malaria◇ claims the lives of 1 million people, diarrhoea kills up to 4 million, tuberculosis◇ 3 million, measles◇ 2 million and acute respiratory infections 5 million. Throughout the Third World◇ there are 200 million cases of schistosomiasis◇, 90 million cases of filariasis◇, 17 million victims of both onchocerciasis◇ and sleeping sickness, 20 million cases of chagas disease◇, besides 12 million cases of leprosy◇ and 20 million people infected with leishmaniasis◇.

In the industrialized countries of the West, better housing, food and education◇, together with safe water◇, had minimized the incidence of most childhood killer diseases by the 1930s. Infant mortality had dropped to 4 per cent and life expectancy risen to over sixty years, all over a thirty-year period. Yet health services are nowhere near fully available. By the end of the 1980s, around 11 million children in the United States had no direct access to a general practitioner. Furthermore, since the 1940s, new problems associated with 'modern society' have arisen. The incidence of heart disease brought on by unhealthy diets and smoking has continued to increase markedly – from 8 per cent of all deaths in 1900 to 25 per cent by the 1950s – and the figure is still rising.

The global community already has the ability and the drugs needed to prevent this level of ill health, but the rich nations are not transferring the technology, health know-how, materials, manpower and money to the Third World to let the poorer nations help themselves, Expenditure on health care projects in the world's poorest countries averages less than $5 *per capita*, whereas in wealthy nations such as Canada, Japan and Western Europe the figure is nearer $400 per person, peaking at $1,900 for each US citizen. The WHO estimates that if health spending in developing countries could be increased by as little as $2 per head, all children could be immunized against the major childhood killer diseases, polio◇ would be eradicated and drugs could be provided to cure all diarrhoeal diseases, acute respiratory infections, tuberculosis, malaria, schistosomiasis and sexually transmitted diseases.

HEATH, EDWARD RICHARD GEORGE (1916–)   British politician and statesman. He entered Parliament in 1950, eventually becoming the first elected leader of the Conservative Party in 1965. He assumed the post of Prime Minister following the general election in 1970. His Industrial Relations Act (1971) caused unrest amongst the nation's miners and his attempts to control inflation ultimately led to an electoral defeat for the Conservatives in 1974. As a result of this he was replaced as leader of the Party by Margaret Thatcher. During his premiership he was a staunch believer in European togetherness and took the United Kingdom into the European Economic Community (EEC) in 1973. He was an influential member of the Brandt Commission◇.

HELSINKI CONFERENCE   The eighty nations attending the 1989 Helsinki Conference agreed to phase out chlorofluorocarbons (CFCS)◇ as soon as possible, and no later than the year 2000. They also promised to phase out the use of other chemicals that damage the ozone◇ layer, such as halons, methyl chloroform and carbon tetrachloride. This decision, taken in view of the growing evidence of damage to the world's stratospheric ozone layer, tightened up the terms of the 1987 Montreal Protocol◇ for protection of the ozone layer which focused on the removal of five major CFCS. The agreement was only a 'declaration of intent', later to be formally recognized at the London Conference in 1990. The undertaking in Helsinki was supported by both China and India, the two major nations which had not ratified the Montreal treaty. The countries attending the Helsinki gathering also agreed to establish a global fund◇ to help Third World◇ nations develop the technology necessary to produce alternatives to CFC. The proposed multi-million dollar fund would be extended to cover research and aid◇ to deal with the greenhouse effect◇.

At a follow-up conference in London in 1990, delegates agreed a more specific timetable for phasing out CFCs by the year 2000. The donor nations also agreed to provide technology and funding to help the non-industrialized countries to develop or produce safe alternative chemicals and laid down specific guidelines for removal of other gases harmful to the ozone layer, such as methyl chloroform, a metal-cleaning agent which will be banned by 2005.

HELSINKI DECLARATION   At the Conference on Security and Co-operation in Europe (CSCE)⌂, an international conference held in 1975, agreement was supposedly reached on co-operation over security, economics, science, technology, environment and human rights⌂. The Helsinki Declaration was signed by all thirty-five European nations which attended the conference, including the Soviet bloc but excluding Albania. The Soviet bloc were later held not to have adhered to the promises in the agreement and ignored appeals made to them at regular CSCE review conferences in Belgrade (1977), Madrid (1980) and Vienna (1986). Several groups were formed in Communist countries, known as 'Helsinki Groups'. These tried to influence their rulers to comply with the agreement, and members were frequently persecuted for their efforts. Following the increasing freedoms seen in Eastern Europe during 1989, a second CSCE Summit was held in Paris in 1990. The nations of Eastern and Western Europe, together with Canada, the United States and Albania (with observer status only), attempted not only to revive the Helsinki Declaration but also to broaden it into a new charter to cover human rights, freedoms and security throughout Europe.

HEPATITIS   Hepatitis is inflammation of the liver and a major cause of human ill health and death. It can be caused by chemical damage or by disease-bearing organisms. There are several forms of the disease and a variety of causal agents.

Hepatitis A is an acute infection caused by a virus usually contracted through eating contaminated food.

Hepatitis B is a chronic viral infection transmitted in blood and other body fluids, and is linked with a major cancer⌂ of the liver. Both are common in the Third World⌂ but can be easily identified using simple screening tests and treated successfully. Vaccines have also been developed which stimulate immunity.

Of all the world's infectious and parasitic diseases, hepatitis B is the most prevalent. It affects at least 2 billion people, and causes 1–2 million deaths worldwide each year.

In 1989 scientists discovered two more viruses, Hepatitis C and D, both of which are bloodborne. In 1990 researchers isolated a fifth virus, Hepatitis E, which is believed to cause one of the most lethal forms of the disease, particularly in the developing countries⌂. It is spread through water⌂ contaminated with faeces from infected people and is estimated to affect up to a million people each year. It is especially prevalent in Asia, with pregnant women most at risk. Around 20 per cent of pregnant women who contract the virus die. Hepatis E is regarded as being only 50–75 years old, having spread from its origins in Burma, India and Nepal throughout much of Asia and into Africa. It primarily affects young adults; this may indicate that an attack does not ensure effective immunity,

HERBICIDES   Weeds are responsible for a 12 per cent reduction in the world's harvests; herbicides are any chemicals used to kill unwanted plants or weeds. There are over 180 basic types, classified according to their mode of action, the plants they are used against, and their persistence in soil. Modern agriculture⌂ depends on their extensive use, but despite the wide range available, two, atrazine and alachlor, constitute about half of all herbicides used. Large quantities of 2,4-D and 2,4,5-T are also used; virtually 90 per cent of all herbicides are used in the industrialized countries.

Contact herbicides destroy only the plant tissue they come into contact with. Translocated herbicides are absorbed by the plant and distributed around internally, killing the whole weed. Many act by interfering with the growth of the weed and are based on plant hormones. An example is 2,4-D, a synthetic auxin which is used to control broad-leaved weeds in cereal crops. Selective weedkillers kill the target plant without harming other vegetation. Other broad-spectrum herbicides, such as paraquat, used to clear wasteland and destroy several species of weeds, are non-selective and kill every plant they come into contact with. These herbicides can be applied for a variety of purposes, including use as defoliants in warfare or to avoid tilling of soil. Some can be extremely harmful, notably Agent Orange (a mixture of 2,4-D and 2,4,5-T) and paraquat, for which there is no cure if poisoning occurs in humans.

Although most herbicides are not persistent, some remain active for up to two years. The majority are not toxic to humans and mammals, although some, such as paraquat and dinoseb, are markedly so. 2,4,5-T contains dioxin⌀ and is thought to pose a major threat to human health. Its use has been severely restricted, particularly in the United States. Fish⌀ are particularly sensitive to herbicide poisoning and insect numbers may also be reduced through use of toxic herbicides. Many are volatile chemicals and must be applied carefully. Spraying must be controlled to avoid aerial drift, and use in hot weather should be avoided to prevent evaporation and shift to non-target crops or fields. Weeds in many areas are developing resistance, as the chemicals tend to be overused.

HIGH YIELD VARIETIES (HYVs)   Modern hybrid strains of crops, mainly cereals, developed through plant-breeding techniques, that produce higher-than-normal yields when grown under specific conditions. By the year 2000, the greatest contribution to increased crop output – no less than 60 per cent – is forecast to come from increases in yields. One of the first widely successful HYVs was IR8, a strain of rice⌀ developed in the early 1960s. It gave high yields but required a vast amount of fertilizer⌀, a common feature of all HYVs. HYVs also tend to have high water⌀ requirements and low resistance to pests, necessitating irrigation⌀ and repeated applications of pesticides⌀ if elevated yields are to be realized. IR8 proved to be extremely susceptible to bacterial blight and rice tungro virus, leading to entire crop losses in several years in some areas. New varieties resistant to these diseases were developed, but proved to be vulnerable to insect attack. Plant-breeders are turning to wild stock to try and breed in resistance. Their task is difficult because varieties generally display resistance to only a few pests and the loss of wild stock is exacerbating the problem.

Use of HYVs in the Green Revolution⌀ led to the proliferation of irrigation schemes in Third World⌀, many of which have been mismanaged. Approximately 55 million hectares, representing almost one-third of cultivated land⌀ in the developing countries⌀, are now planted with HYV of wheat⌀ or rice. Improved wheats were largely responsible for increasing India's wheat production from 11.4 million tonnes in 1964 to 34 million tonnes by 1980. Similarly, in Indonesia, new HYV accounted for rice yields rising from 12.2 million tonnes in 1970 to over 22 million in 1981. In general, cereal yields in developing countries rose by 2 per cent a year between 1961 and 1980 through the use of HYV.

Extensive use of HYV leads to loss of genetic diversity, and they often fail to outperform traditional varieties where inputs of water, fertilizer and pesticides are not available. The shift from traditional indigenous crops to cultivating HYVs for economic gain has also led to socioeconomic problems in several regions of the developing world, including the rise in price of the neglected indigenous food crops and dietary changes.

HYVs are not a panacea for the world's food supply problems. Even using HYVs, only an estimated 25–40 per cent of the potential yield in temperate zones is being realized, whereas the position in the humid tropics is even worse, with only 10–20 per cent of the potential yield actually being harvested.

HOMELANDS   Under the Land Act passed by the South African government in 1913, the right of native black Africans to own land⌀ in the country was removed and 87 per cent of the land was apportioned to whites. The rest, mostly composed of infertile land with virtually no minerals⌀ and the minimum of natural resources⌀, was defined as 'homelands' or 'bantustans'. Several of these homelands are composed of fragmented strips of land rather than a unified area. Every black South African was allocated to one of these homelands and became a citizen, effectively becoming a foreigner in the other parts of the country, although over 11 million black Africans live permanently in white-designated areas of South Africa. The South African government granted these homelands independence, but no other country in the world recognizes them as independent states and they are totally dependent on South Africa for their economic well-being and survival. There are ten such homelands: Bophuthatswana, Ciskei, Gazankulu, Kangwane, KwaNdebele, Kwazulu, Lebowa, Qwaqwa, Transkei, and Venda. The leaders of each, initially selected because of their support for the South African government's plans and bolstered with finance from the Republic, strive to maintain the mantle of 'independence' for their homeland. Some Africans have benefited from the system, not least those who have taken advantage of the fact that thousands of white South Africans flock to the homelands to indulge in the gambling and prostitution which are available there but strictly prohibited on the other side of the virtually non-existent borders.

HUMAN DEVELOPMENT REPORT   An annual series of analyses charting the progress of improvements in living standards seen amongst the world's nations. The reports, produced by the United Nations Development Programme (UNDP)✧, follow the assumption that economic growth does not necessarily equate to human development and improved well-being. First produced in 1990, the reports assess the actual impact of development on human populations, rather than being simple calculations of Gross National Product (GNP)✧ based on purely financial criteria such as income and savings, commodity production and the accumulation of capital. Other factors, such as changes in life-expectancy✧ and access to education✧, are taken into account, in the belief that human development requires sustained growth accompanied by an equitable distribution of income and resources. The initial report discovered that some countries, including Costa Rica, Jamaica and Sri Lanka, achieved a high level of human development with modest levels of *per capita* income. Others, including Brazil, Oman and Saudi Arabia, failed to translate their comparatively high income and rapid economic growth into commensurate levels of human development. Many oil-producing nations, particularly those in the Middle East and North Africa, had failed to translate their wealth into improved living standards for the populace. Despite the use of new criteria, in a league table based on the Index of Human Development the countries of Western Europe, North America and Japan filled the top positions, while sub-Saharan African nations occupied nine of the bottom ten.

HUMAN FRONTIER SCIENCE PROGRAMME (HFSP)

The HFSP is a multi-million-dollar international co-operative scientific research programme. Proposed by the Japanese government in 1987, it follows the lines of the United States's Strategic Defense Initiative (SDI)✧ and Europe's Framework programmes. The HFSP aims to promote, through international collaboration, basic research focused on the elucidation of the sophisticated and complex mechanisms of living organisms. To the fullest possible extent the results of the research will be used for the benefit of all humankind.

The original plan was for Japan, North America and Western Europe to sponsor studies in areas of the life sciences, although Western governments initially suspected the scientific credibility of the programme, viewing it as merely a political gesture. The G7✧ governments which attend the economic Summit of industrialized nations, together with some other members of the European Community✧, finally agreed to participate in a three-year programme, with the bulk of the funding being provided by Japan. The first HFSP project was announced in 1989: a research effort to gain understanding of the human brain and the molecular roots of biology. The Japanese government agreed to finance its first three years at a cost of around $66 million. A headquarters, and the Secretariat to administer grants to research groups around the world and to handle communication between them, was opened in Strasbourg in 1989. In its first year of existence, over 750 applications for funding were received from 22 countries, the USA producing the largest number of applications.

HUMAN RIGHTS   Privileges claimed or enjoyed by every human being, simply by virtue of being human. The concept developed from Roman ideas of 'natural laws'. Since the Second World War, moves have been made to ensure the enforcement of human-rights agreements on a global or international scale, as embodied in the United Nations✧ Charter. Under the UN Universal Declaration of Human Rights, adopted in 1948, people were deemed to have the right to life, liberty, education✧ and equality before the law; to freedom of movement, religion, association, information and to a nationality. Although not a legally binding code, the UN Declaration has spawned various subsequent agreements, such as the Covenants on Civil and Political Rights and on Economic, Social, Civil and Political Rights (1966), which have been accepted as binding by thirty-five countries. Under the European Convention of Human Rights (1950) the Council of Europe founded the European Commission of Human Rights, which investigates grievances and complaints by states or individuals. The findings of the Commission are examined by the European Court of Human Rights, which was founded in 1959 and whose jurisdiction has been recognized by a number of European countries.

HUNGER   Although the total calorie and protein content of food production is more than twice the minimum requirement of the world's population✧, famine and malnutrition✧ remain widespread. During the first half of the 1980s the number of 'hungry' people in the world was reported to have reached 512 million. Every year another 40 million people were being added to the total. This compared to an increase of 15 million between 1970 and 1980. Some 500 million men, women and children were said to be chronically undernourished.

Hunger and starvation are difficult things to quantify. The number of calories needed daily, for example, varies with age, gender and lifestyle – from the 800 needed per day by infants under the age of one to the 3,600 required to sustain an adult male leading an active life. In general, the average person leading a normal life requires around 2,500 calories per day. The Food and Agriculture Organization (FAO) defines undernourishment as a daily caloric intake less than 1.4 times the Basic Metabolic Rate (the energy requirement in a state of fasting and at complete rest), which is around 1,650 calories but varies according to human characteristics, age and environmental conditions.

In the 1960s, half the global population had a daily intake of fewer than 1,900 calories. Despite significant improvements, by 1990 one in ten people still had a daily consumption of 1,900 calories or fewer. Estimates suggest that 950 million people, about 20 per cent of the world's population, still consume too few calories each day to support an active working life. By the year 2000, FAO forecast, food sufficient to provide at least 2,500 calories per day for everybody in the world should be available, but by then there will be 1 billion more mouths to feed than there were at the beginning of 1990. If the forecast is to be fulfilled, food production – or, more correctly, food availability – will probably have to increase by approximately 40 per cent, and a much larger increase will be needed if hunger is to be eliminated.

If food was evenly distributed around the world, everyone would have enough to satisfy their needs. Unequal distribution means that hunger and malnutrition are commonplace. It is found predominantly among those people who lack the cash income to buy – or the land⌂ and inputs to grow – the required amounts of food for them and their families. This is true even in the richest of countries. Within the family, hunger and malnutrition are also found disproportionately, women and children suffering the most; in the United States, 500,000 children are affected by malnutrition, with one child in five living below the poverty line.

The majority of the world's hungry and undernourished people, however, live in Asia, and about 85 per cent of the total live in rural areas. A 1987 United Nations report ranked the hunger and malnutrition problem in Asia as 'one of the world's most serious issues of human welfare'.

International efforts to guard against hunger have been successful to some extent, but are under threat. Rapid depletion of global cereal stocks saw the total fall to well under 300 million tonnnes by 1990, the lowest level since 1981 and substantially below the levels suggested by the FAO as necessary for safeguarding world food⌂ security.

HUNTING   In many developing countries⌂ local people still hunt for survival. Wild animals, mainly mammals, contribute as much as 80 per cent of the diet in Ghana, 70 per cent in Zaïre and 60 per cent in Liberia.

Man began to hunt long before he took up agriculture⌂, but in the industrialized countries hunting has evolved into a pleasure or sporting activity, mostly followed by well-off people. A combination of hunting for economic value of the prey and as a form of recreation is now quite widespread – often the practice of rich foreigners in developing countries, but mostly in developed countries. The better-off have always regarded hunting as a privilege of social status and wealth. Locals have always hunted – mostly for subsistence, not for recreation – but now their actions are often labelled as illegal.

Recreational hunting can range from organized safaris and individual fishing⌂ to a local hunter shooting a potential predator or danger to crops, and it is difficult to quantify accurately. In many developed parts of the world where hunting has been regulated and institutionalized, fees now have to be paid for licences, leases, travel or equipment. In the United States during 1980, 17.4 million people took part in some form of official hunting and a further 42.1 million participated in some form of fishing. These hunters spent $16 billion, almost three times what was spent in 1960. In Western Europe, 2 per cent of the population hunt, and during the early 1980s European Community⌂ hunters spent $3.3 billion hunting 72,000 tonnes of wild game. In the Soviet Union, 3 million people hunt and fish. In Latin America only a very few hunt for pleasure or sport, and they are mostly wealthy landowners. In Asia hunting is banned in Buddhist countries for religious reasons. In the Middle East, recreational hunting is commonplace to the extent that hunters have exhausted their own lands. Maltese hunters are now killing protected birds in Egypt. Rich Saudi Arabian princes now hunt in wildlife reserves in Niger (where hunting was banned in 1964), the Sudan, and other African nations, as well as in Pakistan, where hunting was also severely restricted years ago.

Many African countries now rely on recreational hunting as a valuable source of foreign exchange. Kenya, the Sudan, Tanzania, Uganda, Zimbabwe and others have all closed or restricted hunting due to combinations of political unrest, rampant poaching, and pressure from international conservationists. Yet they all allow specialized hunting for wealthy foreigners. In 1983, Zimbabwe reported earnings of $6.2 million from sport hunting, charging overseas hunters $767 per buffalo, $2,987 per elephant○ and $1,125 per lion.

Hunters were in fact the first to promote the idea of species conservation○, and the game reserve established to satisfy the needs of hunters long preceded any wildlife sanctuary. Today, much of the wildlife conservation activities in the US and elsewhere is funded by the income from hunting licences. In Botswana, the government helps finance the nation's conservation measures from hunting revenues, utilizing an average annual income of around $450,000.

HUXLEY, JULIAN (1887–1975) British scientist and conservationist. Son of a famous British biologist, he was himself a zoologist and scientific administrator who made valuable contributions to the philosophy of science. He was Director of London Zoo and one of the original founders of the World Wildlife Fund (WWF)○, and was appointed first Director of the United Nations Educational, Scientific and Cultural Organization (UNESCO)○, serving in the post from 1946 to 1948.

HYDROELECTRIC POWER Electricity generation using the power of falling water○, often referred to as hydropower. The falling water turns a turbine: electricity is generated with an efficiency of 90 per cent at full load. The higher the reservoir, the less water is needed for the same output. It is a relatively cheap source of power, particularly in mountainous areas where rainfall is high, although these tend to be far removed from the industrial communities where the need for power is greatest. It is also non-polluting and a single 10-kilowatt hydropower plant is estimated to save 21 tonnes of fuel oil○ per year, as well as reducing the release of carbon dioxide○ by 70 tonnes.

Hydropower supplies around 20 per cent of the global electricity supply (or 7 per cent of global energy○), with North America producing more hydroelectricity than any other region. Worldwide, by 1990, a total capacity of 550 gigawatts of hydropower had been installed. Much of the world's potential hydropower remains untapped, with the developing world having by far the largest share of untapped resources. By 1980 Europe and North America had developed an average 47 per cent of their hydropower potential, whereas only a meagre 7 per cent of that in Africa, Asia and Latin America had been exploited. In the unlikely event that all the world's rivers were dammed, an estimated 73,000 terawatt/hours (t/h) of electricity – equal to the output of 12,000 nuclear reactors – could be produced each year (1 terawatt/hour = 10^{12} watt/hours). Experts attending the World Energy Conference concluded that it was feasible to tap 19,000 t/h, compared to the 1,300 t/h produced today.

The large dams○ needed to produce effective hydropower plants carry heavy environmental and social costs, especially in the tropics, but the World Bank○ projects that 223.5 gigawatts of large hydropower capacity will be installed in developing countries○ between 1991 and 1995, with more than half the total in Brazil, China and India. Rivers in the Amazon Basin in Brazil have an energy potential of 100 gigawatts, which would go a long way to reducing the nation's fuel oil import bill.

HYDROPONICS The cultivation of plants without soil using a liquid nutrient solution. An aqueous solution of all the nutrients required for healthy growth is flowed over crop roots. The plants are anchored in an inert medium such as gravel or sand which provides support for the root system. On a commercial scale, the solution is pumped through the system and the nutrient levels are monitored and maintained. Hydroponics allows crops to be grown in arid regions of the world where soil is infertile or toxic, but large-scale hydroponic cultivation is an extremely expensive operation, and this restricts its use. The soil-less cultivation techniques are used commercially only in extremely arid areas, such as parts of California and Israel. However, the system can also be used on a small scale, at low cost, in parts of the world, including cities, where water○ shortages are the norm. In Colombia, in locations where poor women are allowed only twenty minutes every other day to collect water from a communal pump, low-cost hydroponics systems run by women, some operating on rooftops, have been established. During the early days, the rooftop gardens were producing up to 3 tonnes of vegetables per month.

HYDROPOWER See Hydroelectric Power

I

ILLITERACY The exact definition of illiteracy is difficult to establish, but basically it is the inability to read or write. The criterion sometimes employed as a measure is the inability to read or write one's own name. According to the United Nations Educational, Scientific and Cultural Organization (UNESCO)✧, over 25 per cent of the world's adult population✧ is illiterate, and 100 million children have no school in which to learn to read and write. In total, there are well over 1 billion illiterate people in the world. Some 900 million men and women are illiterate, nearly 98 per cent of them living in developing countries✧. Fourteen countries have illiteracy rates higher than 70 per cent, despite over two decades of co-ordinated international action and projects aimed at improving the situation. More than half of today's illiterate people live in India (623 million) and China (229 million). The situation in India, where 57 per cent of adult males are literate compared with only 29 per cent of women, reflects the global position, and UNESCO has adopted the goal of removing the disparity between male and female literacy by the year 2000.

In the past, many mass literacy programmes were introduced into countries without regard to the language. needs or interests of the people concerned. The United Nations World Conference on Education For All held in Bangkok during 1990 – which was also declared International Literacy Year by the UN – saw the launch of a new campaign to improve global literacy. As well as UNESCO, which will be leading the fight against illiteracy, other agencies, including the United Nations Children's Fund (UNICEF)✧, the World Bank✧, and the United Nations Development Programme (UNDP)✧ are also playing a major role in the programme. It is planned to cut illiteracy to 15 per cent of 1990 levels by the year 2000, a task which will require expenditure of $40–50 million during the decade.

INCINERATION High-temperature incineration is potentially the safest means of disposing of most toxic wastes, particularly organic solvents, chlorinated hydrocarbons✧ and oil-based compounds.

Industrial high-temperature burners are used to destroy most toxic wastes, whereas low-temperature incinerators are used to burn municipal waste. Newly developed plasma arc furnaces can destroy toxic wastes – including dioxins✧ and polychlorinated biphenyls (PCBS)✧ – instantly, as they create temperatures four times as hot as the sun's surface. Effective, environmentally sound incineration requires either extremely high temperatures or, where lower temperatures are used, sophisticated equipment such as scrubbers to remove unwanted and dangerous pollutants. Municipal incinerators typically operate at temperatures of around 500°C, too low to destroy plastic materials. Inefficient incinerators are counterproductive as not all the original wastes may be destroyed and secondary, hazardous compounds may be formed if incineration is incomplete. Partially incinerated plastics and PCBS are a source of dioxins, one of the most toxic group of chemicals known. Air borne dioxins are released from municipal incinerators and crematoria. Some of the organic chemicals collected from the emissions of badly operated low- and high-temperature incinerators have yet to be identified – so their impact on human health and the environment is totally unknown.

Properly managed, incinerators can destroy 99.99 per cent of organic wastes, but operational costs are high compared with other waste-disposal options. As a result, incineration is not widely practised; in the United States it accounts for, at most, 2 per cent of the disposal of hazardous waste✧.

In Europe incineration is more extensively used. Rotary kiln incinerators operating at 1,200°C detoxify solvents, oily sludge and organic chemicals. Reclaimed waste oils help to fuel the burners and steam produced can be used for district-heating systems. In Scandinavia, cement and lime kilns are used to burn hazardous wastes. No major dangerous components have so far been detected in exhaust gases and the hazardous wastes used cut the need for fuel for the kilns by up to 30 per cent. Elsewhere, many Western European countries, notably those bordering the North Sea, traditionally carried out the incineration of hazardous chemicals at sea. In 1984, some 100,000 tonnes of highly toxic industrial wastes – mostly organochlorines – were incinerated in the North Sea. However, parties to the North Sea Conference agreed to stop the incineration of toxic waste in the North Sea at the end of 1990. At the same meeting agreement was reached to destroy all of Europe's 250,000 tonnes of highly toxic PCBS by 1999 through high temperature incineration within their country of origin.

INDEPENDENT COMMISSION ON INTERNATIONAL DEVELOPMENT ISSUES (ICIDI)

The task of the ICIDI, better known as the Brandt Commission⌀, was to study the grave issues arising from the economic and social disparities within the global community, and to suggest ways of promoting adequate solutions to the problems involved in the development process and in combating absolute poverty. The publication of its two reports – *North–South: A Programme for Survival* in 1980 and *Common Crisis* in 1983 – attracted widespread public interest. Both documents were important and valuable contributions to international discussion of the problems facing developing countries⌀, and helped to give renewed impetus to consideration of the issues involved amongst the international community. The Commission's membership was made up of statespersons of diverse experience and background, all with acknowledged standing in international economic matters.

INDEPENDENT COMMISSION ON INTERNATIONAL HUMANITARIAN ISSUES (ICIHI)

A think-tank composed of thirty leading personalities from the international community concerned with global development and the welfare of the world's peoples. Created in 1983 and based in Geneva, the ICIHI was established in response to increasing international concern. It addresses itself to questions which are somewhat neglected in the usual political, economic and social debates and fora. Studies have been published on refugees⌀, street children, war⌀, famine, statelessness, disappeared persons, desertification⌀ and deforestation⌀. The ICIHI attempts to sensitize public opinion and that of decision-makers, to the suffering and deprivations of those without the power or opportunity to voice their protest at misconceived development strategies and projects.

INDONESIA/TRANSMIGRATION

The transmigration programme in Indonesia is a government-led initiative to encourage Indonesians from the overpopulated, fertile islands of Java and Bali to resettle on the outlying, underpopulated, tropical-forested islands such as Kalimantan and Irian Jaya. Between 1984 and 1990 the Indonesian government planned to move 750,000 families, representing some 3 million people in all. The programme has come under constant criticism for the devastation that it is causing to tropical forests⌀ on the outlying islands and the social disruption the indigenous people are having to undergo.

Over 10 per cent of the world's tropical forests are in Indonesia, but they are rapidly being eroded by commercial logging and the resettlement scheme and at least 3.3 million hectares of tropical forest have been razed. The Food and Agriculture Organization (FAO)⌀ estimates that around 1 million hectares of Indonesia's forests are now disappearing each year. It costs about $9,000 to move each family to the outer islands, and the bulk of the financing for the project is being provided by the World Bank⌀. At least $650 million has been contributed so far, with the possibility of the investment reaching over $1 billion by the early 1990s. The transmigration programme has become a major cause of conflict in Irian Jaya. Indigenous groups have been fighting for independence since Indonesia annexed the island in 1962. They see the resettlement programme as a threat to their livelihood and culture, while the government view their stance as an expression of opposition to authority.

INDUSTRIAL BIOTEST LABORATORIES

A private laboratory in the United States which was used to carry out safety tests and evaluations on the toxicity of a wide range of products, including pesticides⌀ and irradiated foods. The results from the laboratory were used to determine whether or not these products were safe for use or human consumption. Investigations by the United States government found that the results of hundreds of safety tests had been falsified. Tests on at least 200 pesticides had been misread or faked. As a result, it was concluded that 15 per cent of all pesticides on the global market have been approved for use without proper or reliable testing of their safety for the environment or for human health. Some of the pesticides are in worldwide use, but have still to undergo proper safety checks. A list of the suspect chemicals was produced by the US government; this resulted in some nations, such as Canada and

Sweden, suspending the use of several of those listed, while others such as Britain, took no action.

INFANT MORTALITY RATE (IMR) The number of deaths of children aged 0–12 months, per 1,000 live births in a given year. This includes neonatal mortality – death within the first four weeks of life – from such causes as asphyxia and injuries sustained during the birth process. Neonatal deaths account for two-thirds of infant mortalities in developed countries, primarily as a result of the lack of medical facilities. The IMR is regarded as a measure of social affluence rather than a reflection of the quality of antenatal and obstetric care, and is perceived as a general indicator of overall health status and socioeconomic development.

Despite dramatic improvements over the last twenty years, in the least-developed◇ nations an infant has the risk of dying before reaching its first birthday of about 125 per 1,000. The IMR in Europe represents the lowest regional average, at 13 per 1,000 live births, while Japan, Finland, Iceland and Sweden have the lowest national levels: 6 deaths per 1,000 live births. In the world as a whole, the average IMR in 1990 of 70 per 1,000 live births represented a drop of 31 per cent over the 1970 figure. Infant mortality is often accompanied by a figure for child mortality; this refers to the number of children who die before attaining the age of five. At the end of the 1980s, the global average child mortality rate was 105 per 1,000 live births, down 35 per cent from the 1970 figure.

INFORMAL ECONOMY A term used to describe business enterprises and workers operating outside the officially recognized, mainstream formal sector of a nation's economy. The informal sector is quasi-legal: regulations governing trading licences, minimum wages, property titles and taxes are generally overlooked. Present in most countries and ignored or legislated against by many governments, the informal market is now expanding rapidly throughout the developing world and attracting official recognition and assistance. In seventeen nations in sub-Saharan Africa, some 60 per cent of urban workers are employed in the informal sector, producing 20 per cent of annual output, worth $15 billion. Approximately half of the workforce in Latin America are employed in the informal sector, in what are sometimes referred to as micro-enterprises. The Inter-American Development Bank◇, one of the Multilateral Development Banks (MDB)◇, has established a separate department to co-ordinate its loans and support to these micro-enterprises.

In the past, aid◇ and development programmes have allocated resources purely to the modern, formal sector, usually large-scale industry, often owned by the state, wealthy private individuals or companies working in co-operation with trans-national◇ organizations. The International Labour Office (ILO)◇ reports that the informal sector is much more efficient and cost-effective in promoting development and improving a nation's economic performance. In Latin America every $1,000 lent to the informal sector creates a job, compared with the $12,000 needed in the formal sector. Traditionally, throughout the world, the formal sector has been unable to absorb the rapidly rising number of people seeking work, whereas the informal sector has offered much greater employment opportunities. In India, the informal manufacturing sector has produced twice as many new jobs over the past twenty years as has the formal sector. ILO studies have also found that despite their quasi-legal status, informal enterprises mostly pay taxes in some form, whereas large companies tend to exploit legal loopholes and ultimately pay a smaller proportion of gross income in taxes, as well as enjoying preferential trading arrangements.

INFRASTRUCTURE Goods and services which, while in themselves not normally directly productive, are essential to the functioning of a sound economy. The term encompasses such things as power generation, transport, roads, housing, education◇, health and other social services. The infrastructure in developing nations generally needs either to be installed or improved but loans to develop or improve infrastructure in the Third World◇ often impose a massive financial burden on the recipient country because funds are not given with a view to the long term. At least eighty-five developing nations have received loans from the World Bank◇ for building and developing roads. Most of these roads, particularly those outside urban areas, require a great deal of maintenance – at an estimated cost of $90 billion over the next decade. Road networks, together with the construction of hospitals, schools, and other transport systems, have expanded much faster than the maintenance budgets, and traffic has generally been far heavier than envisaged. When road systems and other newly installed infrastructure is allowed to deteriorate the economic prospects of a nation become seriously impaired, but the costs of maintenance may be beyond reach.

INSECTICIDES Substances used to kill insects, 10,000 species of which are regarded as pests. Originally, strong inorganic poisons, such as arsenic and cyanide compounds, were used; these were not very selective, so they were also toxic to humans and livestock⟷. Natural organic compounds, such as derris and nicotine, were also employed. Synthetic, organic substances, such as DDT⟷, began to be used extensively in 1945. The development of two groups, chlorinated hydrocarbons⟷ and organophosphates, revolutionized the control of insects but caused widespread and long-lasting environmental damage. The full effects of massive applications of chemicals such as DDT, widely used because of their lack of specificity and because they were cheap and easy to apply, are still being realized some forty years later, with traces of DDT appearing in breast-milk in women around the world.

Insecticides are classified according to their mode of action or application. For example, the contact insecticides exert an effect when they come into direct contact with the target insect. Residual pesticides are applied to surfaces that the insect is likely to touch at some stage. Unlike fungicides⟷ and herbicides⟷, which are predominantly used in the developed world, insecticides are used equally as much in developing and industrialized countries. Despite continued and extensive use, however, they have not succeeded in reducing insect damage to agricultural produce, and in the United States alone, insects are estimated to inflict $4 billion worth of damage annually.

Overuse has encouraged insects to develop resistance to many of the synthetic chemicals used to try and kill them. According to Food and Agriculture Organization (FAO)⟷ figures, since the 1940s over 1,600 species of insect have developed resistance to commercially available insecticides, and at least fourteen major insect pests are resistant to all five classes of insecticides. As a consequence – and following the appreciation of the danger of persistent organophosphates – attention switched to insecticides based on natural products, the most successful of which has been pyrethrin extracted from pyrethrum⟷ plants, which is used to formulate a group of insecticides that are non-persistent in the environment as they are degraded naturally.

In nature, plants develop natural insecticides and insects develop resistance in a fluctuating, dynamic equilibrium. The discovery of a variety of natural hormones, pheromones and chemosterilants, all of which could be applied to control insect numbers, was hailed as a breakthrough, as it was considered that insects would be less able to develop resistance. However, resistance to hormone-based insecticides soon appeared and the high toxicity of most chemosterilants restricted their use. Insect growth-regulating compounds and microbial insecticides such as *Bacillus thuringiensis* have not been in wide usage long enough to allow for a complete evaluation, although house flies have already exhibited resistance to *B. thuringiensis*. Insecticide resistance is becoming more and more widespread, and is developing at an accelerating rate. At least 829 species of arthropod pests (insects, mites and cattle ticks) were resistant to insecticides in 1980, compared to only 313 in 1970. During the same period, the number of arthropods resistant to pyrethroid insecticides also rose from 3 to 22.

Problems of resistance and environmental pollution⟷ have led to more widespread use of biological forms of control and to the increasing adoption of Integrated Pest Management (IPM)⟷.

INSTITUTE OF DEVELOPMENT STUDIES (IDS) An independent body established in 1966 at the University of Sussex (UK) as a centre for training and research in the field of development. The IDS aims to promote development, help efforts to improve living standards and alleviate poverty in the Third World⟷, and foster better relations between rich and poor nations. The Institute operates a full programme of publications, research, training and practical assignments dealing with all aspects of development, with an emphasis on those relating to poverty, employment and income distribution anomalies within developing countries⟷ and the inequities within the global economic system. Its training programme includes courses for both overseas and British officials. Senior members of the Institute act as consultants for overseas governments and international agencies. The IDS receives financial support from the British government as well as deriving funds from research grants awarded by public and private sources and through consultancy and course fees.

INSTITUTE FOR EUROPEAN ENVIRONMENTAL POLICY (IEEP) A voluntary international institute, part of the European Cultural Foundation, an independent, not-for-profit centre, based in the Federal Republic of Germany. The IEEP aims to facilitate contact between national parliaments and other institutions connected with environmental policy making in a European context, and to improve overall environmental policy and strategies in Europe. The IEEP undertakes analytical and comparative studies on industrial, agricultural and other topics.

INSULIN A protein hormone, produced in the pancreas that is important for regulating the amount of sugar (glucose) in the blood. Lack of the hormone gives rise to diabetes⌀, a condition which can be treated successfully by insulin injections.

Human insulin, the first genetically engineered hormone, was introduced commercially in 1982 and is now used by over 80 per cent of insulin-dependent diabetics in Europe and the United States. Before its introduction, insulin could be obtained only from the pancreatic glands of cows or pigs. These older and cheaper insulins are still used throughout the Third World⌀. Bovine insulin has three amino acids that are different from those found in human insulin, and some hypersensitive diabetics develop allergic reactions as a result. Porcine insulin is closer chemically to human insulin but still contains one amino acid not found in the human form.

INTEGRATED PEST MANAGEMENT (IPM)
Despite the long-term widespread application of massive quantities of pesticides⌀, 10,000 species of insect are still pests, 1,500 species of nematode damage crop plants, 50,000 species of fungus cause 1,500 diseases, and around 2,000 of the 30,000 species of weed cause large economic losses.

There are four basic methods for controlling agricultural pests: cultivation methods which discourage the build-up of pest populations; control using chemicals; crop breeding and selection of resistant varieties, and nurture of the natural enemies of pests (so-called biological control⌀). The latter includes the encouragement of natural, endemic predators through such measures as the preservation of their natural habitat, and the introduction of either natural or novel predators. Integrated Pest Management is a combination of all these, specially tailored to counter particular pests under locally prevailing conditions. IPM programmes do not attempt to eradicate pests but aim to maintain pest populations below acceptable levels.

IPM incorporates preventive measures such as planting resistant varieties, crop rotations and the destruction of diseased plants to stop contagious diseases from spreading. Early planting can avoid peak pest emergence and the removal of crop residues minimizes damage from the dormant stages of boring insects. Manipulation of water⌀ levels in irrigated areas and intercropping⌀ of insect- and disease-resistant varieties also reduce pest losses. IPM makes full use of improved crop hygiene and protection of natural enemies and does not eschew the use of chemicals. It uses elements of biological control, crop resistance and cultural practices within a comprehensive management plan, seeking to restore ecological equilibrium.

As well as at least 1,600 pesticide-resistant species of insect, 50 species of plant pathogens have developed resistance to chemical fungicides⌀ and 5 species of weed are unaffected by most commonly manufactured herbicides⌀. The control ability of pesticides decreases over time because of the resurgence of pest populations and growing resistance. Using IPM to control agricultural pests overcomes these problems, reducing the cost to human health and the environment in the process. Widespread adoption of IPM techniques could reduce pesticide use by 50–75 per cent worldwide, with no loss in effectiveness.

A form of IPM has been practised in China for centuries: farmers traditionally flood paddy fields early in the season. This kills off the yellow stemborer, a serious rice⌀ pest, without the need for costly and potentially dangerous pesticides. Ducks are also allowed into the rice fields, where they eat planthoppers, leaf-rollers and army worms, all potentially damaging pests. IPM programmes on rice in India cut pesticide use by 40 per cent and improved yields. On Brazilian soya beans IPM programmes caused pesticide use to fall by 85 per cent, on Nicaraguan cotton⌀ the reduction was 35 per cent and on Chinese cotton pesticide use fell by 90 per cent and costs by 84 per cent. In 1986, Indonesia spent $120 million on pesticides. The success of a national IPM programme subsequently introduced led to the Indonesian government banning 57 potentially hazardous pesticides. By applying IPM to control Indonesia's worst rice pest, the brown planthopper, the government is saving $35 million a year. Nicaragua has developed one of the most extensive IPM programmes in the world, mainly focusing on controlling pests of cotton and, in particular, the boll weevil, making extensive use of trap-cropping⌀. Pakistan re-evaluated its subsidies for pesticides and discontinued them in 1980, freeing public funds for research and extension in IPM.

Although successful, IPM depends upon a large amount of information, is extremely site-specific and, to be fully effective, must be implemented in co-ordinated fashion over a large area. It has generally proved economically beneficial, its cost-effectiveness being emphasized by figures published in 1987 showing that farmers in the United States using IPM earned $579 million more in profits than they would have otherwise.

INTEGRATED PROGRAMME FOR COMMODITIES (IPC)

A global trading measure designed to regulate the cost of raw materials (or basic commodities⌀) through the creation and maintenance of international pricing arrangements. The IPC was originally established by the fourth United Nations Conference on Trade and Development (UNCTAD)⌀ in 1976 to determine the scope for new international measures concerning eighteen key commodities of developing countries⌀ and, where appropriate, to establish International Commodity Agreements⌀ to implement the agreed measures. The discussions covered price stabilization – in particular buffer stocking – and 'other measures' such as research and development and market promotion. The major commodity agreements under discussion were those concerning bananas⌀, cocoa⌀, coffee⌀, cotton⌀, vegetable oils, sugar⌀, jute⌀, tea⌀, rubber⌀, oil-seeds, and tropical timber⌀. Two such agreements, those on coffee and cocoa, actually preceded the negotiation of the Programme. A Common Fund⌀ was proposed which would have two accounts, one to finance buffer stocks of these commodities and the other to fund research and development and improve commercial operations. The General Agreement on Tariffs and Trade (GATT)⌀ does not play an active role in the negotiation of these international agreements.

INTER-AMERICAN DEVELOPMENT BANK

A Multilateral Development Bank (MDB)⌀ founded in 1959 to provide assistance to developing countries⌀ in Latin America and the Caribbean by making available low-interest loans. The Bank aims to promote the development of member countries through the provision of both financial assistance and technical co-operation. Membership was originally restricted to the then twenty-one members of the Organization of American States (OAS)⌀, but was extended to Canada in 1972; in 1976 twelve other non-regional members, mostly from Europe, were admitted. The Bank's funds are replenished every four years. In addition to its authorized capital it operates a Fund for Special Operations which provides soft loans⌀. As part of the Bank's Sixth Replenishment (1983–6) a special intermediate financing facility was set up to defray up to 5 per cent per annum interest charges paid by borrowers on loans from the Bank. In 1990, the bank underwent major restructuring. Separate departments were established to expedite activities in six areas considered to be crucial for the region – economic integration, co-financing, export promotion, micro-enterprises, environmental protection, and human resources development.

INTER-AMERICAN INVESTMENT CORPORATION (IIC)

The IIC was established in 1984 as a subsidiary of the Inter-American Development Bank⌀. Its purpose is to promote the region's economic development by encouraging the establishment, expansion and modernization of private enterprise. Priority is given to small- and medium-scale organizations. The IIC also plays a catalytic role in mobilizing additional finance for these purposes from other sources. In most respects it is similar to the International Finance Corporation (IFC)⌀ of the World Bank⌀. The IIC's initial capital stock of $200 million was allocated on the basis of regional countries (55 per cent), the United States (25.5 per cent) and non-regional countries (19.5 per cent). The non-regional members are Austria, France, the Federal Republic of Germany, Israel, Italy, Japan, the Netherlands, Spain and Switzerland.

INTERCROPPING

Intercropping involves the growing of two or more crops in the same field – in sequence, in combination or both. This includes the process of relay cropping, when a second crop is planted between the rows of a first which is nearing maturity. Crop and animal production may also be mixed in even more complex, ecologically sound systems designed to maximize production. This system which has been commonly practised by millions of farmers in the tropics for centuries, may be particularly beneficial to cash- or credit-poor farming families.

Most subsistence farmers prefer to plant at least two crops in the same field, thereby minimizing the risk of total crop failure. In Africa, 98 per cent of the continent's most important legume crop – cowpea – is grown in combination with other crops. In Nigeria, over 80 per cent of the cropland used is given over to polyculture (mixed cropping). A Nigerian farmer may typically grow up to eight crops, including bananas⌀, beans, cassava⌀, melons and yams, in a variety of mixtures. In India, over eighty crops are utilized in mixed-crop combinations.

In many instances, crop mixtures produce more than if the individual crops are grown in monoculture⌀. Where the crop species complement each other in growth rhythms, rooting depths, nutrient uptake and use of water⌀ and light, production is maximized. Intercropping maize⌀ and pigeon pea increases the efficiency of nutrient uptake, primarily as a result of differences in nutritional requirements and root depths between the two crops. The polyculture makes use of up to twice the level of nutrients from the soil than does the monoculture.

Efficiency of use of local resources is one of the major advantages of intercropping. Short-term crops planted with others that take a year or more to mature (such as sugar⌀ cane or cassava⌀) can use land⌀, water⌀ and light while the main crop is still small. Planting early- and late-maturing crops together helps to spread the harvest throughout the year. This can be particularly valuable where there are storage problems. Tall-growing crops can also be combined with short-growing, shade-tolerant crops. Crops such as maize respond to increases in row-spacing by raising the yield per plant but do not increase their leaf area. Consequently, crops such as sweet potato can be raised between the maize rows, thereby increasing and diversifying production with no additional input. Multistorey crop combinations – such as coconut⌀–black pepper–cacao–pineapple – are now becoming more commonplace in the effort to produce as much as possible from a unit of land. Food crops, including maize and sorghum, are now being widely grown in combination with tree crops, the basis of agroforestry⌀.

Intercropping produces increased crop residues, often an important source of fodder for livestock⌀. Residues also return nutrients to the soil, improving tilth and hence the long-term viability of the agroecosystem. In rubber⌀–tea⌀ intercrops, plant debris was found to fertilize the soil to a level that would otherwise have necessitated adding 813 kg/ha of nitrogenous fertilizer⌀ and 65 kg/ha of phosphates⌀.

Intercropping also brings benefits through changes in the microclimate. Rubber–tea intercrops not only have a good rate of return of nutrients to the soil; they can raise stand temperatures by 2°C – limiting the chances of frost damage to the rubber trees – and reduce soil erosion by up to 70 per cent. The increased rooting of mixed crops, usually extending to deeper levels than would be seen with single crops, helps to bind soils together.

Changing from monoculture to polyculture brings a marked shift in water balance – from evaporation from the soil to transpiration from the leaves. The overall water loss (evapotranspiration) is little changed, so water and irrigation⌀ requirements of monocultures and polycultures do not differ significantly. Water requirements prove no impediment to changing from monoculture to polyculture. Effectively, production can be increased without the need for improving water supplies.

The effect of viruses, fungi, weeds, nematodes and diseases can all be diminished merely by planting more than one crop. A study of plant-feeding insects has found that 60 per cent of the species tested were less abundant in mixtures than in monocultures. Several disease-resistant crop combinations have already been identified. Cassava and bean mixes reduce both incidence and severity of powdery mildew on cassava and angular leaf spot on beans. Cowpea viruses are also diminished when cowpeas are grown with cassava or plantain.

Intercropping produces a combination of vegetation which permits little room for weeds either to invade or to survive, and the efficiency with which many crop mixes use light, water and nutrients leaves precious little for the weeds.

Some crops are now being used as 'barrier' crops to protect another. For example, soya beans planted around pigeon pea prevents the passage of hairy caterpillars which thrive on the peas.

Mixed-crop systems meet more of the annual needs of subsistence farming families than does monocropping, and with less risk. In areas where hunger⌀ and malnutrition⌀ are endemic, trials of a sorghum–pigeon pea mixture found that, for a given 'disaster' level, sole pigeon pea crops would fail one year in five, sole sorghum one in eight, but that intercrops would fail one year in thirty-six.

Multiple cropping leads to higher production and consequently to higher farm incomes. Additionally, it improves prospects for poorer farmers in following years. Cereals are a basic subsistence crop, but continuous cereal cultivation depletes the soil. Intercropping cereals with crop legumes⌀, which help to improve soil fertility, can offset this, provide much-needed protein, and improve growing conditions for future crops.

The drawbacks associated with intercropping have – so far – been few and far between. Weeding cannot be mechanized because of the close proximity of crops of different physical characteristics and demands. This is not a major disadvantage because intercropping tends to prevent weed invasion and growth. Nevertheless, any weeding that has to be done has to be carried out carefully and is therefore time-consuming. This problem is compounded by the fact that at present there are very few herbicides⊙ that can be used on mixed crops. In a similar vein, harvesting has to be carried out carefully so as not to damage the standing crops. The main problem of intercropping is that it is labour-intensive. This will act against the practice being adopted in the North⊙ where, in future, it will be most necessary.

INTERGOVERNMENTAL OCEANOGRA-PHIC COMMISSION (IOC) The IOC was established in 1960 as a body within the United Nations Educational, Scientific, and Cultural Organization (UNESCO)⊙ to promote scientific investigation of the oceans through the concerted action of its members. The Commission operates a number of separate programmes, including the Global Atmospheric Research Programme (GARP), Integrated Global Ocean Station System (IGOSS), Association for the Caribbean and Adjacent Regions (IOCARIBE), Training, Education and Mutual Assistance (TEMA) and the Voluntary Assistance Programme (VAP). Based in Paris, the IOC has a membership of 111 nations.

INTERMEDIATE TECHNOLOGY DEVEL-OPMENT GROUP (ITDG) An independent voluntary organization established in 1965. Based in the United Kingdom, the ITDG has an international field of operations primarily focused on the needs and resources of the Third World⊙. The ITDG, whose work is based on and follows the theories of E.F. Schumacher⊙, aims to develop new tools, equipment and methods, or adapt existing ones, in order that, through application of their own skills and resources, the poor in the developing countries⊙ can improve their standard of living. A major consideration in the Group's work is that whatever is developed or produced must be appropriate to local conditions and function adequately within prevailing environmental and socioeconomic constraints. Working with local groups, government agencies and international organizations, the ITDG targets its activities on increasing food supplies, water⊙, building materials, renewable energy⊙ and transport and marketing of goods. The Group produces a number of regular publications, operates a technical inquiry service, and has established two

companies to license and control any tools, machines, patented processes or applications that arise through the ITDG research projects.

INTERNATIONAL ATOMIC ENERGY AGENCY (IAEA) The mandate of the IAEA, which was founded in 1957, is to accelerate and enlarge the contribution of atomic energy to peace, health and prosperity throughout the world, and to apply safeguards over nuclear materials, ensuring that they are used only for their intended peaceful purposes. The Agency, based in Vienna, has two main roles: to act as an international information exchange on civil nuclear applications (a role that, even during the Cold War, embraced the countries from Eastern and Western Europe) and to administer the Nuclear Non-Proliferation Treaty. The IAEA organizes conferences and symposia, publishes conference proceedings, and conducts interdisciplinary research. It has 113 members and a special relationship with the United Nations⊙. In 1990 the IAEA controversially concluded that up to 500 new nuclear power⊙ stations could and should be commissioned worldwide between 2001 and 2010 to provide a quarter of the forecast global energy⊙ needs.

INTERNATIONAL BANK FOR RECON-STRUCTION AND DEVELOPMENT (IBRD) Commonly known as the World Bank⊙, although it is only one wing of the World Bank group, the IBRD was set up by the 1944 Bretton Woods⊙ agreements to facilitate economic and infrastructural reconstruction following the Second World War. Based in Washington, the IBRD is an independent body with its own statutes but has the status of a United Nations⊙ body. It has over 151 members, all of whom must belong to the International Monetary Fund (IMF)⊙.

The Bank is the largest of the Multilateral Development Banks (MDBS)⊙ and finances development projects in member countries by making loans to governments or under government guarantee. All members make a capital subscription in accordance with a formula related to their economic strength. The United States is the largest contributor, with Japan rapidly approaching parity. The Bank's main resources are obtained by direct borrowing in international capital markets and from governments, and from the sale of participations in its loans. It lends on near commercial terms. Since 1948 most of its loans have been directed to secure the economic growth of developing countries⊙: it has also lent for educational and social schemes, including family-planning and resettlement programmes. It plays an important role in appraising and advising on the development programmes of developing countries⌂, and sponsors consortia and consultative groups, as well as producing several influential annual statistical publications. Shares in the Bank are allocated according to contributions, the biggest shareholders getting the most votes on the Executive Board. Loans recommended by the Bank President rarely come to a vote. Each member nation is represented on the Bank's Board of Governors but most decisions are taken by a group of twenty-two Executive Directors. Five are appointed (from France, Germany, Japan, the UK and the USA); the rest are elected from country groupings.

The World Bank group consists in total of the IBRD, the International Development Association (IDA)⌂ and the International Finance Corporation (IFC)⌂, which all use common services and staff and are all responsible to the President of the IBRD, who is always from the United States.

INTERNATIONAL BOARD FOR PLANT GENETIC RESOURCES (IBPGR) Established in 1974 under the auspices of the Consultative Group for International Agricultural Research (CGIAR)⌂, the Board promotes the collection, documentation, evaluation, conservation⌂ and utilization of genetic resources of important plant species. Special attention is paid to those areas where the spread of new varieties may put traditional varieties in danger of extinction⌂. The IBPGR also supports training in various aspects of genetic resources work and provides funding for research projects relevant to its mandate.

Until 1990 the IBPGR operated under a degree of guidance from the Food and Agriculture Organization (FAO)⌂, working with over sixty institutions in the developing world to maintain stocks of germplasm. It is of great value to poorer countries with no resources or facilities to maintain their own stocks. Over 90 per cent of all the germplasm collected by the IBPGR has come from the Third World⌂. However, 40 per cent is now in gene banks in Europe and North America and another 40 per cent is kept in the gene banks of the International Agricultural Research Centres, also part of the CGIAR. The IBPGR itself co-ordinates over forty international plant gene banks. Although around half of these are sited in the Third World, they house only one third of the world's germplasm.

Germ plasm from all banks is said to be freely available, although this has proved not to be the case.

Biological products and processes account for approximately 40 per cent of the world economy and a heated debate over access to and ownership of the IBPGR-collected germplasm has yet to be resolved. In 1984 FAO members adopted, almost unanimously, the International Undertaking on Plant Genetic Resources. Proposed by a group of Third World nations, the resolution proposed that germplasm, including patented seeds, should be viewed as 'common heritage' and therefore be freely available to all. The motion was adopted, with only the United States voicing opposition. Within two years the Undertaking was being ignored, with both industrialized and developing nations restricting the availability of germplasm for either commercial or political reasons.

The IBPGR is under the direct influence of Western governments and donor organizations, whereas Third World countries are able to have a greater say through the FAO. The FAO administers the funds for the Board, but the CGIAR actually provides the Board's funding, with the United States traditionally supplying 25 per cent of the CGIAR's annual budget. The majority of developing countries⌂ would like the FAO to take more direct control of the IBPGR in order to protect their interests.

The 1990 announcement that the IBPGR will be moving to a new base in Denmark – a decision later rescinded – forced many developing countries to express concern that they would lose access to an important collection of germplasm, at a time when the question of 'farmers' rights' and patenting of genetically engineered seeds had not yet been fully solved. In the developed world, plant breeders are allowed to derive substantial economic returns from the results of their work – so-called plant breeders' rights. In the developing world, the farmers who supplied the original germ plasm used by plant-breeders – and who have themselves spent years manipulating crop varieties in the field – receive no return for their labours – hence the call for farmers' rights.

The advent of biotechnology⌂ has led to increasing problems over the patenting of biological material. The value of wild-stock material is increasing as breeding techniques become more sophisticated and because many high-yielding crop varieties (HYVs)⌂ do not have the natural resistance characteristics of indigenous crop varieties. The questions of farmers' rights, ownership of germ plasm, patenting of seeds and freedom of access to plant material are likely to be major issues of the 1990s and are under debate in the United Nations Environment Programme (UNEP)↻, the FAO, General Agreement on Tariffs and Trade (GATT)↻, the World Intellectual Property Organization (WIPO)↻ and the Union for the Protection of New Varieties of Plants (UPOV Convention).

INTERNATIONAL BUREAU OF EDUCATION (IBE)

Founded in 1925 by a group of Swiss educationalists, the IBE became an intergovernmental organization in 1929 and was incorporated into the United Nations Educational, Scientific, and Cultural Organization (UNESCO)↻ in 1969. Based in Geneva, it is concerned with the study of comparative education↻ and aims to act as an international centre of information and educational research, as well as providing a wider exchange of data so that individual countries may be encouraged to profit from the experience of others. It holds regular conferences, including the biannual International Conference on Education, and provides information services, including the Educational Abstracting Service and an Educational Reporting Service. It produces several publications on aspects of educational policy and innovative developments in the field of education, fostering and maintaining an international education library.

INTERNATIONAL CENTRE FOR CONSERVATION EDUCATION (ICCE)

An independent, voluntary organization based in the United Kingdom which aims to promote greater understanding of the natural environment and the concept and methods of conservation↻ in particular, with an emphasis on the situation in developing countries↻. The ICCE provides consultancy, advisory and specialist media services and runs a specialized training programme for conservation educators from developing countries.

INTERNATIONAL CENTRE FOR DIARRHOEAL DISEASE RESEARCH (ICDDR)

Originally established as the Cholera Research Laboratory by the member states of the South East Asia Treaty Organization (SEATO)↻ in 1960, the current name was adopted in 1978. The Centre, based in Bangladesh, undertakes research, training and information dissemination on all diarrhoeal diseases and directly related subjects of nutrition and human health and welfare, with special reference to developing countries↻.

INTERNATIONAL CENTRE FOR INSECT PHYSIOLOGY AND ECOLOGY (ICIPE)

The ICIPE is a specialized international institution located in Kenya and funded by a consortium of donors. It undertakes research in insect science and pest management and also offers high-level training courses for pest-control scientists and technologists from African and other developing countries↻. It also plays a significant role in promoting institution-building and improved scientific performance in Africa as a whole, as well as co-sponsoring conferences and round-table discussions of globally important matters such as farmers' rights. It focuses much of its research on insects and pests of importance in Africa and has developed an effective low-cost trap for tsetse↻ flies.

INTERNATIONAL CENTRE FOR SETTLEMENT OF INVESTMENT DISPUTES (ICSID)

Founded in 1966 under the Convention on the Settlement of Investment Disputes between States and Nationals of Other States which came into force the same year. Sponsored by the International Bank for Reconstruction and Development (IBRD)↻ and located at its headquarters in Washington, the Centre aims to facilitate the settlement of investment disputes between national governments and foreign investors and thereby promote an atmosphere of mutual confidence and stimulate the flow of private international capital. Seventy-two states have signed the Convention, and most have also ratified it.

INTERNATIONAL CIVIL AVIATION ORGANIZATION (ICAO)

A specialized agency of the United Nations↻ which came into existence after its charter, known as the Chicago Convention, had been ratified by twenty-seven governments. The ICAO, established in 1947, now has 159 members. It aims to promote high operating standards, safety, efficiency, and fair competition among international airlines. With headquarters in Montreal, it formulates agreed standards in telecommunications, personnel training, flight procedures, and air-traffic protocols. Its governing council of representatives drawn from twenty-seven of the member countries implements the decisions of the Assembly, its legislative body, and, when necessary, resolves disputes between member nations. The ICAO also devotes itself to developing and evaluating measures to protect international travellers against the hazards of flying, including helping to prevent hijacking and other acts of terrorism. It also plays a role in controlling the transport of dangerous cargo, narcotics and other illegal substances and goods, and improving air traffic flow and navigation.

INTERNATIONAL CLEANER PRODUCTION INFORMATION CLEARINGHOUSE (ICPIC)

A project, launched by the United Nations Environment Programme (UNEP)⋄ in 1990, aimed at persuading business to adopt practices which pose little or no threat to the natural environment. The ICPIC is a computerized information system based in Paris and operates in tandem with a similar system run by the US government's Environment Protection Agency (EPA). Using the data freely available, businesses are encouraged to monitor and re-evaluate entire production processes with a view to limiting pollution⋄. The ICPIC also offers advice on reducing consumption of energy⋄ and raw materials. Users are given access to details of global environmental legislation, including data on policies that have been developed to cope with tougher environmental protection laws. A bibliography is provided which contains information on clean technologies and a directory of useful contacts.

INTERNATIONAL COMMISSION ON RADIOLOGICAL PROTECTION (ICRP)

Originally founded in 1928 as the X-Ray and Radiation Protection Committee; the Commission's present name was adopted in 1950. The ICRP is composed of leading experts from different countries, its major role being to determine what levels of exposure to radiation should be accepted as a global standard of safety. Since the Commission was established, permitted levels of exposure to radiation have been reduced to one-hundredth of the original level. Most nations have established their radiological exposure standards according to the recommendations detailed in a 1977 report known as ICRP 26. In this, the Commission recommends that the dose limit for the general public is 5 milliSievert (0.5 rem) per year. Industrial workers can be exposed to ten times this level.

Over the past decade or so the Commission has been under attack from many prominent scientists and environmental protection groups who have accused it of a lack of scientific judgement and called for tougher standards. The Commission is a self-appointed body of thirteen scientists who oversee specialist working groups. In total, 75 scientists from 20 countries are involved in its work. The ICRP has no proper headquarters and is purportedly free of both commercial and governmental pressures. However, many of the Commission's scientists are heads of national radiation protection agencies and the majority of members come from a background of work or training within the nuclear industry. The ICRP took no steps to help stop atmospheric nuclear weapons testing.

INTERNATIONAL COMMODITY AGREEMENTS (ICA)

Agreements between producers of primary, unprocessed commodities⋄ to regulate their production and sale in order to stabilize prices and conserve supplies. Examples include the Tin Agreement (1956), Coffee Agreement (1962) and Rubber Agreement (1976). Natural factors, such as crop failures and the discovery of new mineral⋄ deposits, cause considerable fluctuations in the supply of commodities, and prices oscillate accordingly. Under the agreements, producers, mainly developing countries⋄, agree to limit production. Minimum prices are fixed, marketing systems established, and buffer stocks built up by purchasing surpluses, which are then sold when prices exceed a specific ceiling in order to stabilize prices.

Several ICAS have been negotiated through the Integrated Commodity Programme of the United Nations Conference on Trade and Development (UNCTAD)⋄. Those concerning tin⋄, rubber⋄, coffee⋄ and cocoa⋄ were designed to stabilize prices, at a level equitable to producers and fair and remunerative to consumers, using buffer stocks and export controls via a quota system. Other ICAS without economic provisions were agreed to cover jute⋄ and tropical timber⋄ and were concerned primarily with research and development. A purely administrative International Sugar Agreement was also negotiated.

By 1990, the system had more or less collapsed and the prospects for those developing countries which rely on a single commodity for the bulk of their export earnings are not good. The Coffee Agreement collapsed in 1989; export quotas were suspended for two years and the price of coffee fell by almost half as a result. The inability of the 74 producer and consumer nations involved in the pact to agree on a new arrangement followed the collapse of the International Cocoa Agreement and the failure of the International Tin Council⋄. The failure of similar commodity pacts on wheat⋄ (1979) and sugar⋄ (1984) consigned both to the free market.

Rubber now remains the only commodity whose trade is governed by an agreement between producers and consumers seeking to bring stability to world prices.

INTERNATIONAL COUNCIL FOR BIRD PRESERVATION (ICBP)

The oldest of the international wildlife conservation◇ organizations, founded in 1922 and originally known as the International Committee for Bird Preservation. It has 66 national sections and 3 continental sections: for the Americas, Asia and Europe. Between them they have 270 national member organizations. The ICBP seeks to achieve the conservation, management and wise utilization of birds and their habitats. Like the Species Survival Commission of the World Conservation Union◇ (IUCN) it operates through specialist teams of experts working on various groups of birds. There are thirteen such groups, all of which collaborate with the IUCN. The ICBP holds quadrennial conferences and works closely with the International Waterfowl and Wetland Research Bureau, which concentrates its activities on preserving ducks, geese, swans and other wildfowl.

INTERNATIONAL COUNCIL FOR RESEARCH IN AGROFORESTRY (ICRAF)

Initiated in 1977 and based in Kenya, the Council is an autonomous, not-for-profit international institute whose principal support is derived from voluntary contributions provided by governments and by international public or private agencies. Its objective is to support, stimulate and co-ordinate on a global basis research in combined land-management systems of agriculture◇ and forestry and to improve land use in developing countries◇ through more efficient use of forests◇ and tree resources.

INTERNATIONAL COUNCIL OF SCIENTIFIC UNIONS (ICSU)

An international non-governmental scientific organization, first founded in 1919 as the International Research Council, which aims to encourage worldwide scientific research and collaborative activities for the benefit of humankind. ICSU plays a major part in initiating, designing, and co-ordinating international scientific research projects. It also acts as a focal point for the exchange of ideas and information. ICSU membership comprises 18 international scientific unions, 65 national members and 17 associate members, drawn from the international scientific research and development community in 71 countries.

ICSU functions through a series of co-ordinated specialized research programmes; the three main ones are the Scientific Committee for Antarctic Research (SCAR), the Scientific Committee for Oceanic Research (SCOR), and the Scientific Committee on Problems of the Environment (SCOPE).

INTERNATIONAL COUNCIL OF VOLUNTARY AGENCIES (ICVA)

Founded in 1962 through the merger of three international non-governmental bodies concerned with refugees◇ and migrants. It aims to provide a centralized reference and co-ordinating centre for voluntary bodies involved in providing assistance to people in need of help from the international community. Based in Switzerland, it has a membership of over 100 national organizations which contribute most of its funds.

INTERNATIONAL COURT OF JUSTICE (World Court)

The League of Nations◇ originally established the Permanent Court of International Justice in 1920 to preside over disputes among member states. Its successor, the International Court of Justice (or World Court), was founded in 1946 as the major judicial body of the United Nations◇. However, the Court, which is meant to maintain international law and peace, is effectively impotent. Over 110 of the UN's members do not recognize its jurisdiction.

Only sovereign states may take disputes to the Court which is based in The Hague, Netherlands. The Court's authority depends solely upon both parties to a dispute consenting to accept its ruling. Nations may also withdraw their acceptance of the Court's jurisdiction, as did France in 1974 to preclude any judgement on a case presented to the Court by Australia and New Zealand which attempted to ban nuclear testing in the South Pacific in accordance with the 1963 Nuclear Test Ban Treaty◇.

The Court consists of fifteen judges, elected by the UN Security Council and the General Assembly. The elections are staggered; five judges are chosen every three years, each judge serving for a nine-year period. In general, the fifteen seats on the Court are distributed in the same way as membership of the Security Council. Consequently, the five permanent members of the Council have always been represented on the World Court bench – except China, which has not nominated a judge since 1967.

INTERNATIONAL DEVELOPMENT ASSOCIATION (IDA)

An affiliate of the World Bank◇ established in 1960 to provide soft loans◇ to help those developing countries◇ with a need for – and ability to use – outside capital but without adequate foreign exchange to service loans on conventional terms. Credits for approved projects are given interest-free for fifty years, with a ten-year grace period. About 137 World Bank members are subscribers to the IDA, but the bulk of the usable resources are provided by some thirty richer members. IDA's resources are replenished at roughly three-yearly intervals. The criterion used to determine which countries are eligible for loans is an IDA-set *per capita* annual income ($805).

INTERNATIONAL DRINKING WATER SUPPLY AND SANITATION DECADE (IDWSSD)

In 1977 the United Nations◇ member states passed a resolution that the 1980s should be proclaimed as the IDWSSD, with the target that, if possible, everyone in the world should have access to a supply of safe water◇ and proper means of sanitation by 1990. The resolution stressed that:

> all peoples, whatever their stage of development and their social and economic conditions, have the right to have access to drinking water in quantities and of a quality equal to their basic needs. Similar considerations apply to all that concerns the disposal of waste water, including sewage, industrial and agricultural wastes and other harmful sources, which are the main task of the public sanitation systems of each country.

Most of the sickness and disease in the developing world is caused by unsafe water and inadequate sanitation. Four million children below the age of five die every year from diarrhoeal diseases stemming from contaminated water and an unsanitary environment.

Officially scheduled to run from 1980 to 1990 – although some IDWSSD projects actually started operating in 1978 – the Decade aimed to supply piped water and latrines to over 1 billion people at an estimated cost of $140 billion. The UN raised less than 25 per cent of the target figure, and a chronic lack of funds and inadequate maintenance continually dogged the programme. By 1985, the UN claimed to have supplied 345 million people in the Third World◇, mostly in rural areas, with a supply of clean water. Over 140 million had new waste-disposal facilities. In all, new or improved water

supplies have been provided to over 500 million people and better sanitation has been provided for 250 million. But in 1987, ten years after the first IDWSSD projects were started, 1.2 billion people were still without access to safe water and 1.9 billion lacked basic sanitation.

According to the independent, UK-based Water-Aid organization, when the IDWSSD started, an estimated 2 billion people were without reasonable access to safe water. At the end of the decade, in 1991, the figure stood at 1.97 billion. But while the overall figure showed hardly any change, over 700 million people did gain safe water supplies during the decade. Results on sanitation were not so good. When the IDWSSD began some 2.6 billion people needed to be supplied with decent sanitation. By 1991 an estimated 3.1 billion still needed this basic amenity, only slightly over 250 million people gaining access to safe sanitation during the decade. The Water-Aid figures for those without safe water and sanitation by the end of the decade were considerably higher than those projected by the United Nations.

By 1991, the official end of the IDWSSD, roughly three people out of every eight in the world were considered to be without access to safe water supplies. Developing countries◇, donors and support agencies, recognizing the IDWSSD's failure to reach its target, decided to extend their deadline to the year 2000 as part of the 'Health For All'◇ concept, but it is unlikely that many developing countries will achieve the goal by the year 2000.

Providing piped water to rural and urban areas has had its costs, environmental and economic. New pumps and cleansing systems soon corroded or simply broke down, never to be repaired, as no maintenance costs were built into the majority of IDWSSD projects. Access to and increasing use of water has boosted consumption, leading in some instances to the depletion of supplies. In India increasing water use has led to groundwater◇ depletion in Maharastra, actually raising the number of villages without proper water supplies from 112 in 1975 to 23,000 in 1985.

The main constraints preventing attainment of the IDWSSD targets have been financial. Spending during the early part of the decade was around $6 billion annually. This rose to between $8 and $10 billion by the end of the 1980s, but was simply not enough. The UN Children's Fund (UNICEF)◇ calculates that for an outlay of $11 billion per year, and adoption of low-cost, appropriate technologies, 80 per cent of the global population could have access to safe water and adequate sanitation by the year 2000.

INTERNATIONAL ENERGY AGENCY (IEA)

The IEA is an autonomous, twenty-one-member body, established in 1974 within the framework of the Organization for Economic Co-operation and Development (OECD)⟡, to implement an international energy⟡ programme. All members of the OECD except Finland, France, Iceland and Portugal are also members of the IEA. The Agency was set up following the decision of the Organization of Petroleum Exporting Countries (OPEC)⟡ to increase oil⟡ prices dramatically. Among its basic objectives are co-operation among participating countries to reduce excessive dependence on oil through energy conservation⟡ and the development of alternative energy sources. Members also agree to share their oil in emergencies. The IEA also aims to promote co-operation with other oil-consuming countries, together with oil-producing countries, with a view to stabilizing international energy trade as well as the development of a rational management programme for use of world energy resources in the interest of the global community.

INTERNATIONAL ENVIRONMENTAL EDUCATION PROGRAMME (IEEP)

Established in 1975 by the United Nations Environment Programme (UNEP)⟡ and the UN Educational, Scientific, and Cultural Organization (UNESCO)⟡, the IEEP is the only major worldwide programme for environmental education⟡. Set up in response to a recommendation from the 1972 Stockholm Conference on the Human Environment⟡, the IEEP was developed on the premiss that the understanding of the fundamentals of environmental problems lie to a great extent in social, cultural and economic factors and cannot be solved by technology alone. The programme is interdisciplinary, implemented in both the formal and informal education sectors, and encompasses all levels of education, including that for the general public.

INTERNATIONAL ENVIRONMENT BUREAU

A specialized agency of the International Chamber of Commerce established in 1986 and funded by contributions from companies from around the world. The Bureau aims to promote more efficient management of the environment so that sustainable economic growth can be achieved. It serves as a 'trans-industry clearing house' for the latest technological developments and environmental expertise and provides assistance on environment and development, particularly to commercial enterprises in developing countries⟡.

INTERNATIONAL FINANCE CORPORATION (IFC)

Established in 1956 as an affiliate of the World Bank⟡ to assist the progress of less-developed member nations. The IFC encourages and provides investment in, and loans to, productive private enterprises and financial intermediaries, such as investment and development banks, where sufficient capital is not otherwise available on reasonable terms. Finance is always provided in association with funds from private investors, but does not compete with private capital. The IFC has 133 members and a total authorized capital of over $1 billion.

INTERNATIONAL FUND FOR AGRICULTURAL DEVELOPMENT (IFAD)

The IFAD was founded as the result of a resolution adopted by the World Food Conference⟡ in 1974 which urged the creation of a new fund to finance agricultural development projects in the developing countries⟡, making resources available on concessional terms. IFAD is mandated to mobilize resources for agricultural development and to focus its activities on food production, the rural poor, and the creation of employment for the landless poor so that they can purchase their basic needs.

The Fund began operating at the end of 1977 as the thirteenth specialized agency of the United Nations⟡, with an initial funding of $1 billion. Membership is in three categories: those countries of the Organization for Economic Co-operation and Development (OECD)⟡, nations belonging to the Organization of Petroleum Exporting Countries (OPEC)⟡, and the developing countries. Member countries of the OECD and OPEC provide the bulk of IFAD's funds. There is a governing council and an executive board, each membership category having equal voting power in these bodies. Financing is provided by IFAD via loans and grants, mainly on the basis of project identification and following appraisal by co-operating international funding institutions. The low-interest loans to small farmers and the rural poor put seeds, fertilizer⟡ and tools into the hands of those who previously derived little benefit from outside assistance. Although IFAD is widely recognized as an effective agency and one that helps those most in need, replenishments of its Fund have regularly fallen below expectations: the latest replenishment was $523 million less than the $750 million sought.

INTERNATIONAL INSTITUTE FOR EDU-CATIONAL PLANNING (IIEP)

Established in France by the United Nations Educational, Scientific, and Cultural Organization (UNESCO)✛ in 1983 to serve as a world centre for advanced training and research in educational planning. Its purpose is to help member states of UNESCO in their social and economic development by enlarging the fund of knowledge about educational systems and the supply of experts in their field. While administratively a part of UNESCO, the Institute's special statutes ensure its intellectual autonomy.

INTERNATIONAL INSTITUTE FOR EN-VIRONMENT AND DEVELOPMENT (IIED)

An international, independent Non-Governmental Organization (NGO)✛ based in London and Buenos Aires. The IIED undertakes research and analytical studies on a wide range of environmental and development subjects, including environmental planning, human settlements, Antarctic✛ resources and the use and production of energy✛, forestry✛ and agriculture✛, paying close attention to the environmental and socioeconomic impact in every case. It primarily works in the area between the United Nations✛ and other international institutions, national government agencies and NGOs, aiming to provide a two-way flow between policy-makers and those seeking to influence policy, through the collection, collation and dissemination of information. It owns, in partnership with the World Wide Fund for Nature (WWFUK), a publishing house called Earthscan Publications Ltd. This editorially independent company publishes books which fall into the areas of IIED interest for a wide audience throughout the world. IIED also co-operates closely with other international environmental development agencies, regularly participating in joint ventures with the World Conservation Union✛ (IUCN), World Resources Institute (WRI)✛ and United Nations Environment Programme (UNEP)✛, amongst others.

INTERNATIONAL IRRIGATION MAN-AGEMENT INSTITUTE (IIMI)

The Institute, located in Sri Lanka, concentrates on finding solutions to all manner of problems connected to or arising from irrigation✛ and carries out its operational activities through a network of programmes attached to existing institutions and research centres throughout the world. The IIMI is funded by contributions from donor governments and international organizations.

INTERNATIONAL LABOUR OFFICE (ILO)

First convened in 1919, when it was affiliated to the League of Nations✛, the ILO became a specialized agency of the United Nations✛ in 1946. It strives to improve working conditions and living standards for the world's population✛. Based in Geneva, the 150-member ILO advocates a world labour code to protect the interests of workers, supports labour research projects, monitors labour legislation around the world, and provides technical assistance to developing nations. The ILO's main emphasis is on the adoption of international minimum labour standards and the development of methods for regulating labour conditions on an international and equitable basis. It is also especially concerned with the conditions under which young persons and women are employed, the protection of workers outside their native countries, the activities of transnational✛ companies, industrial relations and human rights✛. In 1977 the United States withdrew from the ILO on the grounds that it had become dominated by political interests, but rejoined in 1980.

INTERNATIONAL LAW

Present international law evolved from the regulations drawn up to govern relations between the European Christian States in the sixteenth and seventeenth centuries. Formation of the League of Nations✛, later replaced by the United Nations✛, allowed any state to become a member and so gain access to the global legal framework.

The scope and range of international law has been extended to cover the realms of the open ocean, Antarctica✛ and space. In reality, however, international law cannot be successfully enforced unless it coincides with the aims of one or more of the predominant major powers, as shown by the examples of the Korean War✛, France's decision not to recognize an International Court of Justice✛ (World Court) ruling that would have halted nuclear testing in the South Pacific, and the United States' refusal to accept a judgement against it in a case brought by Nicaragua.

Third World⟳ countries, home to the majority of the world's population⟳, are supposed to be free and equal partners in determining world development, but in effect they have little opportunity to press for their rights or seek redress for unjust treatment. International institutions such as the United Nations and the World Court are the only fora at which disputes among nations can be settled, but the major powers have the right to veto some decisions in the UN system and to reject any pronouncements by the World Court that are not to their liking.

Public international law, also called the Law of Nations, is administered by the World Court, and is based on (1) natural law or laws recognized by sovereign nations; (2) codified agreements between states, including international conventions; (3) customs followed in practice; (4) the writings and opinions of respected legal figures.

Private international law, also called Conflict of Laws, determines the laws of which country should apply and in which courts jurisdiction lies. This can prove crucial for the fair treatment of claimants – as shown by the Bhopal⟳ case – and highlights the inequalities that exist within the differing legal systems prevailing in the world.

With the growing recognition of transboundary pollution⟳, exports of hazardous waste⟳, and the exploration and exploitation of Antarctica, space and the high seas, all 'common heritage', there is increasing interest in the role and importance of international law, and the 1990s have been declared the Decade of International Law by the UN General Assembly.

INTERNATIONAL MARITIME ORGANIZATION (IMO)

The IMO, one of the specialized United Nations⟳ agencies, was set up in 1959, originally as the International Maritime Consultative Organization, with headquarters in the United Kingdom; it adopted its new name in 1982. The IMO has 133 member nations and one associate member. Its main objectives are the development of internationally acceptable measures for the improvement of safety at sea, the prevention of marine pollution⟳ from ships, and means of improved navigation for maritime traffic. The IMO facilitates the drafting and adoption of international conventions, protocols, codes of practice and recommendations. It also arranges seminars and workshops, and produces publications and conference proceedings. As part of its general programmes the IMO also provides technical assistance, mainly in the form of expert personnel and training facilities and mainly for the less-developed countries.

INTERNATIONAL MONETARY FUND (IMF)

The IMF is a specialized financial agency of the United Nations⟳ established under the 1944 Bretton Woods⟳ agreements. Operational since 1947, the Fund now has 151 members. It seeks to promote international monetary co-operation, the growth of world trade and foreign exchange stability, and to smooth multilateral payments and financial arrangements amongst member states. It also provides financial assistance to member countries in balance-of-payments difficulties. The ultimate goal is to provide a secure global financial base that would allow a liberal trading system and avoid the protectionist policies which contributed to the global economic Depression in the 1930s.

Part of the IMF system broke down in the 1970s, when members opted for floating rather than fixed currency-exchange rates. Having originally operated its accounts in US dollars linked to gold⟳, the IMF switched to the use of special drawing rights (SDR)⟳ as a standard unit of account in 1972. These are valued in terms of a weighted collection of major currencies. Countries receive SDR in proportion to their national quotas, and these may be exchanged for foreign currency to buy imports. Members contribute resources in accordance with pre-set national quotas which also determine the amount a member may withdraw. Each member country contributes to the Fund in both gold and national currency. The higher a member's contribution, the greater its voting rights.

The IMF provides stand-by loans to members with balance-of-payments problems; the amount of the loan is limited by the quota system. The Fund uses its resources to provide foreign currency to member countries. 'Drawings' of foreign currency are temporary, they must be paid back, usually within five years. If large amounts are withdrawn there are conditions; the recipient country has to give undertakings to the Fund about the economic policies it intends to pursue and follow advice given on the measures needed to resolve its financial crisis.

The operations of the IMF are theoretically non-discriminatory: the same rules apply to all members. However, the Fund operates special facilities for developing nations, including a Compensatory Financing Facility which makes additional resources available to compensate for unexpected shortfalls in export earnings for commodities⟳, and a facility which makes financing available for the setting up up of buffer stocks agreed under International Commodity Agreements⌾.

A review of the IMF's articles of agreement in 1976 recommended the setting up of a Trust Fund, using part of the IMF's gold reserves, which would be used to subsidize the use of IMF facilities by the poorer developing countries⌾. The IMF is primarily the lender of last resort to Third World countries, since its loans are usually accompanied by strict conditions. IMF conditionality takes several forms. Debtor governments must reduce expenditure, in particular so-called 'non-productive' costs such as welfare, education⌾, health and food subsidies. Real wages must also be reduced. The IMF frequently insists on the expansion of export cash crops⌾ at the expense of food grown for internal consumption. Countries are also often asked to devalue their currency to boost exports and limit imports. The IMF strictly forbids debtor nations to set up import quotas or any system whereby manufactured goods from the industrialized world could be stopped. Arms spending⌾ is also not restricted.

The IMF is based in Washington and the head is always a European. Power resides, in theory, with twenty-one executive directors. In fact, control is exercised by the major quota nations – France, Japan, the Federal Republic of Germany, the United Kingdom and the United States, which has the largest quota.

INTERNATIONAL ORGANIZATION FOR STANDARDIZATION (ISO)

An organization, founded in 1946 and based in Geneva, which is concerned with establishing, controlling and making uniform all international scientific, industrial and commercial standards of measurement and design. All the national Standards Institutes from around the world are members.

INTERNATIONAL ORGANIZATION OF CONSUMER UNIONS (IOCU)

An independent international institution, founded in 1960, that links the activities of well over 100 national consumer organizations in more than 50 countries worldwide, drawn from both the developed and developing world. Based in Malaysia and the Netherlands, the IOCU aims to promote worldwide co-operation in consumer protection, information dissemination and education⌾ through establishing or strengthening information networks, arranging international seminars and holding a regular world congress. It also initiates investigative research and encourages the comparative evaluation of products

and services, acting as a clearing house and international information centre and advocate for consumers in global fora such as the United Nations⌾.

INTERNATIONAL PLANNED PARENTHOOD FEDERATION (IPPF)

Formed in 1952 and based in London, the IPPF is a grouping of independent Family Planning Associations from around the world. It encourages national associations to provide family-planning services, helps them to maintain high clinical standards and train personnel, and assists in initiating educational programmes. The IPPF supports programmes in over 120 countries; its funding is provided by voluntary contributions from governments, international organizations, and the private sector. The Federation acts as a consultant to the United Nations⌾ Economic and Social Council and other specialized UN agencies; this allows it to participate in discussions within the United Nations system as an accredited authority on the subjects within its competence.

INTERNATIONAL PROGRAMME OF CHEMICAL SAFETY (IPCS)

An international co-operative exercise set up in 1980 to monitor and oversee the safe use and management of industrial and agricultural chemicals. The system, founded following an initiative proposed by the International Labour Office (ILO)⌾, World Health Organization (WHO)⌾ and United Nations Environment Programme (UNEP)⌾, brings together 27 countries and 67 participating institutions concerned with all aspects of risk evaluation and maintenance of chemical safety standards. There are over 7 million known chemicals, and thousands of new ones are discovered or introduced into common use every year. At least 80,000 in regular use are potentially hazardous to humans. The IPCS reports on the effects of these different chemicals on the human body, sets guidelines for exposure limits, proposes means of standardizing measurements used in exposure and toxicity testing, and provides information on methods for coping with chemical accidents or the treatment of poisoning.

INTERNATIONAL REFERRAL SYSTEM FOR SOURCES OF ENVIRONMENTAL INFORMATION (INFOTERRA)

A network of national information centres established by the United Nations Environment Programme (UNEP)⌾, following the call at the 1972 Stockholm Conference⌾ for an international mechanism to facilitate the free exchange of environmental information. INFOTERRA is a decentralized system designed to promote smooth and free interchange of data on environmental matters. Each member country compiles a register of research institutions and centres of excellence willing to share expertise in environmentally related areas of interest or concern, such as energy⟳, agriculture⟳, wildlife and pollution⟳. An international directory includes details from the individual national registers, so listing all experts available for consultation. INFOTERRA handles well over 10,000 requests annually, the majority from developing countries⟳. The system is available to governments, industry and researchers in 129 nations, covering 98 per cent of the world's population.

INTERNATIONAL REGISTER OF POTENTIALLY TOXIC CHEMICALS (IRPTC)

A centralized register of data on hazardous chemicals, located in Geneva and run by the United Nations Environment Programme (UNEP)⟳. The IRPTC is basically concerned with providing information to enhance chemical safety. It disseminates data on chemicals, locates and publicizes the lack of knowledge in specific areas and encourages research to fill any gaps identified. The network of contributors to the register includes 115 national correspondents sited in over 100 countries, as well as many national and international organizations, chemical companies and governmental institutions.

INTERNATIONAL TELECOMMUNICATION UNION (ITU)

A specialized agency of the United Nations⟳, founded in 1934 to promote international agreement on standards of use and development of telecommunications systems. Originally established in France in 1865 as the Union Télégraphique Internationale, its name was changed after amalgamation of the International Telegraph Convention and International Radiotelegraph Convention.

ITU's aims are to maintain and extend international co-operation for the improvement of rational use of telecommunications, to promote the development of technical facilities, to improve efficient use of all telecommunications systems, and to increase their usefulness and make them generally available.

Based in Geneva, with 166 member nations, the ITU makes recommendations on the regulation and use of radio frequencies and general operating procedures, and helps to establish telecommunications

systems in developing countries⟳, including radio, telegraph, telephone, television and space communications.

INTERNATIONAL TIN COUNCIL (ITC)

The International Tin Council was set up to promote consumer–producer co-operation with regard to tin⟳ and to implement the International Tin Agreement, one of the International Commodity Agreements (ICA) set up to help stabilize the worldwide price for individual commodities⟳.

The Tin Agreement effectively kept the base price for tin artificially high, encouraging surplus production and placing a steadily increasing financial strain on the Tin Council, which was charged with buying and stockpiling surpluses in order to maintain global prices. In 1985 the ITC collapsed when it ran out of money to continue its price-support role in the glutted market, leaving gross debts of almost $1.5 billion. Its collapse caused the world tin price to fall by 57 per cent, devasting the industry in high-cost producer countries such as Bolivia and the United Kingdom.

Following lengthy and expensive legal action, during which Tin Council member governments argued that they were not responsible for the Council's debts, in 1990 the twenty-two governments involved finally agreed to pay $300 million in compensation to the organization's creditors. The money was an out-of-court settlement against net claims of over $840 million, mainly from brokers on the London Metal Exchange.

INTERNATIONAL TRADE CENTRE (ITC)

Jointly sponsored by the United Nations Conference on Trade and Development (UNCTAD)⟳ and the General Agreement on Tariffs and Trade (GATT)⟳, the ITC was set up in 1964 at the request of developing countries⟳ to help them improve their exports and overseas trading opportunities. Based in Geneva, the ITC aims to identify export market opportunities, to make export marketing and promotion techniques better known, and to draw atttention to organizations that can assist in export production. It also operates a training programme. The ITC advises governments of developing countries on their national trade promotion strategies and how to upgrade institutions and services. It also works to rationalize import operations on a global scale as well as providing advice on how best to adapt products for sale abroad.

INTERNATIONAL TROPICAL TIMBER ORGANIZATION (ITTO) An agreement reached in Geneva in 1983 to set up an international organization to promote the production, sustainable utilization and conservation⊙ of tropical timber⊙ led to the formation of the ITTO. After several years of indecision and heated debate, the ITTO was finally sited in Japan with the mandate to oversee the implementation of the International Tropical Timber Agreement. It is composed of representatives from producer and user countries and from the timber industry. Becoming operational in 1987, the ITTO forum accounts for 75 per cent of the world's tropical forests⊙ and more than 95 per cent of international trade in tropical timber. This is based on the membership tally of 45 nations, although 70 nations took part in the original negotiations to establish the organization.

The ITTO objectives are broadly classified as:

- the promotion of research and development in all aspects of forest management and wood⊙ utilization;
- the promotion of reforestation⊙ and forest management;
- the promotion of timber processing in the producer countries;
- the establishment of a system of market information to help boost trade.

The promotion of sustainable management is central to all these objectives. Research projects will focus on the fuller and more efficient use of forest resources, manpower development and constraints on the development of industries in the producing nations. The ITTO's task is not to halt the trade in tropical timber but to promote it and, at the same time, ensure that it becomes viable in the long term.

Global trade in tropical timber amounts to around $8 billion annually, with virtually all wood being derived from natural forests which are inadequately managed. The amount of wood entering global trade from plantation sources is comparatively negligible. The ITTO aims to see that all timber comes from 'sustainably managed' forests by the year 2000. Of the 828 million hectares of productive tropical forests around the world, significantly less than 1 per cent is currently being exploited without irreversible damage.

By 1990, research carried out by the ITTO had identified precious little sustainable forest management in the tropics. In the Caribbean and Latin America only 75,000 hectares in Trinidad and Tobago are being managed properly on a sustainable basis. Investigation of the position in six African countries found no sustainable forestry. Examples of sustained timber management were found mainly in Asia and Australia. The Mae Poong forest in Thailand and management systems in parts of Malaysia were put forward as specific examples of sustainable forestry, together with logging operations in Queensland, although conservationists disputed these findings.

INTERNATIONAL UNDERTAKING ON WORLD FOOD SECURITY Adopted by the Food and Agriculture Organization (FAO)⊙ in 1974: governments agreed to ensure that adequate amounts of basic foodstuffs, especially cereals, were always available. The goal was to avoid or overcome acute food shortages⊙ through a co-ordinated system of national food reserves. Over eighty nations subscribed to the undertaking; most reserve stocks are maintained in a few industrialized countries.

INTERNATIONAL UNION FOR THE CONSERVATION OF NATURE (IUCN) *See* World Conservation Union

INTERNATIONAL WHALING COMMISSION (IWC) Established in 1946 under the International Convention for the Regulation of Whaling, which came into force two years later, the IWC governs whaling⊙ operations by nationals of contracting governments. Its aim is to promote the whaling industry on a sound and sustainable basis, for the industry's benefit. It regulates the hunting⊙ of twelve species of great whales and the minke whale and does not attempt strictly to curtail or prevent whaling altogether. The IWC sets catch and size limits and specifications for whaling gear. In the effort to conserve stocks and develop an orderly and functional industry, the IWC, through a specialized Scientific Committee, encourages research and collects and disseminates information concerning whale numbers and whaling operations. It meets annually to decide upon regulations which are usually adhered to by member governments. The IWC has been the main focal point and forum for the international anti-whaling protest. In 1988 it introduced a moratorium on whaling, effective until 1991, which most nations respected, although some whaling was carried out under the guise of being for 'scientific' purposes. Its existence and effectiveness came into question in 1989. With several members years behind in paying their dues, the resultant dire financial straits forced the IWC to announce that its scientific budget would be cut by over one third and member-nation fees would increase by 50 per cent.

INTERNATIONAL WHEAT COUNCIL (IWC)   Established in 1949. The Council's work is concerned with the co-ordination and implementation of the Wheat Trade Convention and the International Wheat Agreement. Original objectives included furthering international co-operation in connection with world wheat⬦ problems, promoting expansion of international trade in wheat and wheat flour, contributing to the stability of the international wheat market, and providing a framework for the negotiation of provisions relating to wheat prices and the rights and obligations of members in respect of international trade in wheat. The IWC was also made responsible for administering the Food Aid Convention (FAC)⬦.

INVISIBLE TRADE   The wealthier, developed countries receive a considerable amount of income from various services they provide, such as transport, tourism⬦, banking and insurance, together with other income from interest, dividends, and returns on investments. This is known as invisible trade. In 1987, the total of this trade topped the $1 billion mark; the United States (17.4 per cent of the total) and the United Kingdom (12.5 per cent) were the leading beneficiaries.

IRON   A metallic element that has been known and used since prehistoric times. It is an essential component in the human body, forming haemoglobin in the blood which facilitates the supply of oxygen. Iron is the fourth most abundant element in the earth's crust. Iron ores are widely distributed throughout the world, the most commercially important containing 25–60 per cent iron. Virtually all mined ore is used to make pig iron. A broad range of ferro-alloys are manufactured from pig iron, including steel – a general term used to describe alloys which can be forged, made into sheets, drawn into rods or wires, or machined. Iron is widely used in toolmaking, construction, shipbuilding and car manufacture.

Annual global production of iron ore is in excess of 990 million tonnes; the Soviet Union (246 million tonnes), China (163 million tonnes) and Brazil (157 million tonnes) are the largest producers.

IRRIGATION   The artificial watering of land⬦ to enable crops to be grown. Irrigation is one of the most productive known forms of agriculture⬦. The world's irrigated cropland already produces 40 per cent of total global crop yield. Of the 1.5 billion hectares (ha) of cropland in the world, between 220 and 270 million are irrigated, although 500 million could potentially be so.

Irrigation may involve the flooding of whole fields, as in rice-growing, or the water⬦ may be run in channels between the crop. High-pressure sprinklers may also be used where the land is undulating. Traditional low-cost systems linked to seasonal changes in river level, using earth banks, channels and lifting devices, are still used by farmers throughout the world. Modern methods are frequently large-scale, expensive and involve the construction of dams⬦, artificial reservoirs, canals, and pumping systems. Between 1950 and 1970 the irrigated area of the world doubled, but the industrial-scale irrigation projects responsible for this are expensive – up to $10,000 to irrigate a single hectare of land in some areas. The irrigated land is usually given over to the production of cash crops⬦ to raise the foreign exchange to repay the installation costs.

Irrigation has been practised for centuries and has been in continual use in parts of Africa, Asia and South America. It remains predominantly a Third World⬦ phenomenon, with 75 per cent of all irrigated land still found in developing countries⬦. By 1985, developing countries had invested an estimated $250 billion in irrigation. By 2000, another $100 billion will have been spent. China has well over a third of all the world's irrigated land, with India fast approaching this figure, the Indian authorities having committed $20 billion to irrigation development over the past thirty-five years. However, a tenth of the world's irrigated land is badly waterlogged, a similar amount is affected by salinization⬦, and productivity on these lands is falling. Almost 10 million ha of irrigated land in India alone are reportedly in danger of becoming infertile.

Estimates for the amount of irrigated land worldwide showed a rise from 163 million hectare in 1968 to 271 million by 1985, with projections that the total would be be approaching 400 million hectare by 2000. However, such are the problems associated with irrigation that by early 1990 forecasts had been revised drastically downwards, and it was thought that only 228 million ha of land around the world were being irrigated and that this would rise to only just over 270 million ha by the end of the century.

Irrigation is extremely water-intensive. Over the world as a whole, 1,620 cubic kilometres of water are used for irrigation each year, but well over 2,200 are withdrawn. Consequently, around one-quarter to one-half of all water withdrawn for irrigation is lost during storage or transport. Although water withdrawals continue to increase, by 1991 a decreasing proportion of withdrawals were being used for irrigation. The 63 per cent of water being used in 1991 was projected to decline further: to 55 per cent by the year 2000.

Excessive or ill-managed irrigation not only wastes large amounts of water, it can also leach out soil nutrients, create problems of salinization and threaten public health. Good management is essential if irrigation is to be successful, and especially so if it is to be of long-term value. Traditional systems have evolved safe methods of avoiding most of the major pitfalls of artificially watering land. In pursuit of maximum production and short-term gain, modern schemes are large-scale and perennial, with three crops often being raised on the same piece of land. Drainage is rarely adequate – primarily because it is expensive to install good drainage systems – and salinization is common. Salinity problems reduce productivity on 7 per cent of the world's irrigated land, nd hundreds of thousands of hectares are lost each year as a result of salt build up – as much land as is brought into production. By 1990, 30–40 per cent of the world's irrigated cropland was thought to be either waterlogged or suffering from excessive salinization.

Irrigation projects provide new and permanent niches for the vectors of diseases such as malaria⟡, filariasis⟡ and schistosomiasis⟡, which are a danger to public health. Agricultural pests, insects and mammals such as rats thrive in newly irrigated areas where year-round supplies of food suddenly become available.

With these and other problems, such as those arising over the costs of installing and maintaining machinery and disputes over water rights, it is generally recognized that irrigation projects are now most likely to succeed where fallow periods are observed and management is left up to local communities – irrigation schemes should be small-rather than large-scale.

ISLAM The Muslim religion originated in Arabia in the seventh century and was spread by Arab conquests 1,300 years ago. Its scriptures are in Arabic, detailed in the Koran, which is the basis of Islamic belief, practice and law. Its holiest shrine is at Mecca in Saudi Arabia, to which hundreds of thousands of Muslims make the pilgrimage every year. The majority of the world's 850 million Muslims are not Arabs: more live in India, Pakistan and Bangladesh than in all the Arab countries combined. The Islamic Conference, an organization of governments founded in 1969, has forty-five member states. Most of them agree on such issues as the Palestine Arab struggle against Israel and the Afghan guerrillas' action against Soviet occupation forces. The war between Iran and Iraq caused deep divisions, and several nations are concerned about the way in which countries such as Libya, Iraq and Iran have tended to use religious appeals to increase their sphere of influence.

Within the Muslim religion there are deep sectarian divisions, primarily between the Shias and the Sunni. Over the past fifty years there has been an accelerating adoption of the Islamic faith by black Americans in the United States seeking an alternative to the racist attitudes of many white Christians.

ISLAMIC DEVELOPMENT BANK (ISDB) Established in 1974 and operational since 1975. The stated purpose of the Bank is to foster the economic development and social progress of member countries and Muslim communities, individually as well as jointly, in accordance with the principles of Islamic law [*Shariah*]. The ISDB is involved in project loans, equity participation, leasing operations, profit-sharing schemes, foreign trade financing and provision of technical assistance to member states. Although the Bank concentrates its activities on its poorer members, projects can be undertaken in any member state. The Bank operates under Islamic law so does not charge interest on loans, but does make a service charge based on administrative costs.

IVERMECTIN Ivermectin is a persistent, broad-spectrum, anti-parasitic drug of enormous potential to the livestock⟡ industry as well as for improving human health. It is now routinely administered to cattle, horses, sheep and pigs in many countries around the world; it provides effective control of gastrointestinal and respiratory-tract nematode worms, warble fly, mites, ticks and lice.

Research has shown that treatment with Ivermectin causes cattle to produce manure containing significant amounts of the drug. The dung so produced exhibits strong insecticidal properties and eliminates the dung-degrading insect community, thus preventing cowpats from being decomposed. Animals can produce dung with insecticidal properties for at least fifteen weeks after first receiving treatment. The manufacturers of the drug refuted these observations claiming that hundreds of millions of cattle in seventy-seven countries have been treated with Ivermectin, yet no complaints had been received during the first six years after the drug was licensed for use.

Another dimension was added to the controversy as a result of the Onchocerciasis Control Programme in West Africa. Onchocerciasis⟁, or river blindness, is caused by a parasitic worm transmitted to humans via the bite of a blackfly. It was found that Ivermectin effectively kills the parasitic forms of the worm in the human body, without causing any internal damage. All previous therapeutic drugs caused dangerous side-effects due to the build-up of dead parasitic material in the body. With full backing from both the World Bank⟁ and World Health Organization (WHO)⟁, safety evaluations have been carried out in France and the drug is being used in West Africa, although humans cured of onchocerciasis following treatment with Ivermectin may produce excrement that does not biodegrade normally.

IVORY Close-grained, whitish calcified tissue forming the tusks of elephants⟁ and walruses and the teeth of hippopotamus, used to make piano keys, name seals, jewellery, knife handles and ornaments. In Japan the market for piano keys and name seals alone accounted for eighty tonnes a year by the late 1980s; the country bought around 130 tonnes of ivory each year. Plastics have now replaced ivory for many purposes, but ivory is easy to carve and polish, has a unique feel, and remains a valuable and much-sought-after commodity, even though the main source, the African elephant, is threatened with extinction⟁.

In 1989 the African elephant was moved from Appendix II to Appendix I of the Convention on Trade in Endangered Species of Flora and Fauna (CITES)⟁ for a trial period of two years, effectively preventing trade in ivory. The ban became operational in 1990. Botswana, Burundi, China, Malawi, Mozambique, South Africa, Zambia and Zimbabwe announced their intention of taking a reservation on the elephant, meaning that they could continue trading ivory with other countries doing the same or with countries who are not CITES members, although China acceded to the global ban in early 1991. In 1990 the United Kingdom announced a six-month reservation allowing Hong Kong – a British colony and the hub of the legal and illegal ivory trade – to reduce its 670 tonnes of stockpiled ivory, much of it illegally acquired.

As the elephant has gradually been removed as a source of ivory, attention has been switched to alternative animals and the walrus is now being hunted and illegally exploited.

IVORY SUBSTITUTES Ivory nuts offer one of the best alternatives to animal ivory⟁. These nuts – more correctly called corozo nuts – are the fruit of the ivory palm tree, which is native to the valleys of the Andes in South America. Before the development of the plastics industry, these hard white nuts were used to manufacture buttons. Research is also under way to manufacture a material that resembles proper ivory more closely, in terms of both quality and texture, than currently available plastic subsitutes.

Tusks from mammoths discovered in the frozen Soviet tundra are reportedly being used as an elephant-ivory substitute. The 10,000–20,000-year-old tusks polish faster than elephant⟁ tusks and have an interesting grain and colour. New ceramic ivory is being developed; this is better than plastic substitutes because it is porous and so feels more like real ivory. In 1990, Japanese scientists announced the development of a new ivory substitute made from dairy products – a mixture of eggshells and milk – and titanium dioxide; its capillaries make it more absorbent than other substitutes. Failure to absorb water⟁ has been the major failing of all other substitute materials.

J

JOJOBA   Jojoba (*Simmondsia chinensis*) is a drought⌕-resistant, long-lived, evergreen desert shrub. It is native to the south-western region of the United States, growing wild only in the Sonoran Desert. The seeds are composed nearly 50 per cent of a liquid wax which has properties very similar to spermaceti, the oil obtained from sperm whales. Spermaceti is unique amongst animal oils in that it contains a liquid wax, which makes it useful as a starting point for the manufacture of high-pressure lubricants. No mineral oils, animal oils or vegetable oils can act as a suitable substitute for spermaceti – except the liquid wax from jojoba seeds. Jojoba oil is a clear, odourless liquid with molecules slighter larger than those in spermaceti, which improves its pressure lubrication properties. It can be used for all the applications for which sperm whale oil has traditionally been employed. The shrub can be put to other uses. Browsing animals feed freely on the foliage, and the leaves can be used for cattle fodder in times of shortage. The seeds contain a toxic substance, simmondsin, so they must not be fed to livestock⌕, even to the extent that meal produced from the seeds – which contains 30 per cent protein – must be carefully treated so that all the toxin is denatured. The global demand for jojoba seeds now exceeds 5,000 tonnes annually, and plantations are appearing throughout the world. The plant does not require irrigation⌕, grows well in poor soil, and needs little or no fertilizer⌕, thus enabling marginal and semi-arid land⌕ to be used for cultivation. Disadvantages are few. The flowers are wind pollinated, and bushes will not yield a crop for three to five years. Planting, caring for seedlings and harvesting are also all labour-intensive operations. Jojoba is now being cultivated on arid land in Australia, India, Israel, South Africa and in several parts of the United States, although few, if any of these plantations are yet producing oil on a commercial scale.

JUTE   Jute is the world's most significant fibre crop in terms of quantity. It is the most widely used of all the vegetable fibres, ranking second only to cotton⌕ in global production of natural fibres. Jute is mainly grown in India, Pakistan and Bangladesh, where the main centres of production are in the Ganges–Bramaputra delta. Jute is mainly being grown on small farms which rarely exceed 10 hectares in area.

Varieties of two species are used for commercial cultivation. *Corchorus olitorius* is native to Asia and parts of Africa, whereas *Corchorus capsularis* is indigenous to southern China, from where it was introduced into India and Bangladesh. Three to five months after sowing, jute plants mature and develop stems up to 5 m high. The average yield is around 1,100 kg per hectare (ha) for *C. olitorius* and about 1,700 kg/ha for *C. capsularis*. Fibre yield is approximately 6 per cent of the green weight of the stem. The fibre is stronger dry than wet. After harvesting, the stems are retted. This is a process of fermentation and decomposition in which the tissues surrounding the fibres in the bark are disintegrated or loosened by submersion in water⌕. The loosened fibres are peeled from the stem by hand and beaten until they are separated; this is known as ginning or decortication. The fibre is spun and woven and is commonly used for making sackcloth – jute sacks do not tear even when ripped by a sharp object. Young leaves of the jute plant can also be eaten as a green cooked vegetable and the leaves are also believed to have medicinal properties.

About 3 million tonnes of jute are produced annually; India (1.2 million tonnes), Bangladesh (728,000 tonnes) and China (580,000 tonnes) are the major producers.

K

KAKADU NATIONAL PARK The Kakadu National Park, in Australia's Northern Territory, has been designated as a reserve for Aborigines⟡ and included on the World Heritage Convention⟡ list. Land in the Park is steeped in legend and aboriginal folklore. Several sacred sites and traditional cave paintings trace 20,000 years of continuous human occupation, detailing the passage of time in the area and providing a crude record of many animals that have since become extinct. The Park also contains one of the world's richest uranium⟡ deposits and has long been a battleground between Aborigines, conservationists and mining interests.

Mining is not permitted in National Parks in Australia, and as the Kakadu Park was being established – in three stages since 1979 – pressure to allow more and more mining increased annually. The mining lobby succeeded in having the Stage Three National Park proposal reduced to encompass only 65 per cent of the originally planned area. The remainder was labelled a 'conservation zone', which is open to commercial mining exploitation. One mine is currently operating and two more are planned, although the government has not granted export licences to the new operations. In total, a further 300 applications for exploration leases to prospect for uranium have been lodged. As the pressure to allow mining increases, controversy over the Park and the plight of the Aborigines continues. The Australian government revoked an original agreement to allow mining on land⟡ in the Park and deferred a final decision on the matter until 1991. Even if mining is allowed, it will now be limited to a 37 sq km area rather than the 2,252 sq km area originally designated.

KALA-AZAR The visceral form of the disease leishmaniasis⟡, caused by the parasitic protozoan *Leishmania donovani*. The parasite is transmitted to humans by the bite of sandflies. It invades the cells of the lymphatic system, spleen and bone marrow. Symptoms include enlargement and subsequent lesions of the liver and spleen, coupled with anaemia⟡, a low white blood cell count, and irregular fevers. The disease, also known as dum-dum fever, causes a darkening or blackening of the skin – 'kala-azar' is a Hindi word meaning 'black disease'. Millions of people are at risk of infection – the disease occurs in Africa, Asia, the Mediterranean region, and South America. Detection is relatively difficult and may necessitate analysis of bone marrow samples and blood tests. Drugs containing antimony are used to treat this potentially fatal disease, but the drugs themselves are extremely dangerous.

KALIMANTAN East Kalimantan in Indonesia was the scene of one of the greatest ecological catastrophes of recent times. In the spring of 1983, following an unusual drought⟡, a fire broke out in the local rainforest⟡. The fire was thought to have started while landless peasants were clearing land⟡ for cultivation. It burned for ten months, destroying over 2 million hectares of tropical rainforest along with countless species of wildlife. Wood⟡ worth $7 billion was destroyed, and the smoke created by the fire was so dense that it eventually forced the cancellation of flights into Singapore, more than 1,000 km away. The Indonesian government originally did not report the fire and attempts were made to cover up the scale of the disaster. Torrential rains finally put the fire out, but as the charred ground could not absorb the water, floods⟡ followed the flames.

KANGAROOS In their native Australia, kangaroos are widely regarded as pests. Every year the Australian government sanctions the killing of over 2 million, although they are not legally recognized as having 'pest' status. Both the skins and the meat enter world trade. The meat is used for human and pet food⟡; the skin is traditionally used worldwide in the manufacture of sports shoes and other sports accessories and equipment. Around 75 per cent of kangaroo meat exported from Australia goes to Asian and Pacific countries whereas 90 per cent of all skins commonly end up in Europe, a trade which the European Parliament has taken steps to stop.

Japan and the United States are also major consumers of kangaroo skins. Kangaroo leather is believed to have more strength-for-weight than any other leather. It is also very fine and does not scuff, making it ideal for sports footwear, whose users need suppleness allied to strength. The true status of kangaroo populations in the wild remains unknown.

Reproduction rates in several species of kangaroo actually go up in times of drought⬦ or environmental stress and the dingo, a wild dog which is the kangaroo's only natural enemy, has suffered a marked decline because of farming activities in Australia. Yet of the seven types of kangaroos hunted, population models for four have never been established. One species, the whip tail, may be in extreme danger of extinction⬦ due to indiscriminate hunting⬦.

KAPOSI'S SARCOMA   A malignant, cancerous tumour usually arising in the skin. It is common in Africa, where it occurs in a mild form, but rare in the Western world, except in people suffering from Acquired Immune Deficiency Syndrome (AIDS)⬦, where a more virulent form develops. The tumour starts as a purplish, flat and painless lesion, usually on the legs and ankles. It evolves slowly but may then spread quickly to other parts of the body. It becomes life-threatening when it invades the lower intestines, the lungs and, more significantly, the bones, which crack, leaking calcium into the blood and causing great pain. Radiotherapy is the treatment of choice, but chemotherapy may be of value.

KAUNDA, KENNETH DAVID (1924–)
Zambian statesman. Originally trained as a teacher, he joined the Northern Rhodesian African National Congress in 1949. In 1958 he founded the more militant Zambia African National Congress and was imprisoned for subversion. On his release two years later, he became President of the United National Independent Party. In 1964 he was chosen as the first Prime Minister of Northern Rhodesia, then first President of Zambia following independence. In 1973 he introduced one-party rule. Although a follower of Gandhian philosophy of non-violent opposition, he lent strong support to liberation movements operating in Rhodesia/Zimbabwe and Mozambique.

KENAF   Kenaf *Hibiscus cannabinus*, originated in tropical and subtropical parts of Africa but it is now widely grown as a fibre crop in India, Pakistan, the Soviet Union and Thailand. Annual global production is in the region of 1 million tonnes. Kenaf is an annual plant, reaching 4 m in height, and is much more hardy and tolerant to poor soils, drought⬦ and floods⬦ than jute⬦. The crop is harvested three to five months after sowing. Yields vary from an average of 246 kg per hectare (ha) in the field to 1,500 kg/ha on experimental farms. Kenaf stems are submerged in water, and the fibres are separated from woody parts of the stem and soft tissue. The fibre produced is similar to jute – coarser and less supple, but more resistant to rotting. Kenaf fibres have traditionally been used for weaving into ropes, making fishing nets, sacks, and cloth. The seeds contain 20 per cent oil which can be extracted and used as a lubricant or for lamp-oil, as well as for making soap, linoleum, paints and varnishes. Kenaf is now being grown commercially for the production of pulp to make newsprint⬦ and paper⬦.

KESTERSON RESERVOIR   A reservoir, built in California in 1971, which lies on a major migration route for several species of birds, causing it to be declared a National Wildlife Refuge. The reservoir quickly became badly contaminated by runoff drainage water⬦ from nearby irrigated farmlands. Pesticides⬦, salt and toxic wastes added to the pollution⬦, but of greatest concern was selenium, a natural element, essential in small doses but highly toxic in larger quantities. Birds at the reservoir had already begun to show signs of genetic damage and deformity before the US government's Department of the Interior ordered the reservoir to be closed in 1985. Directives were also given to local farmers to improve their drainage practices and reduce pollution levels in the effluent emanating from their farms. Plans to clean up the reservoir have not been finalized in detail, primarily because of the costs involved. To remove all contaminated silt alone would cost an estimated $150 million.

KEYNES, JOHN MAYNARD (1883–1946)
British economist who proposed that financial crises and high levels of unemployment could be avoided by adjusting demand through government control of credit and currency, basically surmising that unemployment can best be alleviated by increased public spending. He is regarded as the most important economist of the twentieth century, responsible for the field of 'macroeconomics'. His ideas may be contrasted with those of monetarism. He attended the Versailles Peace Conferences in 1919 as a representative of the British Treasury, and in 1944 he was

the chief British negotiator at the Bretton Woods Conference⌀. He published two major works, *The Economic Consequences of Peace* (1919) and *General Theory of Employment, Interest and Money* (1936).

KISSINGER. HENRY (1923–)   American politician and diplomat, born in Germany. In the administration of President Richard Nixon, he was assistant for National Security Affairs from 1969 and Secretary of State from 1973 to 1977. He undertook secret missions to China and the Soviet Union which led to détente, President Nixon's visits to both countries and a general improvement in East–West relations. He also negotiated truces between Syria and Israel in 1974, and shared the Nobel Peace Prize with Le Duc Tho for their negotiations for a peaceful settlement to the Vietnam War⌀. He also played an active part in trying to find a lasting solution to the Arab–Israeli conflict and to the Angolan and Rhodesian crises. In 1983 President Reagan⌀ appointed him to head a bipartisan commission on Central America, and he continued to play an active role in diplomacy and peace initiatives.

KOLA   Kola is an evergreen tree, closely related to cocoa⌀, native to West Africa and grown primarily in the forests⌀ of the region. The tree is valued for its stimulating properties, which are mainly due to the caffeine and theobromine content of its nuts. These chemicals are potent stimulants and are believed to reduce both thirst and hunger⌀. The tree occurs in many varieties, but only two are widely cultivated: *Cola nitida*, which is commercially important, and *Cola acuminata*, which has significant social and ceremonial uses in West Africa.

Kola trees are robust, producing dense foliage. Raised from seeds, they take five to ten years to begin bearing fruit. Production is then more or less continuous until the tree dies. The economic life of a kola tree may be anywhere from seventy to a hundred years, but most die after about thirty. Kola pods are generally harvested at monthly intervals; each pod contains six to ten nuts. The pods are opened and left with the fruit in the open, where the pulp begins to ferment. The nuts are then cured to reduce the moisture content. For commercial purposes, they are used to produce a range of drinks and beverages.

Kola is also used for medicinal purposes: in laxatives, as a heart stimulant and as a sedative. Dyes extracted from the tree are used to colour cotton⌀ and other textiles; the bark is also reputed to have several medicinal properties, and the wood⌀ is of good quality for boat-building and furniture-making. Nigeria is the world's leading producer of kola nuts, the chewing of which is widespread and commonplace throughout West Africa and Sudan, where kola nuts are also important in marriage ceremonies as symbols of friendship and fertility, as well as being used in birth, initiation and death ceremonies.

KOREAN WAR   An indecisive conflict between Communist and non-Communist forces in Korea between 1950 and 1953. In 1948 two Korean states were established on either side of the '38th parallel'. Growing friction led to forces from the North invading the South in 1950. The United Nations⌀ condemned the invasion, and sixteen member nations sent troops to support South Korea. The Soviet Union was boycotting the UN meetings at the time and could not use the power of the veto to block the UN involvement. The North Koreans were supported by Chinese forces. Northern Communist forces were driven back across the 38th parallel but refused to negotiate a peace, provoking the UN allies to advance into the North. Several peace negotiations were held from 1951 onwards, but some 5 million people had died during the conflict before a truce was signed in 1953. A peace conference held in 1954 brought no concrete agreement but the border on the 38th parallel was reintroduced, although South Korea did not take part in the ceremony.

KORUP   Africa's oldest rainforest⌀, located in Cameroon, the site of a conservation⌀ project which was originally regarded as a model to be followed in rainforests around the world. Designed to allow and promote the sustainable use of rainforest land⌀, the project is being run in the newly created Korup National Park by the Cameroon government and the Worldwide Fund for Nature (WWF)⌀, with financial backing from the British government and the European Community⌀. The project aims to protect an area of rainforest and its 8,000 species of plants and animals, while providing the local human population⌀ with land in a surrounding 'buffer zone' area from which they can derive their living.

Korup is unique in that it is home to a quarter of the world's primate species and half of the plant varieties found in Africa's tropical forests. It also has poor soils, even for a rainforest. Under these conditions, plants and animals have produced a diverse range of chemicals which could have medicinal and industrial uses. Over 90 such substances have been identified, 38 of them new to science.

Cameroon law forbids people to live in National Parks. The local population, who suffer from a high level of malnutrition⌕ and hunt local wildlife – including the endangered forest leopard – to supplement their diet, were thus moved to more fertile land in the buffer zone around the core conservation area. They are being helped to grow fruit trees, medicinal plants and food crops, and begin livestock⌕ production. A road was also planned to facilitate the transport of produce to markets. However, since the project began, concessions have been given to international logging companies to work in areas surrounding the Park, and the government has begun to allow logging in the buffer zone originally designated for agricultural purposes. Although specific directions have been given for logging, including which tree species can be harvested, the guidelines are being ignored, and split or damaged wood is being burnt. In 1990, the World Bank⌕ also reported that the logging companies involved, predominantly of European origin, had yet to pay 70 per cent of the taxes they owe.

KRILL Shrimplike omnivorous marine crustaceans, 8–60 mm long, of the order *Euphausiacea*, of which 82 species have been identified. Periodically, swarms of the dominant species, *Euphausia superba*, form in certain regions of the world, notably in the southern reaches of the oceans around Antarctica⌕. Huge swarms comprising over 2 million tonnes of krill and covering 450 sq km have been observed. During the day they appear as giant red patches on the ocean. At night their bioluminescent organs create a shimmering light.

Krill contain a great deal of protein and form an important food source for 20 species of fish⌕, 3 species of seals, over 50 species of birds, including penguins, and – most significantly – 5 species of Baleen whales. Following the whaling⌕ moratorium, whaling nations quickly began to exploit krill as a source of protein for both humans and livestock⌕. Harvesting increased dramatically: from 4 tonnes in 1962 to 520,000 tonnes in 1982 with the Soviet Union accounting for 90 per cent of the total catch. Overfishing may be the reason behind the steady reduction in catches since the peak in the early 1980s: from 233,800 tonnes in 1983 to 128,329 in 1984, and falling gradually since. The biomass⌕ of krill is estimated to be of the order of 250–600 million tonnes and, given its rapid yearly increase, a harvest of tens of millions of tonnes could theoretically be maintained annually.

KWASHIORKOR A form of malnutrition⌕ arising from a diet severely deficient in protein. It mainly occurs in young children under the age of five, although it is not common in children during the first three years of life. The related disease, marasmus⌕, is a deficiency of not only protein but also carbohydrate and fat. Both can occur together, lessening the infants' chances of survival.

Kwashiorkor, which was identified as a clinical condition in the 1930s, develops soon after breastfed⌕ babies are weaned on to an inadequate family diet. It occurs commonly in poor countries – especially parts of West Africa, where the normal diet is lacking in protein and families are large. The name kwashiorkor itself is derived from a Ghanaian word meaning 'the sickness that the elder children get when the next baby is born'. Affected children fail to grow, are apathetic, and have swollen stomachs and ankles, sparse body hair, diarrhoea and enlarged livers. Associated gastric infections are also common: the slightest infection usually proves fatal for a child suffering from this condition. Recovery is rapid and complete when children are supplied with a healthy, nutritionally balanced diet.

KYSHTYM The site of a nuclear weapons production plant in the Soviet Union, at which a serious accident occurred in the late 1950s. A concrete storage tank containing 160 tonnes of high-level radioactive waste⌕ from the production of nuclear weapons exploded following the failure of a crude cooling system. The explosion, kept secret until 1976, was revealed to have been almost as serious as the nuclear accident at Chernobyl⌕. It is believed that 20 million curies of radiation were released by the Kyshtym accident, as against 50 million curies at Chernobyl. Soviet scientists now admit that a total area of 15,000 sq km were contaminated as a result of the explosion, mostly by the long-lived radionuclide strontium-90. Most of the radioactivity⌕ was associated with large droplets and so fell to earth close to the site of the explosion. Although 600 people were evacuated from the worst-hit areas within the first week, it took between eight and twenty-two months to evacuate the final total of 10,000. Most of the land⌕ contaminated has since been returned to agricultural use but about 20 per cent has been set aside as a 'radioecological' reserve.

L

LAKE BAIKAL Lake Baikal in Siberia is the world's largest freshwater lake, holding 80 per cent of the Soviet Union's water⊃ and 20 per cent of the global total. Only the seventh largest lake in terms of surface area, covering 34,000 sq km, it is up to 1.6 km deep in places giving it a greater volume of water than any other lake. The lake is the world's oldest, revered by Buddhists as a holy sea and a spiritual entity. It also boasts a rich diversity of wildlife including 1,550 species of animals and 1,085 species of plants, of which 1,000 are found nowhere else.

Up until the middle of this century the lake was noted for the purity of its water, but that is no longer the case. The Baikalsk paper⊃ pulp factory, built nearby, has been discharging polluting effluent into Lake Baikal for several decades. Despite the fact that environmental concern led to the installation of a waste treatment system at the factory in 1969, effluent continues to be discharged. The result has been the creation of a pollution⊃ zone covering over 60 sq km. In addition, the smoke emitted from the factory's chimneys has polluted 2,000 sq km of surrounding Siberian forests⊖ and almost 35,000 hectares of fir trees have been badly damaged by the airborne pollution. The Seleninsk pulp and cardboard plant, based on the nearby Selenga River, which started to build a water-recycling system at least twelve years after it was supposed to have had the project completed, is also contributing to the ever-increasing pollution of Lake Baikal.

Environmentalists had hoped that the Baikalsk factory, which produces around 200,000 tonnes of cellulose fibre annually, would be closed by 1991 but it will continue operating at least until 1993, and probably beyond that. Effluent from the factory is now to be carried away by a new 71-km pipeline (costing $155 million) and discharged into the River Irkut. Although this may reduce pollution levels in Lake Baikal somewhat, conservationists fear that the scale of the investment in the new pipeline will mean that the factory will not close after all.

LAND There are an estimated 13 billion hectares (ha) of land above sea level. Of this, approximately 1.5 billion ha are being used to grow crops. This total has increased by 9 per cent since 1965, whereas the land set aside for pasture – currently 3 billion ha – and the world's forests⊖ and woodland – now covering some 4 billion ha – have both been declining steadily over the same period. In all, well over 8 billion ha of land are capable of being agriculturally productive to some degree, or around 1.6 ha per person – a ratio that is fast diminishing. The geographic variation in land exploitation is substantial and may be an indicator of future food shortage problems. In Asia, an estimated 82 per cent of cropland is already being cultivated, whereas in Africa, where 30 per cent of the continent's 840 million ha of land is potentially cultivable, only around one-third is being cultivated.

Since 1945, as a result of the burgeoning human population⊃, the *per capita* amount of cropland has fallen by one-third. An increase in cropland of only 2.7 per cent between 1976 and 1986 did not keep pace with the expansion of the human population and continuation of this state of affairs is projected to result in a further fall to around 0.17 hectares *per capita* by the year 2000. Under present conditions, every human requires about 1 ha of fertile land to ensure an adequate diet.

LAND EQUIVALENT RATIO (LER) An agricultural science term denoting the amount of land⊖ planted in monoculture⊃ that would be needed to achieve the same yield produced by a mixture of crops. Researchers in Mexico found that 1.73 hectares (ha) of land would have to planted with maize⊃ to produce the same amount of food as 1 ha planted with a mixture of maize, beans and squash. Measured in terms of total biomass⊖ produced, the ratio increased to 1.78 ha of monoculture. In terms of total food production, in the majority of cases where comparisons have been made, mixed cropping, or intercropping, outperformed monoculture.

LANDLESS LABOURERS In 1988, a report from the Food and Agriculture Organization (FAO)⊖ warned that 'landlessness is emerging as possibly the single largest agrarian problem in the developing world'. Over 75 per cent of the world's population⊃ live in the developing countries⊖, mostly in rural areas. Although at least 70 per cent of these people are small-scale farmers, living off the land◇, many of them are effectively landless – a term which describes a variety of situations. Millions of subsistence farmers in the developing world do not own any land, not even the land where their family dwelling stands. These people form the majority of the agricultural wage-labour force. Tenant farmers and sharecroppers work the land owned by someone else, usually a wealthy landowner, paying rent in the form of a large proportion of the crops they raise. Slash-and-burn◇ cultivators follow a nomadic lifestyle, clearing a plot of land that has no apparent owner and growing crops until it becomes unproductive, then moving on to repeat the procedure elsewhere. Many other people are deemed to be functionally landless. They may have title or customary rights to land, but it is either insufficient to support them and their families or their land tenure may be rendered worthless through other factors, such as the absence of water◇ rights. Similarly, women have no legal right to land in some countries but continue to work it when their husbands are away or have died.

In many countries where landholding patterns derived from colonial days still persist, small-scale farmers are excluded from vast tracts of land, usually the best and most fertile, even though small-scale operations produce much more per unit of land than commercial or industrialized farming systems◇, primarily because subsistence farmers cultivate every available area and devote more time to the task. In developing countries◇ the shift in land-tenure patterns favouring large farms is often the result of deliberate government policy geared to developing cash crops◇ for the export market. Meanwhile population◇ pressure, coupled with a limited supply of even remotely productive land, causes overuse and destruction of already marginal areas. Cultivable land shrinks as generations of misuse deplete its productive capacity. Fallow periods may be reduced through sheer necessity, with the soil becoming more and more exhausted and less productive in the process, swelling the total of landless. The average size of family-held units is shrinking throughout the developing world. In Africa and Asia during 1970, farms averaged just under 2 hectares (ha); by 1980 this had shrunk to around 0.5 ha, the size declining with increasing population numbers.

Landless and functionally landless farmers tend to have no collateral and are therefore denied access to the credit needed to buy the seeds, fertilizer◇ and equipment required to improve their situation. The result is a downward spiral into deepening poverty and malnutrition◇ or migration in search of paid employment, causing a massive drift from rural to urban◇ centres, or movements, legal or otherwise, across international borders. Many low-paid, unskilled labourers in the developing world are from a landless, rural background.

LAND USE Current patterns of land use and rate of expansion in global agricultural output will not be able to satisfy the future demands of the world's expanding human population◇. Questions also remain over whether improvements in agricultural productivity can be achieved, or even if current production levels can be maintained in the long term.

Agricultural sustainability and what level of land use is viable are difficult to define. Long-term viability is often sacrificed in the desire for short-term yield increases and profits. But productivity without sustainability is mining. Natural resources◇ have finite limits and can easily be exhausted. Modern, intensively managed agricultural systems differ from natural ecosystems◇ in many ways, but there is one basic rule that cannot be ignored – the productivity of a system cannot be maintained if an unacceptable burden is placed on the natural-resource base. Yet in pursuit of agricultural productivity improvements, farmers all over the world are depleting renewable resources such as soils, water◇, and forests◇ at ever-increasing and alarming rates.

Fertile land◇ is an essential requirement for any agricultural system, yet of the 1.5 billion hectares (ha) of land in the world being farmed, only about half is considered to be really suitable for agriculture◇. Paradoxically, at a time when more agricultural production is needed, vast quantities of valuable land are being destroyed by human activities. More and more marginal land is being cultivated, often by impoverished people using inappropriate methods and technology. Overgrazing◇ and overcultivation are leading to serious soil erosion◇. This problem is most acute in the tropics, where a combination of increasing slash-and-burn◇ operations and the timber industry is denuding the land of forests◇ and vegetation. Rain erodes unstabilized soil when it falls at a rate greater than 25 mm per hour. In the tropics, at least 40 per cent of rainfall exceeds this rate. Consequently, removal of vegetation simply worsens an already unacceptable rate of soil loss.

The lessons of the past are being ignored. Against the 1.5 billion ha used for crop production, nearly 2 billion have been ruined. According to a joint report from the United Nations Environment Programme (UNEP)⌀ and the Food and Agriculture Organization (FAO)⌀, around 6 million ha of cultivable land are being lost each year through soil degradation. If present trends continue, newly cultivated land will not replace that being lost. The FAO's *Agriculture: Toward 2000* report calculated that soil- and water-conservation⌀ measures would need to be extended to 25 per cent of all farmland by the end of the century, and that flood⌀ control will need to be extended to 20 million ha, merely to achieve equilibrium. It is estimated that these measures will cost in the region of $2.5 billion – which is unlikely to be forthcoming.

Worldwide, in addition to land lost to erosion, desertification⌀ and toxification, a further 8 million ha are lost annually to non-agricultural conversion through the building of roads, houses and urban⌀ sprawl. In the United States, 10 million ha of land are designated as military training areas. Consequently, agricultural production is having to rely more and more on fossil fuels⌀ to produce the agricultural chemicals essential to maintain soil fertility on the farming land that is still being used – a situation which cannot last indefinitely.

The need to regulate land use is generally being ignored. Japan is the only country in the world that has drawn up a nationwide land-use plan. A zoning programme, introduced in 1968, divided the entire country into Urban, Agricultural, or Other land. In 1974 the scheme was refined to include Forests, National Parks and Nature Reserves. This policy has helped a nation of 120 million people living on predominantly mountainous land, in an area smaller than the US state of California, to become not only self-sufficient in food but also an exporter of the main agricultural product, rice⌀.

LANGUAGE The chief mechanism by which humans communicate with each other. It is estimated that there are 4,000 different languages spoken in the world today, many more have died out with the passing of time and peoples. Of the so-called 'mother-tongue' languages, over 1 billion people have Chinese as their mother tongue, although Chinese is more correctly a collection of eight different languages. Of the other languages, 350 million people have English as their mother tongue, 250 million Spanish and 200 million Hindi. In reality, English proves to be the world's most common language as it is the 'official' language in a variety of countries throughout the world, including several

nations that speak another mother tongue. With regard to official languages, English covers 1.4 billion people, Chinese 1 billion, Hindi 700,000 and Spanish 280,000.

LASSA FEVER A little-understood, severe viral disease endemic in West Africa, first described in 1969 and named after the village in Nigeria where it was originally discovered. It is a rare but fatal disease, transmitted to humans by certain species of rat. After an incubation period of one to three weeks, headache, high fever and acute muscular pains develop. There is often difficulty in swallowing. Death, mainly from kidney or heart failure occurs in well over 50 per cent of cases. Treatment with blood plasma obtained from patients who have recovered is the best available therapy.

LATIN AMERICA Name applied to those countries of Central and South America lying south of the United States, including Mexico and those islands of the West Indies where a Romance language is spoken. Spanish is the most widely used language, although Portuguese is spoken in Brazil, the largest country, and French in Haiti and French Guiana.

There are several major regional development organizations which encompass these countries, including the Central American Common Market⌀ (1960), the Andean Group⌀ (1969) and the Latin America Integration Association (1980). The Latin American Economic System (SELA) was established in 1975, but this incorporates many Caribbean states that are not, strictly speaking, part of Latin America. The Latin American region has a population⌀ of 450 million and a Gross Product of around $1 trillion.

LATIN AMERICA FREE TRADE ASSOCIATION (LAFTA) Formed in 1960, LAFTA aimed eventually to remove all restrictions on trade among member countries – Argentina, Brazil, Bolivia, Chile, Colombia, Ecuador, Mexico, Paraguay. Peru, Uruguay and Venezuela. Based in Uruguay, the Association worked to promote full-scale regional economic development in Latin America. It made relatively little progress towards forming an effective common market, and in 1980 the same eleven states agreed to relaunch their efforts, replacing LAFTA with the Latin American Integration Association, which was devoted to achieving the same end.

LAW OF THE SEA   The United Nations Convention on the Law of the Sea (UNCLOS) was adopted in 1982. The Convention was signed in Montego Bay by 131 countries, following the participation in discussions of 150 states. It contained many provisions that were originally decided at the earlier Geneva Conferences on the Law of the Sea held in 1958 and 1960, in which only 86 nations took part.

Around 70 per cent of the earth's surface is covered by the seven seas, most of which are 'open' sea and not under any country's legal jurisdiction. These seas contain vast natural resources including fish⇨ stocks and seabed minerals⇨. The 1958 and 1960 discussions failed to provide solutions to the myriad problems, but lengthy negotiations between 1974 and 1982 finally resulted in the new code of sea law. The Convention authorized a twelve-nautical-mile territorial limit extending seawards from every country with a coastline, the area being officially under the jurisdiction of the country concerned. It also adopted the concept of a 200-nautical-mile Economic Exclusion Zone (EEZ)⇨ extending further seawards in which the nations concerned could exercise control over the exploitation of certain natural resources⇨. Under this system, a third of all oceans are brought under the jurisdiction of coastal states, including the whole of the Caribbean and the Mediterranean. Tiny developing island states in the Pacific Ocean can now control sea zones of over 335,000 sq km.

The Law of the Sea upholds the right of free navigation within all EEZs and gives nations stronger powers to regulate pollution⇨ or exploitation of natural resources in their waters. In its 320 articles, maritime zones are defined, legal rights are assigned, and the mechanism for settling international disputes is provided for. Minerals on the deep ocean floor or outside EEZs are declared common heritage. Because of the question of rights to exploit these deep-sea mineral resources, the Federal Republic of Germany, the United Kingdom and the United States withheld their signatures. The treaty has been virtually becalmed because of this opposition to the formation of an 'authority', as proposed, that would oversee the mining of the seabed and distribute any proceeds among the landlocked and least-developed nations⇨. By 1990 only 44 nations had ratified the Convention, 16 short of the required minimum, although much of the treaty has already become customary international law.

LEAD   A poisonous, heavy, dense, soft, blue-grey metal, used by humans since prehistoric times. It is extremely resistant to corrosion, malleable, and has been used extensively since Roman times for pipework, especially for plumbing. Lead was later found to be a powerful neurotoxin, even in trace quantities, and lead pipes, particularly in soft-water areas, have become a major problem and have gradually been replaced by plastic piping. Lead is used for a variety of industrial purposes: in shields to protect against X-rays, in batteries, the manufacture of ammunition, paints, glass production, and as an additive to petrol. Airborne lead from petrol fumes is toxic, particularly to young children. The addition of tetra-ethyl lead to petrol in order to prevent engine knocking has caused universal distibution of lead in the environment, with extremely high concentrations in the atmosphere and dust in urban areas with dense traffic flows. Acute lead poisoning causes diarrhoea and vomiting, but poisoning is more often chronic, characterized by abdominal pain, muscle pains, anaemia⇨ and nerve and brain damage. Lead retards the development of young children, impairing mental function and learning ability.

Around 5.7 million tonnes of lead are produced each year for industrial purposes; the United States (1.1 million tonnes), the Soviet Union (800,000 tonnes) and the United Kingdom and the Federal Republic of Germany (400,000 tonnes each) are the major producers.

LEAF PROTEIN   The leaves of a huge variety of crops and other plants are discarded or used as a mulch. Where food – or, more significantly, protein – is scarce, leaves could fulfil a normal daily requirement, especially in parts of the world where protein shortage causes widespread malnutrition⇨ and hunger⇨. Treated leaves, from the majority of plants that are safe to eat, can provide a protein-rich curd containing significant amounts of Vitamin A, iron⇨ and calcium. In addition, a fibrous material useful for livestock⇨ fodder, and a sugary whey which can be fermented to alcohol or used as a fertilizer⇨ to supply potassium and phosphorus, are also produced. The use of as little as 15–20 g (dry-weight) of leaves per day could provide a cost-effective part of a child's nutritional requirements, and large-scale leaf-protein production projects are already established and under evaluation in Bolivia, Ghana, India, Kenya and Sri Lanka.

LEAGUE OF ARAB STATES *See* Arab League

LEAGUE OF NATIONS An international organization, the first of its type, created in 1920 with the purpose of achieving and maintaining world peace. Following the conclusion of the First World War, the League's covenant was incorporated in the postwar peace treaties. The failure of the United States to ratify the Treaty of Versailles meant that the USA was excluded from the League. The League settled minor disputes and organized international conferences, but in the 1930s it was unable to deal effectively with major international incidents such as the Japanese intervention in China and Italy's intervention in Ethiopia. Germany withdrew in 1933 and the organization gradually became less effective in reaching its objective. After the Second World War, the League was superseded by the United Nations◇.

LEAST DEVELOPED COUNTRIES (LLDC)
A category adopted by the United Nations◇ to describe low-income, commodity-exporting developing countries◇ with low industrial bases. During the 1960s, twenty-four countries were identified as having particularly severe long-term constraints on their development, originally based on calculations of three basic criteria: *per capita* Gross Domestic Product (GDP) of $100 or less (at 1970 prices), manufacturing that contributed 10 per cent or less of GDP, and 20 per cent or less of literate people aged fifteen or over. These twenty-four were recognized by the United Nations Conference on Trade and Development (UNCTAD)◇ in 1968 as LLDCS, the category being adopted by the UN General Assembly in 1971.

At a special Conference on the Least-Developed Countries held in 1981, donor nations pledged to double their official development assistance to countries classified as LLDCS. By 1989, the category had been extended to encompass forty-two countries: Afghanistan, Bangladesh, Benin, Bhutan, Botswana, Burkina Faso, Burundi, Cape Verde, Central African Republic, Chad, Comoros Islands, Djibouti, Equatorial Guinea, Ethiopia, the Gambia, Guinea, Guinea-Bissau, Haiti, Kiribati, Laos, Lesotho, Malawi, Maldives, Mali, Mauritania, Mozambique, Myanmar, Nepal, Niger, Rwanda, Samoa, São Tomé e Principe, Sierra Leone, Somalia, Sudan, Tanzania, Togo, Tuvalu, Uganda, Vanuatu, Western Samoa, Yemen and Yemen Democratic Republic.

In 1990, the Second UN Conference on the Least Developed Countries was attended by 150 nations. Delegates were informed that promises made by donor countries in 1981 to double their official development assistance had not been upheld: only eight nations had reached the target. The Conference adopted a new Programme of Action which placed greater emphasis on human rights◇, the need for democratization and privatization, and the potential role of women in the development process, and stressed the need for stringent population control policies as a fundamental factor in promoting development. Under the plan, donor nations were given four distinct levels of commitment for increasing their official aid to the LLDCS. A review of the criteria used to designate nations as LLDCS was also called for, together with the possible inclusion of Namibia within the group.

LEECH Leeches are blood-sucking annelid worms related to the common earthworm. Most species live in ponds or streams, only a few living in moist vegetation on land◇. Two suckers, one either end of the animal's body, enable the worm to crawl and attach itself to its host. Leeches are capable of taking in massive blood meals of up to nine times their body weight. Having done so, they may not feed again for months afterwards; satiation can last up to a year.

Hundreds of years ago, blood-letting using leeches was a common treatment for most human ailments. This practice was discarded long ago as being unhygienic and unscientific, a viewpoint that has now been disproved. Leeches produce chemicals in their saliva which act as agents capable of counteracting several diseases and human conditions, including atheriosclerosis, thrombosis and cancer◇. The most powerful known anti-coagulant, hirudin, was discovered in leeches over a hundred years ago. Amongst the other agents found in leech saliva are two fibrinase enzymes: one causes blood clots to break up; the other is thought to be effective in breaking up atherosclerotic plaques which build up in arteries, leading to arterial disease and heart attack. Extracts from leech saliva have been found to interfere with the growth of lung tumours and prevent cancerous cells from grouping together.

Doctors performing intricate and delicate microsurgery are now relying on leeches to ensure that their work is successful. After transplanting or reattaching human limbs or tissue, the blood supply at the site of the operation often fails to return to the heart: it builds up and has a secondary effect of restricting the flow of fresh blood. This combination results in the death or decay of the tissue. Virtually all attempts by doctors, using both chemical and mechanical means, to prevent necrosis once circulation problems have arisen have failed. Surgeons are now using leeches to remove blood from the operation site and keep blood flow regular. Leeches suck up the blood, and in so doing provide enough time for capillaries to grow across the sutures and normal blood circulation to resume.

Biotechnologists have already devised means of producing hirudin in large quantities, and the world's first leech farm has been established in the United Kingdom; over 30,000 animals are kept to try and satisfy the growing demand.

LEGIONNAIRE'S DISEASE A bacterial infection of the lungs caused by the bacterium *Legionella pneumophilia*. The disease was first identified and named after an outbreak of 182 cases at the American Legion Convention in Pennsylvania in 1976. Thirty of those infected died. Symptoms appear after an incubation period of about a week. Diarrhoea, malaise and muscle pain are followed by fever, a dry cough, chest pains and hallucinations. Proper functioning of the kidneys is also impaired. Antibiotics provide the most effective therapy.

Since its discovery the disease has been predominantly reported from developed countries, notably in North America and Western Europe, where it is closely linked to hot-water and air-conditioning systems. The bacterium is airborne, transmitted through humidifiers and respirators filled with tap water⌁ as well as through showers, baths, and health spas. Air flow and temperature are important factors in determining its rate of spread. The disease generally strikes males around the age of fifty; those who smoke and drink immoderately are at greatest risk. In 1989 the World Health Organization (WHO)⌁ predicted that the disease will soon emerge in the Third World⌁ as more and more developing countries⌁ urbanize and develop facilities to cater for their growing tourist⌁ industries.

LEGUMES Leguminous plants are found throughout the world, the greatest variety in the tropics and subtropics. With up to 18,000 species of herbs, shrubs and trees, the Leguminosae family is the third largest of all flowering plants and one of the most important to humans. Of all the legume species known, fewer than twenty are cultivated

extensively. Those in common use include groundnuts⌁ (peanuts), soya beans, various peas and beans, and several varieties of clover. In addition, several species of leguminous trees are widely used.

The most important feature of the legumes is their ability to convert nitrogen⌁ gas from the air into inorganic nitrogen compounds which enrich the soil and can be readily used by plants. Their nitrogen-fixing contribution can be vital for maintaining soil productivity over long periods, and a leguminous crop can add up to 500 kg of nitrogen to a hectare (ha) of land⌁ annually. Today, cultivated legume crops still add more nitrogen to the soil worldwide than do man-made fertilizers⌁. In Australia over 100 million ha of land have been brought into cultivation thanks to the use of leguminous plants.

Legume seeds, in the form of beans or pulses, are second only to cereals as a source of human and animal food. Before the potato⌁ was introduced into Europe, beans constituted the bulk of the diet of the poorer classes on the continent. Nutritionally, they contain 2·5 times as much protein as cereal grains. Legume seeds remain major sources of food for millions of people in Latin America, the Indian subcontinent and throughout Asia. Despite the fact that there is a chronic protein deficiency in virtually every developing country, the Green Revolution⌁ has concentrated on cereals and not directed much research into increasing the yield of pulses.

Several species of leguminous trees, such as acacia⌁ and leuceana, are also able to fix nitrogen, and with their protein-rich foliage, pods and seeds, and a general hardiness and tolerance of drought⌁ conditions, they prove extremely beneficial in providing fodder for livestock⌁ in arid regions. In addition, they are fast-growing, stabilize soil, provide fuelwood⌁ and lumber, improve soil⌁ fertility and are the source of a number of gums which enter international commercial trade.

LEISHMANIASIS A parasitic disease, prevalent throughout the tropics and subtropics, caused by microscopic protozoans (*Leishmania spp.*) and transmitted to humans via the bite of sandflies. Symptoms range from anaemia⌁ and enlargement of the liver and spleen to ulcerative growths that eat away the nose, mouth and throat. The disease occurs in two principal forms. The most dangerous, visceral leishmaniasis, (known as kala-azar⌁), damages the internal organs, principally the spleen and liver. The less hazardous cutaneous leishmaniasis affects the skin tissues, causing open sores or ulcers. Cutaneous leishmaniasis itself occurs in several different forms, depending on the region and the species of *Leishmania* involved. Most cutaneous forms self-cure after skin lesions have erupted, giving the patient a degree of immunity to the particular form of the disease that caused the original infection. In parts of Israel and the Soviet Union, where leishmaniasis is endemic, people are deliberately infected so that the site of skin lesions can be controlled to avoid disfigurement or loss of productivity, and so that immunity is developed.

There are 12 million cases of the disease in most parts of the world at any one time. Around 350 million people, spread throughout eighty countries, are at risk of infection, and over 400,000 new cases are reported each year. The visceral form is treated with drugs containing antimony, which destroy the parasites but are themselves dangerous.

LEPROSY A chronic disease, particularly prevalent in tropical countries, which appears in several distinct forms, affecting the skin, mucous membranes and nerves. It is caused by a bacterium, *Mycobacterium leprae*, which is related to the organism that causes tuberculosis⟲. Leprosy is contracted only after close contact – usually sneezing or skin contact – with an infected person. After infection, the incubation period is usually one to three years, and symptoms appear slowly.

Lepromatous leprosy is a contagious, steadily progressive form of the disease characterized by the development of widely distributed lumps on the skin, thickening of the skin and nerves and, in severe cases, muscle weakness leading to deformity and paralysis. The eyes, bone and muscle may also be affected. Tuberculoid leprosy is a benign, often self-limiting form causing discolouration and disfiguration of patches of skin, usually associated with a localized loss or diminution of nervous sensation in affected areas. Indeterminate leprosy is a form in which skin manifestations represent a combination of these two types.

Leprosy can be controlled, but not cured, with potent sulphone drugs. Dapsone, developed in the 1940s, stopped bacteria from multiplying but did not kill them, and dapsone-resistant strains of leprosy soon began to appear. Treatment with a mixture of dapsone and two potent anti-leprosy drugs, rifampicin and clofamizine, proved to be a cure, even for those suffering from dapsone-resistant strains – over 850,000 people have been cured so far.

There are an estimated 11 million cases of leprosy in the world. The disease occurs in 121 countries, with 1.6 billion people potentially at risk of contracting it. Improved living standards have eradicated leprosy from most developed countries but it is still prevalent in parts of the Third World⟲ and in groups of people in developed countries, such as the Australian Aborigines⟲, whose living conditions are still poor.

LIE, TRYGVE HALVDAN (1896–1968) Norwegian politician and international civil servant. He was appointed the first Secretary-General of the United Nations in 1946 and served until 1952. During his term of office he dealt with the Arab–Israeli War and the UN involvement in the Korean War⟲. Opposition from the Soviet bloc to his policies on armed support for the South Koreans eventually led to his resignation.

LIFE EXPECTANCY Most commonly published figures refer to life-expectancy at birth. This indicates the number of years newborn children should live, subject to the mortality risks prevailing for the cross-section of their national population⟲ at the time of their birth. Modern medicine, science and technology all act to prolong life; consequently, life-expectancies for most people in the world have been steadily increasing. By the end of the 1980s, on average, a newborn child could expect to live to sixty-one, whereas one born ten years earlier could expect a lifespan of only fifty-five years. These figures compare with an average life-expectancy in 1950 of a mere forty-six years.

The success brought about by improvements in living standards, health services and medical advances over the last forty years has been reflected in rising life-expectancies everywhere. Even in the Third World⟲, average life-expectancy has risen from forty-one to almost sixty years.

The place of an individual's birth has a profound influence on determining the age of death. To indicate the stark differences between the developed and developing worlds: average life-expectancy in Africa is now slightly over fifty-one years, whereas a child born in Europe is likely to live to the age of seventy-four. Globally, longevity still tends to correlate with *per capita* income levels. Inhabitants of poor countries have the shortest life-expectancy; those in Afghanistan, Ethiopia and Sierra Leone at between thirty-six and forty years, have the lowest. Those in rich nations have the longest: life-expectancy in Japan is highest, at around seventy-seven years.

LIMITS TO GROWTH A report, published in 1972, which provided an in-depth investigation of the relationships between human populations and the resources they depend on for their survival. Sponsored by the Club of Rome and the Massachusetts Institute of Technology, the investigation involved the construction and operation of a series of computer models which, based on historical data and application of current knowledge, allowed a projection of future events.

Limits to Growth warned of the finite nature of the earth and depicted the varying possibilities and problems of supporting a rapidly increasing human population⬦. The somewhat gloomy forecasts prompted a wave of debate about the future. The publication of the report itself followed a decade or more of growing awareness of natural ecology⬦ and the place of the human race within this system, all of which appeared to stem from pictures taken by the Apollo space missions in the mid-1960s which graphically depicted the earth as a 'spaceship' whose inhabitants were obviously dependent on limited supplies of naturally occurring resources. Consumerism, technology, profit, growth and power began to be seen as negative factors which served to destroy and degrade these natural resources⬦ and threatened the long-term survival of natural environmental equilibria. Four major models, incorporating several gross assumptions about variations of critical components, formed the bulk of the report, with results extrapolated to the year 2100:

Standard run – Model run with no major changes in the physical, economic or social relationships that govern the prevailing world system. The population eventually drops, mainly due to food shortages and depletion of resources.

Stabilized population and capital – Assuming that population numbers and capital can be controlled, a temporary stable state is reached, but resources continue to be depleted.

Reserves of natural resources doubled – Increased supplies and availability of resources allow for greater industrialization. However, this causes pollution⬦ to rise, leading to an increased death rate⬦, decreases in food production and an overall depletion of resources.

Equilibrium – When pollution control, recycling, systems for protecting and restoring infertile soils, lower industrialization and other measures are included, together with a stabilization of global population and capital, a balanced state is produced which can exist far into the future.

Although *Limits to Growth* aroused considerable controversy by producing a basic conclusion that 'the limits to growth on this planet will be reached in the next hundred years', detailed work published later in a little-heralded Technical Report showed that it was possible to choose parameters through which population, capital and material growth could all be sustained well past the year 2100.

LINGUA FRANCA Any language which serves as a medium for communication between speakers with different native languages. It may be a hybrid between two languages, such as 'pidgin' English, or an established language such as French, which was once the lingua franca of diplomacy. The name originates from the language used by Mediterranean traders in the Middle Ages.

LIVESTOCK Humans have domesticated animals for thousands of years. Cattle were first used as livestock over 5,000 years ago and are still seen as a means of making poor or fragile land⬦ productive. By 1990 there were an estimated 1.26 billion head of cattle in the world, an increase of 6 per cent over the 1975 number. Of the total, 19 per cent were spread throughout South America and 12 per cent on the continent of Africa. However, there are more cattle in India than in any other country, some 15 per cent of the global total. The cattle in India, which are sacred, are used either as draught animals or for hide.

Although a wide variety of animals are reared as a source of food, clothing or other materials, cattle, sheep, goats, pigs and chickens are by far the most common. By the end of the 1980s an estimated 1.6 billion sheep and goats, 800 million pigs and 8.6 billion chickens were being reared around the world. The impact of these animals on the environment and the costs of feeding them are high.

Sheep are notoriously damaging to poor soils as they trample and destroy vegetation while feeding, and million of hectares of land in Australia have suffered as a result. Goats browse on small shrubs and low trees, eating virtually all forms of vegetation, including several that many other animals will avoid. Yet, of the world's goats, one in three are found in the fragile conditions on the African continent, where they contribute greatly to the desertification⬦ process. In many parts of Africa, where livestock is viewed as a symbol of wealth and social status, nomadic herders traditionally employed elaborate calendars and rotation systems in order to graze millions of livestock on marginal land without destroying its productivity. The creation of artificial international borders, coupled with increasing pressure on land and growing human numbers, have resulted in the livestock being raised intensively on smaller tracts of land, with consequential over-grazing◇, increasing the threat of desertification.

Livestock-rearing deprives people of significant amounts of valuable agricultural land and food. When the costs of cereals, traditional food for live-stock in Europe and North America, rose during the 1970s, vast quantities of cheap cassava◇ were imported from Asia as a replacement. Huge tracts of land in Thailand and elsewhere were devoted to growing the crop in the process. In the United States, of the massive quantities of grain produced – over 1,000 kg for every person – 86 per cent is fed to animals.

The Netherlands is faced with a more pressing livestock problem. It is one of the most densely populated countries in the world: 14 million people share 40,000 sq km with 17 million cattle and pigs. The livestock produce more than 100 million cu m of manure every year, the disposal of which has yet to be satisfactorily tackled.

LIVESTOCK FOOD *See* Pet Food

LOCUST
A term which covers a dozen or so species of orthopteran grasshoppers which have swarming or migratory characteristics. These insects have two distinct phases, a solitary phase and a migratory phase. When environmental conditions are favourable, immature forms of the insects, called nymphs or hoppers, crowd together. As numbers increase, the solitary-phase insects change their behaviour, shape and colour and mature into the gregarious, locust-like migratory phase. These changes are under hormonal control and are triggered by the stimulus of overcrowding. Migratory locusts form swarms which travel over long distances and devour all crops and vegetation upon which they settle. The most damaging variety is the migratory locust (*Locusta migratoria*) found throughout Africa and South-West Asia.

When conditions are right, prodigious swarms assemble. One of the largest ever recorded covered 1,000 sq km and contained an estimated 40 billion locusts which ate up to 80,000 tonnes of food per day. This swarm had a biomass◇ equivalent to that of about 1 million people. Swarms can move 3,000 km in a month devastating vast tracts of land◇, and as much as 20 per cent of the earth's land surface,

mostly in Africa and the Middle East, are period-ically invaded by these swarms. The desert locust is known to invade fifty-seven countries covering 29 million sq km. In a single day, an average swarm can eat the amount of food, of all vegetative types, that would feed an estimated 400,000 people for a year. Every year locusts cause $10 billion worth of damage to crops in Africa and the Middle East.

In 1988, the worst locust outbreak for thirty years swept across northern Africa and penetrated as far afield as Iran, southern Europe and the Caribbean islands. Over $200 million was spent on anti-locust operations, spread mostly over 14.5 million hectares of land in Africa.

For a week or so before they take to the air, locusts 'march' along the ground, covering 1–2 km per day and consuming all the vegetation as they proceed. It is during this period that control measures can best be applied. Attempts at control depend on international co-operation and a very early warning system is essential, for once the young hoppers have undergone the final moult into winged adults, it is too late to prevent swarm damage.

LOMÉ CONVENTION
A convention first signed in 1975 in the capital of Togo which pro-vided for trade concessions between the countries of the European Economic Community◇ and forty-six nations from African, Caribbean and Pacific (ACP) countries◇. Later conventions were signed in 1979, 1984 and 1989, encompassing a series of comprehensive aid◇ and trade agreements. The Lomé Conventions succeeded the Yaoundé◇ Conventions. They create duty-free access to the Community for a wide range of goods and tropical products from ACP states, the first creating special arrangements for bananas◇, sugar◇ and rum. Part of a $3.2 billion aid package from the European Community (EC)◇ was also set aside to finance a System for Stabilization of Exports (STABEX)◇ scheme, designed to compensate ACP nations for loss of export earnings on key commodities.

Lomé II offered improved access to the EC for different ACP agricultural products, extended STABEX and established SYSMIN, a similar system of compensation to assist those ACP states whose eco-nomies were heavily dependent on mineral exports. New provisions on industrial and agricultural co-operation were also included. New commitments on the promotion and protection of investments and recommendations on the fishing◇ industry were also made. Sixty-four ACP states signed Lomé II.

Lomé III further liberalized trade regimes and included provisions to improve the effectiveness of EC aid, of which $8 billion was made available.

Under Lomé IV, which took two years to negotiate, measures to alleviate drought⌀ and curb desertification⌀, and social and cultural projects, were strongly emphasized. Lomé IV will run for ten years compared with the previous Conventions, which had five-year terms, and will have a total $13.7 billion of EC aid over the first five years. The aid package comprises $12.3 billion from the European Development Fund (EDF)⌀; $1.3 billion for structural adjustment⌀; $1.7 billion for compensating ACP nations for loss of export earnings; and $1.4 billion from the European Investment Bank (EIB)⌀. Although Lomé IV also provided for improved access to the EC market for ACP products, debt⌀ problems to be met by converting some loans into grants, special environmental considerations, such as the prevention of exports of toxic and radioactive⌀ waste to the ACP states, contained stronger human rights⌀ elements and specific mention of the need to fight apartheid⌀, it fell short of the ACP countries' expectations. By 1990, the Dominican Republic and Haiti had joined the sixty-six existing ACP signatory nations, and Namibia was expected to become the sixty-ninth member of the grouping following its independence.

LONDON DUMPING CONVENTION (LDC)
Common name for the Convention on the Prevention of Marine Pollution by Dumping of Waste and Other Matter. Originally signed by fifty-eight countries in 1972, the LDC – basically a United Nations⌀ body – came into force in 1975 and has since been ratified by sixty-three countries. The international treaty covers all the world's oceans with the main aim of controlling the dumping of waste from ships and aircraft. Under the Convention, a 'black' list of compounds has been drawn up, the dumping of the materials on this list being strictly prohibited. The list includes organohalogen compounds, mercury⌀ and related compounds, cadmium⌀ and related compounds, persistent plastics, crude oil⌀, fuel oil, high-level radioactive waste⌀ and materials produced for chemical warfare⌀ or biological warfare⌀. The 'grey' list covers materials that may be dumped, but only with special permission; these include lead⌀, copper⌀, arsenic, fluorides⌀, and pesticides⌀ or their by-products.

Several clashes have occurred over the dumping of nuclear waste and incineration⌀ of wastes at sea. In 1985, twenty-five nations, led by Australia and New Zealand, voted for an indefinite ban on the dumping of nuclear waste, although several, including France, the United Kingdom and the United States, opposed the idea. Between 1946 and 1952, vast quantities (46 petajoules) of packaged, unpackaged and liquid nuclear wastes were dumped at over fifty sites in the northern sectors of the Atlantic and Pacific oceans, predominantly by the United States and Britain.

In late 1989 the scope of the LDC's worldwide moratorium was extended to include the dumping of decommissioned nuclear submarines at sea – the original moratorium did not apply to warships. The following year members agreed to stop the disposal of all industrial waste at sea, despite initial opposition from France, the Soviet Union, the United Kingdom and the United States. From 1995 onward, the dumping of all forms of industrial waste will be prohibited outside territorial waters. It will be the responsibility of member countries to prosecute their own ships which transgress the Convention. The LDC has also put into motion plans to establish a global mechanism to control land-based emissions, which are responsible for 80 per cent of marine waste, even though the Convention's remit covers only dumping from ships.

LOVE CANAL
The location of a landfill chemical disposal site near Niagara Falls in the United States, first excavated in the late nineteenth century as part of a hydroelectric⌀ and industrial project. It was used by the Hooker Chemicals and Plastics Company as a dumping site for its wastes: over 43,000 tonnes of chemical waste, much of it carcinogenic, from the early 1940s to 1952.

In 1953 the site was sold to the local Board of Education for $1, on condition that the company was absolved of any future liability that might arise from the chemicals in the dump. A school was built, followed by a housing estate. In 1977 the dump was found to be leaking. Air, soil⌀ and water⌀ around the site were all heavily contaminated with carcinogenic and toxic chemicals⌀. In 1978, the New York State Commissioner for Health ordered the evacuation of 240 families; the dump site was fenced off and declared a Federal Disaster Area.

Two years later the pollution⌀ was found to be worse than first feared and more families were evacuated. Other sites nearby are now known to be leaking, threatening the water supply for 6 million people. These include the Hyde Park site, which contains the largest deposit of dioxins⌀ in the world. There are estimated to be over 200 other hazardous sites in the Niagara Falls area containing 8 million tonnes of toxic and hazardous waste⋄.

People living on the Love Canal site suffered from unusually high incidences of diseases and genetic damage, which could be attributed to exposure to toxic and carcinogenic chemicals. By 1990 around $14 billion dollars' worth of lawsuits had been lodged against Hooker by the families and next of kin of those worst affected.

LOW-INCOME COUNTRIES (LICs) A category within a system used to rank countries based on an analysis of national economic performance. The initiator of the system, the World Bank⋄, originally defined any country with an annual *per capita* Gross National Product (GNP)⋄ of less than $400 as a Low-Income Country. The exact figure and number of countries varies with annual updates.

The World Bank divides all nations with a population⋄ in excess of 1 million into a number of categories based simply on economic performance. In 1989, developing countries⋄ were divided up into Low-Income (less than $500 *per capita* GNP) and Middle-Income⋄ countries. The latter were split into 'lower' ($500–2,199 *per capita* GNP) and 'upper' (more than $2,200 *per capita* GNP). Forty-two nations were in the Low-Income category, 37 in the Lower-Middle-Income, and 17 countries made up the Upper-Middle-Income group.

The Organization for Economic Co-operation and Development (OECD)⋄ uses slightly different criteria for its ranking system, classifying any country with a *per capita* income of less than $700 as Low-Income. The OECD Middle-Income grouping is also split into 'lower' and 'upper', the former having a *per capita* GNP of $700–1,300 and the latter a *per capita* GNP of over $1,300.

LUXEMBOURG ACCORD A landmark agreement by members of the European Community⋄, proposed by France and finalized in 1966, under which any decision by the Council of Ministers could be vetoed by any member nation whose national interests were deemed to be jeopardized.

M

MACADAMIA A genus of bushy, evergreen trees native to Australia, the hard-shelled fruit of which contains a single edible seed called the Queensland or macadamia nut. These nuts are highly nutritious, have a good flavour and command the highest price of any nut used for confectionery purposes. There are ten species but only two are cultivated commercially, the most common being *Macadamia ternifolia*. Harvesting is simple and extremely labour-intensive: nuts are collected by hand once they have fallen to the ground. The quality and flavour of the nut, which contains high levels of both water⋄ and oil, deteriorate rapidly unless harvesting, processing and packing are carried out promptly and properly. The nutshell is extremely hard and has to be cracked by machine: nuts are first sorted for uniformity of size before being rolled between stainless steel drums. Kernels can be eaten raw but are usually exported in roasted form in vacuum-packed cans or jars. As macadamia trees take seven years to reach full productivity – annual yields can reach up to 100 kg per crop – cultivation is a long-term investment. Traditional areas of production such as California, Florida and Hawaii in the United States, and parts of Australia, have been joined by new plantations in East Africa and South and Central America. In many regions macadamia trees are now being inter-cropped with other cash crops⋄ such as coffee⋄.

MAHAWELI DEVELOPMENT PROGRAMME The Mahaweli Development Programme is an ambitious water-resource development project involving irrigation⋄, agriculture⋄ and hydroelectric power⋄ generation. Based in Sri Lanka's dry zone in the northern, central region of the island, the Programme covers 39 per cent of the whole island, 55 per cent of the dry zone and 60 per cent of the undeveloped land⋄ surface. The projects in the

Mahaweli Programme were planned to be phased over a thirty-year period. About a dozen major projects are involved; all are interlinked but can be constructed independently. The original total cost of the Mahaweli Programme was put at over $1 billion. The Victoria high-arch dam⇔ and the Kotmale hydropower scheme are its most expensive projects. The Programme itself has attracted funding from a variety of sources: the United Kingdom provided a grant of around $180 million for the Victoria scheme and Sweden agreed to contribute around $320 million to the Kotmale project. Water⇔ from both projects will be diverted from the Mahaweli ganga at Minipe. At least 135,000 hectare (ha) of previously uncultivated downstream land will be made usable by irrigation. Well over 130,000 families will have to be moved under the Programme's agricultural resettlement⇔ component. Estimates suggest that if all the projects go ahead as planned, 1.5 million people, almost 10 per cent of the nation's population, will need to be resettled.

The successful completion of Project 1 – the Polgolla/Bowatenna Diversion Complex – provided sufficient water to enable 50,000 ha of existing irrigated land in the Anuradhapura, Matale, Polonnaruwa and Trincomalee districts to be cultivated fully.

Outbreaks of several water-related diseases, notably Japanese encephalitis, have become commonplace since the water-resources development projects began. Between 1977 and 1984 the costs of the scheme trebled, and now the Sri Lankan government must find an amount equivalent to three years' total export earnings to pay for the venture.

MAIZE An annual cereal grass, *Zea mays*, also called corn. Cultivated as a grain crop in Central America for 3,000 years, maize is now the staple food for over 220 million people in eighteen countries in parts of Latin America and Africa. In terms of world production, it is the third most important cereal crop after rice⇔ and wheat⇔. Maize has a lower protein content than wheat and is a poor source of some vitamins. People who depend heavily on maize as a source of food are particularly at risk of developing pellagra, a vitamin-deficiency disease characterized by dry, scaly skin.

In poor countries maize is used to produce meal or porridge-like dishes; in developed countries it is eaten directly as a vegetable or processed into flour, sweetcorn or breakfast cereals. In Europe and North America maize is used extensively as an animal feed, when both the grain and leaves are eaten. Corn oil, starches, alcohol, dextrin and adhesives are all industrial products from maize.

Maize is one of the most variable of all commonly cultivated crops: some varieties mature in a little over two months, while others take more than a year. Ideally, maize requires temperatures of over 19°C, although it will tolerate a range from 14 to 30°C. A rainfall of 600–1,200 cm is needed and should be evenly distributed throughout the growing season.

The development of hybrid strains allowed yields to be raised dramatically, although only when agricultural inputs were available. In the 1930s, 1 per cent of the maize grown in the United States was hybrid, now it all is. The USA traditionally produces between one-third and one half of the world's maize harvest, although the bulk of it is used for livestock⇔ feed.

Globally, 405 million tonnes are produced each year; the United States (145 million tonnes), China (74 million tonnes) and Brazil (24 million tonnes) are the world's largest producers.

MALARIA An infectious disease caused by the presence of parasitic protozoa of the genus *Plasmodium* in the red blood cells. The disease, transmitted via blood-sucking female *Anopheles*⇔ mosquitoes, is confined to tropical and subtropical areas and is the most prevalent and devastating parasitic disease to afflict the human race. A hundred million new cases are reported annually, and about 270 million people are believed to be infected by the malarial parasite. Between one and two million people die every year from malaria and its complications and over 2.1 billion people, almost half the world's population⇔, are at risk of contracting the disease.

When a mosquito⇔ bites, parasites are injected into the bloodstream and migrate to the liver and other organs, where they multiply. After an incubation period which ranges from eight days to ten months, parasites return to the bloodstream and invade the red blood cells. Their rapid multiplication ruptures the cells, causing a bout of fever, shivering and sweating. Anaemia⇔ also develops, and the patient may suffer from nausea. When the next batch of parasites is released, symptoms reappear. The intervals between bouts of fever vary with the different types of malaria. Severe forms cause liver and kidney failure and brain and lung complications which prove fatal. Malaria can be cured at low cost by specific anti-malarial agents if treatment is given during the initial stage of the blood infection. Preventive and curative treatment relies on drugs such as chloroquine and proguanil. Widespread programmes to control the mosquito vectors, mostly using the pesticide◇ DDT◇, have been implemented since the 1930s. Many have been mismanaged and have allowed the mosquito to develop resistance to DDT and similar chemicals. Despite malaria's impact on global health and the vast amount of resources being used in the fight to combat it, the situation has shown little improvement over the past twenty years, mainly because development projects tend to ignore public health aspects or economic losses caused by disease.

The peak period of malaria transmission to humans often coincides with the period of greatest agricultural activity, and major agricultural development projects lead to the creation of new foci for the disease. Large-scale forest clearance, particularly in Amazonia◇, has led to a marked rise in the number of malaria cases, while extensive, ill-managed irrigation◇ schemes throughout the tropics have produced numerous ideal breeding sites for malarial mosquitoes. Favourable conditions for the spread of the disease are also being created in the slums which are found in all big cities in the tropics. No cost-effective, safe alternative has been found to replace the hazardous DDT in anti-malarial spraying programmes, many of which have ceased operating, and the disease is on the increase as a result. Safe, effective and cheap drugs were developed to provide protection against malaria, but these are now proving ineffective as the parasite responsible for the severe form, *Plasmodium falciparum*, has become resistant to them.

Although the development of a potentially viable vaccine is under way, the mosquitoes' resistance to insecticides◇ and of the parasites to commonly available drugs means that 470 million people are living in areas – mainly in tropical Africa – where no specific measures to control the disease are being taken.

MALE DOMINANCE   Throughout the world men have an overwhelming dominance over women in positions of power and authority – in politics, industry, the judiciary and in regulatory roles within scientific and international bodies. In terms of work, men, especially those in developed countries, have better access to paid employment because it tends to be organized on the assumption that the availability of male workers will not be restricted by childbearing or domestic responsibilities. In the developing

world◇ boys tend to enjoy better access to education◇ than girls, as learning is not deemed of great importance for girls in certain societies. As a result, by far the greatest number of the illiterates◇ in the world are female. Life, in biological terms, reverses the tables, for in general, women live longer than men, no matter where in the world they are born or grow up. Although some of the biological bases which explain this phenomenon have been identified, the exact reasons why females should generally be able to cope with life's stresses and strains better than men have yet to be fully explained.

MALNUTRITION   A condition caused by an imbalance between what an individual eats and what is required to maintain health. The word literally means 'bad feeding' and can result from eating too little, but may also imply dietary excesses or an incorrect balance of basic foodstuffs such as protein, fat and carbohydrate. A deficiency (or excess) of one or more minerals◇, vitamins or other essential ingredients in the diet may arise from the inability to digest food properly as well as from actually consuming an unbalanced or inadequate diet.

Imbalance in the diet causes a variety of nutritional disorders, including beri-beri◇, rickets, and pellagra. Chronic underfeeding or lack of nutrients leads to listlessness, immune-system damage, impairment of mental functioning and eventually death. Excess eating leads to obesity and heart disease, and hastens death.

Malnourished people are usually more open to infection and prone to develop other diseases or general ill health. In 1980 an estimated 340 million people were either not getting enough to eat or had a seriously imbalanced diet. Now, it is estimated that there are between 450 and 750 million severely malnourished people around the globe. At least 20 million underweight babies are born each year simply because their mothers are malnourished. Poverty is the most immediate indirect cause of this malnutrition in both the developed and developing worlds. In 1985, 20 million people in the United States were suffering from hunger-based malnutrition. What money is available to most low-income families is frequently spent on food of dubious nutritional quality.

Malnutrition is commonly applied to describe the condition caused by underfeeding in the developing world. In the Third World◇, 60 per cent of deaths in children under the age of five are related to malnutrition caused by underfeeding. In such cases, doctors measure the degree of malnutrition by assessing a child's weight in relation to height. A moderately malnourished child is one whose weight is 70–80 per cent of the norm for its height. Children who weigh less than 70 per cent of the norm are labelled severely malnourished. A child who weighs less than 60 per cent is unlikely to survive without immediate treatment.

MALTHUS, THOMAS ROBERT (1766–1834) British clergyman and economist, famous for his theories on human population⟳ and food supply. In his *Essay on the Principle of Population*, (1798) he argued that population control was one of the major problems facing the world, since people increased in a geometrical ratio whereas food increased only in a mathematical ratio. He concluded that the human race was doomed to remain living at starvation level if population was not regulated. Later in life, he proposed that 'moral restraints' – sexual abstinence and delaying marriage – could keep human numbers from increasing too quickly and, failing that, birth control⌂ should be introduced and made compulsory.

MAN AND THE BIOSPHERE (MAB) The Man and the Biosphere Programme, launched under the auspices of the United Nations Educational, Scientific, and Cultural Organization (UNESCO)⟳ in 1971, aims to help advance and develop scientific knowledge with a view to the rational management and conservation⌂ of natural resources⟳. It follows on from work and ideas arising from the International Biological Programme (IBP), which began in 1964 and concluded in 1974. The IBP was a natural-science project and did not adopt a comprehensive approach, taking little account of social and economic data in its study of ecosystems⌂. In contrast, the MAB identifies and studies specific ecosystems⌂ but focuses on the problems of peoples and their needs, incorporating research, personnel training and dissemination of information at all levels, from governments and policy-makers to the general public.

The MAB operates through a series of National Committees established by member states of UNESCO. Over 100 countries have formed such committees, which usually comprise scientists from universities or national research institutions alongside representatives of other private or public bodies concerned with environmental research or management. An International Co-ordinating Council of thirty members, elected by the UNESCO General

Conference, is the main policy-making body. It meets every two years to review and assess progress. By 1988, 269 biosphere reserves⌂ incorporating a variety of ecosystem types and covering 143 million hectares had been established under the MAB programme.

MANDELA, NELSON ROLIHLAHLA (1918–) South African lawyer and politician. A black South African, he practised law in Johannesburg and became a leading member of the African National Congress (ANC)⌂. He was arrested and tried for treason, a lengthy process which lasted from 1956 to 1961. Although aquitted, he was retried in 1963, found guilty of sabotage and plotting to overthrow the government, and sentenced to life imprisonment. His wife, Winnie, subsequently became a prominent spokesperson for the ANC and was placed under house arrest several times before the slight relaxation of government controls, coupled with mounting international pressure for the release of Nelson Mandela, brought about the easing of restrictions. Mandela was eventually released from custody in 1990 and continued his efforts to see the dismantling of the apartheid⌂ system and introduction of majority rule in South Africa through dialogue with the South African authorities and government leaders from around the world.

MANGROVES Trees and shrubs which form dense thickets and low forests⌂ on coastal mudflats, salt marshes and estuaries throughout the tropics. Mangrove forests or swamps are characteristically low and dense with a tangle of underwater and aerial roots, although some mangroves can grow as high as 50 m. Mangroves cover at least 14 million hectares and are typically associated with river mouths where the water is shallow and the sediment levels are high. They form one of the most diverse of all ecosystems⌂ and provide a unique habitat⌂ for 2,000 species of fish⌂, invertebrates and plant life.

Mangrove forests or swamps are a valuable source of timber, pulpwood, fuel and charcoal. The timber from the trees is impervious to water⟳ and resists attack by marine worms and the trees also provide raw materials used in making dyes, glues, rayon and tannin. Mangrove swamps are home to several species of food fish⌂ and act as breeding-grounds for a variety of commercially important fish, shrimps and crab stock, as well as being a source of other useful products such as wax and honey. Mangroves are able to play an important role in desalinating sea water and are one of the major factors in stabilizing shorelines. In addition, they are able to reclaim land⌂ from the sea. A network of horizontal roots anchor trees to soft mud, trapping sediment. This enables the land to progress seaward, and mangroves can move seaward at a rate of up to 100 m a year.

Throughout the tropics mangroves are under threat from clear-cutting, charcoal production, sand and shale mining, and land reclamation for agriculture⌂ or aquaculture⌂. Massive felling in Asia is taking place to provide exports to Japan as wood⌂ chips destined for paper⌂ pulp and rayon manufacture. The Sunderbans in Bangladesh, the most extensive mangrove forest in the world, is being threatened by a reduced flow of water in the River Ganges⌂ due to various dam⌂ projects and irrigation⌂ systems.

MANILA DECLARATION   At a meeting of the Group of 77⌂ in Manila in 1976, developing countries⌂ adopted a set of directives concerning an equalization of global trading systems. The Declaration was intended to guide negotiators in discussions, principally at the fourth United Nations Conference on Trade and Development⌂ (UNCTAD IV) trade negotiations, but also in other international fora. The declaration represented an important basis for the consensus resolutions finally adopted by UNCTAD IV.

MAN-MADE RIVER PROJECT   The Great Man-Made River Project, devised by Libyan authorities and being carried out with the assistance of the United Nations Educational, Scientific and Cultural Organization (UNESCO)⌂, will provide coastal regions of the country with water⌂ for agriculture⌂, industry and municipal uses. Vast quantities of water lie in a massive aquifer under the Sahara Desert in the southern part of the Libyan Arab Jamahiriya. The severe climate, poor soil⌂ and lack of manpower in the region make use of the water resource difficult. The authorities have decided to extract water from the aquifer and transport it 2,400 km north to the littoral regions where soil is better, workers are available, and the water can be used efficiently. The ten-year scheme is estimated to be costing $6.3 billion. The first phase of the project, due for completion by mid-1992, involves the construction of two well fields, each producing 1 million cubic metres (cu m) of water daily. This water will be transported through a four-metre-diameter pipeline over a distance of some 1,600 km. It is intended that the scheme will eventually be supplying 5.5 million cu m of water per day to the costal zones.

MAO ZEDONG (Mao Tse-tung) (1893–1976)   Chinese statesman. Born into a peasant family, he became a Marxist politician and helped form the Chinese Communist Party in 1921. He proclaimed the People's Republic of China in 1949. As Chairman of the Communist Party, he provided the pattern and direction for the development of the country through the Great Leap Forward of 1959 and the Cultural Revolution of 1966. His philosophy and thinking behind these events were published in the famous 'Little Red Book'. Although he stepped down as Party Chairman during 1958 on the grounds of ill health, he reappeared with a greater following during the Cultural Revolution.

MARASMUS   A nutritional disorder resulting in severe wasting in infants: body weight falls below 75 per cent of that expected for the sufferer's age. Affected infants look 'old', pallid and apathetic. They lack skin-fat and have subnormal temperatures. The condition usually results from gross shortages of energy and protein and may be due to malabsorption, incorrect feeding, vomiting or diarrhoea. Severe diseases of the heart, lungs, kidneys or urinary tract, or bacterial or parasitic infections in tropical climes, are also causal factors. Treatment by gentle nursing and the provision of nourishment and fluids in gradual steps is sufficient to return sufferers to adequate health.

MARCOS, FERDINAND EDRALIN (1919–89)   Highly controversial Filipino politician. During his studies at law school in 1939 he was convicted of murdering a political opponent of his father, but eventually managed to secure his own acquittal. During the Second World War he was a guerrilla fighter and managed to survive terms in prison camps. He became President of the Philippines in 1965 and his regime, backed by the United States government, became increasingly repressive. His presidency was continually threatened by civil unrest and claims of financial irregularities and fraud. In 1972, he declared martial law and assumed dictatorial powers. Although the controls were ostensibly relaxed during the early 1980s, the murder of opposition leader Begnino Aquino Jr, allegedly with government complicity, caused further unrest and the controls were tightened again. He was eventually overthrown in 1986 and went into exile in Hawaii, being replaced as Head of State by Mrs Corazon Aquino, widow of the murdered opposition leader. During his leadership, the country ran up a $3.9 billion dollar debt amid many claims that the Marcos family had fraudulently diverted vast sums overseas – money which, upon his overthrow, the Philippine government took steps to retrieve.

MARITIME LAW The world's oceans and seas cover 362 million sq km, almost 72 per cent of the earth's surface. They contain enormous natural resources⟳ of incalculable wealth and, not surprisingly, disputes over ownership have been common. Many people regarded the seas as *res nullius*, meaning that they were no one's property and could be appropriated by the first comer. Others were of the opinion that the seas were *res communis*, belonging to all humankind, and should be exploited for the benefit of all.

Since the formation of the United Nations⟳, there have been several attempts to establish an international agreement governing the oceans and the resources they contain. The International Law Commission established by the UN in 1950 struggled for seven years to produce articles which were put before the UN Conference on the Law of the Sea (UNCLOS) in 1958. Four conventions were adopted: on the high seas; on the territorial sea and the contiguous zone; on the continental shelf; and on fishing⟳ and conservation⟳ of resources. These conventions proved totally inadequate, and less than a quarter of the world's nations became parties to them.

In 1970 the UN passed a Declaration which treated the seabed and ocean floor as 'the common heritage of mankind'; this initiative eventually led to a third Conference on the Law of the Sea in 1973. This gave birth to the Convention on the Law of the Sea⟳, adopted in 1982, which governs peace and security, the allocation of natural resources, navigation, access and transport, and scientific research in the marine environment. It also provides a framework for resolving any disputes.

MARPOL CONVENTION Common name for the International Convention for the Prevention of Pollution from Ships, which was adopted under the auspices of the International Maritime Organization (IMO)⟳ in 1973 and became operational in 1983. The Convention contains five annexes, the first two of which, covering pollution⟳ by oil⟳ and from

noxious liquid substances, came into force in 1983. The last three are optional and need at least fifteen countries, whose combined merchant fleets total at least half of world tonnage, to ratify them in order to make them effective.

In 1988, Annex V finally came into force. It is designed to prevent pollution by the dumping of refuse, which includes all kinds of victual, domestic or operational waste and applies to all kinds of ships. The annex bans the dumping of all plastics, including fishing⟳ gear made from synthetic materials, but allows the disposal of other forms of refuse under controlled conditions. In 'Special Areas', such as the Mediterranean, Baltic and Black Seas, food wastes only can be dumped into the sea, and then not within 19 km of land⟳. All contracting parties to the Convention are obliged to provide adequate waste-disposal facilities for shipping within their ports.

Annex III, dealing with hazardous waste⟳ in packaged form, and Annex IV, covering sewage disposal, have still not received the level of support required to bring them into force.

MARSHALL, GEORGE CATLETT (1880–1959) American general and statesman, Army Chief of Staff and strategic adviser to the President of the United States during the Second World War, in which he played a major role in overseeing the build-up of US forces. After the war he served as Secretary of State (1947–9) and devised the European Recovery Programme (more commonly known as the Marshall Plan), in which the United States undertook to provide economic aid⟳ to Europe to repair the structural and economic damage inflicted by the war. He was awarded the Nobel Peace Prize in 1953.

MARSHALL ISLANDS Group of thirty atolls in the northern Pacific used by the United States as a nuclear weapons testing site from 1946 to 1958. The largest atoll, Kwajalein, was the first Pacific territory captured by United States forces in the Second World War. It is still a US military base and is used to test missiles capable of carrying nuclear warheads. From 1947, the Marshall Islands were part of the Trust Territories of the Pacific Islands, theoretically under the control of the United Nations⟳ Trusteeship Council, but administered by the USA under the Trust Territory⟳ System. The islands have sought independence but have opted to retain an association with the USA which is still officially responsible for their defence, enabling the

US military to maintain a presence. Subsistence farming is the main occupation; US military services provide the wages for most of the island's employees.

From 1946 onward, over seventy atmospheric tests of nuclear weapons were carried out in the islands, including the atomic bomb testing on Bikini atoll. The local islanders were evacuated to accommodate the testing or because their homes had become too radioactive but were allowed to return, often resettling on atolls that proved unable to support them. Many are now dying or have died from cancer⌀. The inhabitants of Rongelap⌀ were reportedly exposed to fallout from the 'Bravo' hydrogen bomb test for three days before being evacuated. Many suffered from radiation sickness. They were allowed to return to the atoll three years later despite high radiation levels. Natives of Bikini were allowed to return in the late 1960s, but less than ten years later they were again evacuated as they were found to be consuming dangerous levels of radioactive chemicals in their food. US officials now admit that it may be as long as fifty years before the Bikini atoll will be fit for human habitation.

MATERNAL MORTALITY Every year 500,000 women around the world die during pregnancy or childbirth. Well over 90 per cent of these deaths occur in the Third World⌀, where rates of maternal mortality for poor rural women are usually over 200 times higher than they are for women living in deprived areas in Europe and North America.

Lack of medical attention, poor education⌀, housing, sanitation and nutrition in Third World countries contribute to make the average lifetime risk of women dying as a result of complications arising from pregnancy between 1 in 25 and 1 in 50. This compares with a risk of between 1 in 4,000 and 1 in 10,000 for a woman in the developed world.

In the developing countries⌀, death during childbirth accounts for around 25 per cent of deaths of women of childbearing age, In the United States, the figure is less than 1 per cent. According to the World Health Organization (WHO)⌀, maternal death rates per 100,000 live births average 640 in Africa, 420 in Asia and 270 in Latin America. For the developed world as a whole, the figure is thirty.

Haemorrhage, infection, toxaemia, obstructed labour and unskilled abortion account for 75 per cent of all maternal deaths in the Third World. The vast majority of women in the developing countries deliver their offspring at home, with fewer than half of the births being attended by personnel who have undergone medical training. Despite the best efforts of primary health care⌀ workers and 'traditional birth attendants', many of these fatalities will be avoided only when the medicinal materials and specialist medical attention available to women in the developed world become widely disseminated.

MAXIMUM SUSTAINABLE YIELD An estimate of potential used to determine levels of hunting⌀ or fix limits for culling. When a population of animals has expanded to the limits of its space and food supply it tends to remain stable, exhibiting a fairly consistent age structure, with surviving young replacing those that die. The rate at which the greatest number of animals can be removed, without lowering the population or jeopardizing reproduction and replacement rates, is known as the Maximum Sustainable Yield. It is calculated on the assumption that if a proportion of a population is killed, the increased availability of food for the remainder can result in an increase in the surviving young or increased breeding rates. This assumption is not always correct and the calculations take no account of a number of other factors, such as the role of social structures and behaviour in animal populations. The concept can also be applied to crops and forests⌀ to determine harvesting levels.

MEASLES A highly infectious viral disease (rubeola), especially common in children. Measles is acquired through personal contact and spread by infected people coughing and sneezing. It tends to appear in epidemics every two to three years. After an incubation of eight to fifteen days, patients develop severe catarrh and small spots inside the mouth, and a raised, blotchy red rash, lasting three to five days, appears on the body. During the whole of this time the patient remains infectious. Those infected are susceptible to pneumonia and middle-ear infections. Complete recovery may take two to four weeks, after which time the body develops a persistent immunity. A vaccine has been developed which, when given to young children, provides effective and lasting immunity, although evidence is beginning to emerge to indicate that the vaccine used worldwide may not always provide lifelong protection. A single dose is necessary and should be given to children twelve to fifteen months after birth. Virtually every unprotected child will develop the disease.

In the developed nations of the world measles is not serious, although complications may develop, especially when the disease is contracted in adult-hood. In the Third World⬦, however, the disease has become a major killer, particularly of small children. Every year an estimated 67 million cases occur, resulting in over 2 million deaths. All Third World children contract the disease, usually before the age of three, much earlier than in the industrialized world. Due to the greater risk of infection and the lower level of general health, children in developing countries⬦ need to be immunized earlier – nine months after birth – for the vaccination to be effective. Measles is the main killer disease among the six vaccine-preventable diseases⬦ targeted under the World Health Organization's Expanded Programme on Immunization (EPI)⬦.

MEDITERRANEAN ACTION PLAN In 1975 the Mediterranean Sea was so polluted and in such a deplorable state that the countries bordering it decided to take concerted action to try and improve it. The Barcelona Convention for the protection of the Mediterranean was signed by seventeen of the eighteen nations involved; only Albania failed to lend its support. As a result, under the guidance of the United Nations Environmental Programme (UNEP)⬦, a multi-billion dollar Mediterranean Action Plan was drawn up under the Regional Seas⬦ Programme.

In 1988, 80 per cent of bathing beaches were regarded as safe for swimming as opposed to 65 per cent before the Convention came into force. Despite the claims made by those involved in the plan, conservation groups such as the Worldwide Fund for Nature (WWF)⬦ maintain that the Mediterranean is still 'clouded with pollution'. Every year 2.5 million tonnes of fossil fuel waste⬦ and heavy metals, plus 300,000 tonnes of phosphates⬦, reach the Mediterranean, and 90 per cent of human wastes from coastal settlements is dumped untreated into the sea.

A major part of the problem is the effect of the tourist⬦ industry as, on average, over 100 million people, representing around one-third of the world's tourists, visit the Mediterranean each year. UNEP has warned that if trends are not reversed, by 2030, 95 per cent of the coastline will be urbanized and the Mediterranean Basin will have to support 500 million inhabitants together with 200 million tourists. At an international conference called in 1990 to discuss the parlous state of the Mediterranean, attended by the seaboard nations and international development agencies, delegates adopted the Nicosia Charter. Under this broad agreement, the

World Bank⬦, the European Investment Bank (EIB)⬦ and the European Community (EC)⬦, committed themselves to provide $1.5 billion immediately to help towards cleaning up the Mediterranean Basin by the year 2025.

MEDITERRANEAN FRUIT FLY Commonly known as the Medfly, this pest is native to Europe where it causes extensive damage to orchards in the countries bordering the Mediterranean Sea. In 1975 it reached the North American continent: its sudden appearance in California threatened the state's $16 billion agricultural industry. Repeated eruptions of the fly since then have been fought by the state authorities using the insecticide⬦ malathion. In combination with corn syrup to attract the fly, the mixture was controversially sprayed at night from helicopters. Critics questioned the safety of malathion and local residents complained that their houses, vehicles and streets were covered with a sticky coating of syrup. The appearance of the Medfly in other states with important fruit-producing areas, such as Florida, coupled with general public outcry against aerial spraying, has led to the development of a control programme based on the release of sterile male flies: the Californian authorities have ceased aerial spraying and are relying on the release of 400 million sterile male flies per week to keep the Medfly in check.

MERCURY Mercury is a unique metal that is liquid at normal temperatures. It is usually obtained from cinnabar, a sulphide of mercury that is found sparsely in volcanic rock. The few known reserves of cinnabar may well be exhausted by the end of this century. Mercury has a number of uses. In metallic form it is used in switches and thermometers. Mercury compounds are widely used in anti-fungal agents and as detonators for explosives, whilst mercury amalgams are extensively used for dental treatment.

Mercury is highly toxic, and in large doses fatal. Its toxicity is heightened because it can be swallowed, inhaled and is easily absorbed through the skin. The human body cannot excrete it, so it has a cummulative effect. Inorganic mercury compounds are toxic, organic ones more so. The latter are soluble in fat and build up in body tissue.

In humans, mercury damages the kidneys and the nervous system; the symptoms are an increasing tremor in the limbs and loss of mental faculties. Historically, mercury poisoning has been relatively commonplace. In the nineteenth century, mercury nitrate was used in the manufacture of top hats. Hat-makers exposed to mercury developed tremors and mental damage, giving rise to the phrase 'mad as a hatter'. Most cases of mercury poisoning now arise through the inhalation of mercury vapour, usually in people who work in mercury-refining factories or those who are involved in the manufacture of instruments, such as thermometers and barometers, which contain mercury. There have also been several examples of mass poisoning. Mercury-based compounds are commonly used on seeds to prevent moulds from growing, and hundreds of people in Iraq were poisoned when dressed seeds were cooked and eaten. In Japan, effluent discharged from a mercury-refining factory polluted an area of coastal waters: the fish◇ become contaminated with mercury. This concentrated in the food chain and the inhabitants onshore, who consumed large quantities of locally caught fish, developed Minamata disease◇.

Modern technology and its applications have meant that mercury poisoning has become a matter of some urgency. Warning labels will have to be put on all batteries which contain mercury, according to a 1988 Swiss law. The legislation also places severe restrictions on the amount of mercury each battery can contain. This legislation is just one example of the growing campaign in Western Europe aimed at reducing mercury pollution.

The French government has launched a national scheme to combat mercury pollution◇ arising from mercury-containing miniature batteries used to power watches and cameras. These batteries. although small, are discarded *en masse* as they quickly run out. The battery casing quickly corrodes, releasing mercury into the subsoil and groundwater◇, where it breaks down into methyl mercury, the causal agent of Minamata disease. France's National Waste Collection and Disposal Service is encouraging people to discard their used batteries in collection boxes, thousands of which are being distributed around the country. The boxes will be emptied by volunteers and the mercury recycled. Similar battery-collection points are becoming commonplace in other European countries.

New ceramic materials are being produced which make excellent tooth fillings, focusing attention on the effect on the body of the mercury in dental amalgam. The amalgam reputedly binds the mercury and does not allow it to enter the body in a dangerous form. In Sweden, health authorities have advised dentists not to use amalgam when filling cavities in the teeth of pregnant women, although there has always been a lack of any conclusive evidence that the amalgam could be responsible for mercury poisoning of the fetus or of adults.

Annual global production of mercury is in the order of 6,400 tonnes; the Soviet Union (1,900 tonnes), Spain (1,300 tonnes) and the United States (900 tonnes) are the main producer nations.

METHANE A colourless, odourless flammable gas, the main constituent of natural gas◇, used as a fuel and as source of other chemicals. It is a natural trace gas in the atmosphere and plays a significant part in influencing the processes of global warming◇ and ozone◇ destruction.

In the stratosphere methane acts as a sink for chlorine atoms, thereby reducing the depletion of the ozone layer. Methane's contribution to global warming◇ is twofold. As well as being a greenhouse gas◇ itself, when stratospheric methane oxidizes it causes water vapour – which is also a powerful greenhouse gas – to increase. Atmospheric methane concentration increased markedly around 150 years ago and has continued to do so at a rate of 1 per cent per year – faster than the build up of carbon dioxide◇. Since pre-industrial times, levels of atmospheric methane have more than doubled. Molecule for molecule, methane is twenty times more effective at warming the globe than carbon dioxide.

The total amount of methane added to the atmosphere annually has been estimated at about 550 million tonnes, about half of which comes from anthropogenic sources. Most, about 90 per cent, is removed through oxidation, just over 50 tonnes remain airborne. Temperate and boreal forests◇ play an important role in absorption. Temperate forest soils absorb 9.3 million tonnes of methane annually, almost four times that absorbed by tropical forests. Clearing forests for agriculture◇ and the use of nitrogenous fertilizers◇ sharply reduce the amount of methane absorbed by soils. The concentration of atmospheric methane in the northern hemisphere is almost 10 per cent higher than that in the southern hemisphere.

Methane is produced naturally from a variety of sources. Decaying organic matter under shallow water produces methane which bubbles up in marshes and bogs. Other principal sources are the enteric fermentation in ruminant animals, production by termites◇, and the burning of biomass◇. Biogenic sources, livestock◇, paddy-field cultivation and termites are all increasing in line with agricultural expansion. Termites release between 5 and 150 million tonnes of methane into the air each year as they digest vegetative matter. They prefer tropical grasslands, and as tropical forests are replaced with pasture, so termite numbers increase. The global livestock herd is estimated to vent between 50 and 100 million tonnes annually, to go with the gas produced by 5 million sq km of marshland and 1.5 million sq km of rice◇ paddies which release an estimated 150 million tonnes. Over the past two decades, landfill waste-disposal sites have also been found to produce significant quantities of methane, releasing around 70 million tonnes a year, more and more of which is now being used as a fuel.

By 1990, human activities were increasing the amount of methane in the air at a rate fifty times faster than at any other time in the past 160,000 years. The Intergovernmental Panel on Climate Change (IPCC) concluded that a 15–20 per cent reduction in emissions was required merely to stabilize atmospheric concentrations at 1990 levels by the year 2000.

MICROBIOLOGICAL RESOURCES CENTRES (MIRCENS)

A network of six regional institutions, set up under the auspices of the United Nations Educational, Scientific, and Cultural Organization (UNESCO)◇ and the United Nations Environment Programme (UNEP)◇, which act as centres of excellence for microbiological research and development. Micro-organisms are essential components in all biological cycles and are particularly important in decomposing natural materials and fertilizing soil◇. They also have enormous potential with the advent and progress of biotechnology◇. The MIRCEN system is based in the developing world, with centres in Brazil, Egypt, Guatemala, Kenya, Senegal and Thailand. Focal projects include the conversion of crop wastes into fuel or fertilizer◇, the development of biological fertilizers based on nitrogen-fixing bacteria, and the development of micro-organisms for pest control or the decomposition of man-made pollutants.

MICRONESIA, FEDERATED STATES OF,

A group of islands in the Pacific Ocean comprising Ponape, Truk, Yap and Yosrae. Together with the Marshall Islands◇ and the Republic of Belau◇, they made up the Trust Territory◇ of the Pacific Islands, administered by the United States on behalf of the United Nations◇. Self-government was achieved in 1979, with the USA retaining official

responsibility for defence of the islands and providing grant-in-aid for several years.

MIDDLE-INCOME COUNTRIES (MICs)

A term used to describe those developing nations that are relatively far along the development path – as measured by economic performance – such as Brazil, Chile, Republic of Korea, Kuwait, Malaysia, Mexico and Yugoslavia.

The Organization for Economic Co-operation and Development (OECD)◇ categorizes developing countries◇ into four main groupings: the Low-Income Countries (LICS)◇, the Lower-Middle-Income Countries (LMICS), the Upper-Middle-Income Countries (UMICS), and the Central and Eastern European Nations. There are 67 LMICS, including the 42 nations defined by the United Nations as Least-Developed Countries (LLDCS)◇. The main criterion for classifying a country as an LIC is a 1987 Gross National Product (GNP)◇ of under $700 *per capita*. Nations with a 1987 *per capita* GNP of between $700 and 1,300 – a total of 28 – are classified as LMICS, while any country with a *per capita* GNP of over $1,300 – some 60 in all – is placed into the UMIC group. The OECD's Central and Eastern European grouping, which traditionally comprised Bulgaria, Czechoslovakia, the German Democratic Republic, Hungary, Poland, Romania and the USSR, is having to be totally revised following the momentous happenings in the region during the late 1980s.

The World Bank◇ classifies 42 nations as Low-Income, 37 as Lower-Middle-Income and 17 as Upper-Middle-Income, although it does not include nations with a population◇ of fewer than 1 million. Most of the severely indebted nations with debts◇ outstanding to commercial banks are Latin American countries in the Middle-Income group. Together, they account for around 44 per cent of the $1.3 trillion Third World◇ debt◇. Most debt-reduction plans, including the Brady Plan, have been aimed at reducing the debt of UMIC countries and savings of $22 billion have so far been realized, easing the pressure on commercial banks in the process. However, these countries' debt repayments still account for 26 per cent of their export earnings.

MILLET

Millet is a name applied to several distinct, small-grained cereal species which grow in a range of environments, warm and moist or cool and dry. Of all the commonly grown cereal crops, bull rush millet, *Pennisetum awesccanum*, is one of the most drought-resistant and can withstand much hotter conditions with far less rainfall than maize⌀. However, it is relatively low-yielding and particularly prone to attack from granivorous birds. Several species of millet are cultivated in areas of the world with extremely low rainfall – 500 mm per year – and a short rainy season. The grain grows well on poor soils and is the staple crop for many of the world's rural poor in arid regions.

Together with sorghum, millet is the staple food for almost 200 million people in thirteen arid countries in East and West Africa. Millet is often intercropped with sorghum, groundnuts⌀ or cowpeas in these areas to avoid total crop loss in adverse conditions. At harvest time, the heads of the plant are usually cut off by hand. Yields of 450 kg per hectare are average. Millet is used to make flour or cooked and eaten in the form of meal, and can be fermented to produce alcoholic drinks. Around 32 million tonnes are produced worldwide each year; India (9 million tonnes), China (6 million tonnes) and Nigeria (4 million tonnes) are the world's leading producers.

MINERALS Minerals are the materials which form rocks, and the term can also be loosely applied to any material that is extracted from the earth. They provide the oil⌀, coal⌀, metals, fertilizers⌀ and building materials upon which the industrial processes of the world depend. Many industrialized countries are forced to import minerals from primary producer countries, most of which are in the Third World⌀. The European Community⌀ imports over 95 per cent of its chromium⌀, copper⌀ and manganese, and Japan imports all its bauxite⌀ and over 90 per cent of its chromium, copper, iron⌀ ore and nickel. Many Third World countries are dependent on mineral exports to finance their economies. Conflict of interest between the producer nations in the Third World and consumer nations in the industrialized world was highlighted by the oil crisis in the 1970s – oil being a mineral fuel. As the producer nations do not process or use a great proportion of the minerals they produce for industrial purposes they do not get as much benefit as they could from their natural resources⌀, and stand to lose out most when reserves start to run out. Rates of consumption fluctuate, but the average American is thought to use some 1,600 tonnes of various materials in a lifetime, fifty times more than a native of India.

If present global levels of consumption are maintained, all easily accessible deposits of at least thirteen minerals will be exhausted within the next fifty to a hundred years. Supplies of industrial diamonds will stop in ten years, and reserves of lead⌀, tungsten, tin⌀, zinc⌀, mercury⌀ and silver will have been used up before 2040. Reserves of some minerals, such as bauxite⌀ and phosphate⌀, however, will last for over a thousand years at current usage levels.

MINAMATA DISEASE A form of mercury⌀ poisoning that killed forty-three people in the Japanese town of Minamata (referred to occasionally as Minimata) between 1953 and 1956 and injured many more. The disease was contracted by eating fish⌀ contaminated with dimethyl mercury from the effluent of a local plastics factory. Regular eaters of fish began to absorb and accumulate relatively large amounts of mercury in their bodies causing the disease to develop. Symptoms include tremors, impaired hearing, severe anaemia⌀ and bone deformities – and in acute cases brain damage, delirium and paralysis, and eventually death. Many women in the area gave birth to deformed babies, and the incidence of genetic deformities also rose.

MONOCROPPING *See* Monoculture

MONOCULTURE Cultivation of large tracts of land⌀ to produce a single variety of crop which is harvested all at once, the process being regularly repeated. It is a typical feature of modern large-scale agriculture⌀ as it reduces labour costs and maximizes machine use and marketing efficiency. The alternative is intercropping⌀ (or polyculture), the growing of a mixture of crops, either together or in sequence.

Monocropping has a number of major disadvantages compared with polyculture. Cultivation of a mixture of crops minimizes pest problems, and insures against drought⌀, frost and other natural phenomena that may curtail production. Where one crop is lost, the other crops increase the chance of at least some harvest. Intercropping can also make use of complementary crop mixes which help to increase yields. Mixing crops of different height, shading area, and microclimate requirements helps to increase soil⌀ cover to prevent loss of soil and moisture. With monocropping, all these are lost.

In monoculture systems, devices for maintaining soil⌀ fertility, structure, and moisture, and reducing susceptibility to pests, are enfeebled or discontinued. More nutrients, pesticides⌀ and water⌀

have to be added. Monoculture systems predominantly use high-yield variety (HYV)⌀ crops, increasing the need for these inputs. Although monoculture is responsible for recent growth in agricultural production, it has made agro-ecosystems more and more artificial, unstable and prone to collapse.

MONSOON A seasonal large-scale reversal of winds in the tropics, occurring chiefly as a result of differential heating of the oceans. It is best developed and exerts greatest effect in India, China and South-East Asia. The word, derived from the Arabic word *mawsim*, is now commonly used to describe the intense rainfall that generally accompanies this wind reversal.

MONTREAL PROTOCOL An agreement reached by representatives from most of the world's governments who gathered in Canada in 1987 to review the problem of the destruction of the ozone⌀ layer and the role of chlorofluorocarbons (CFCs)⌀ and as a follow up to the Vienna Convention for the Protection of the Ozone Layer⌀, two years earlier. The Protocol was signed by thirty governments, together with the European Community (EC)⌀, although some CFC-producing Third World⌀ countries, notably China and India, withheld their agreement until assurances on compensation or technology transfer for production of alternatives were agreed.

The Protocol, which came into force in 1989, aimed to see CFC production frozen at 1986 levels. It also called for consumption to be reduced gradually: to 20 per cent below 1986 levels by 1994, reaching 50 per cent by 1999. It was also suggested that the production and consumption of halons should be frozen at 1986 levels, this target to be met by 1992. Halons – gases commonly used in fire extinguishers – also harm the ozone layer.

At a follow-up conference in London in 1990, the industrialized world finally agreed to fund the transfer of technology to the developing world to help meet CFC phase-out timetables. Delegates agreed to set up a special fund worth $240 million over three years. They also decided to implement much stricter and swifter moves to curb CFC production and use. They agreed to implement a 50 per cent reduction by 1995, 85 per cent by 1987, and move towards a total removal of the chemicals by the year 2000. The German government announced plans to eliminate CFCs by 1995. Restrictions on halons – which are up to ten times more destructive than CFCs – were also

tightened. They are now to be phased out by the year 2000, with a 50 per cent cut by 1995, except for essential firefighting and medical uses where there are no suitable alternatives. Two other key ozone-depleting chemicals – carbon tetrachloride and methyl chloroform – were added to the Protocol during the London meeting. Carbon tetrachloride use is to be ended by 2000, with an 85 per cent cut by 1995, while methyl chloroform production will be completely phased out by 2005.

Developing countries⌀ will be allowed up to ten years' leeway on the phasing out of certain chemicals. The London meeting also confirmed that the Protocol will be reviewed in 1992, rather than the originally agreed date of 1994.

MOSQUITO A small fly belonging to the family *Culicidae*. There are about 3,000 species, exhibiting an almost worldwide distribution. Mosquitoes are especially abundant in the tropics and are viewed by many as man's greatest pest. In most species, males feed on plant juices, while females, who must have a blood-meal to reproduce, bite and suck the blood of animals, often transmitting serious diseases to humans and livestock⌀ in the process. Mosquitoes are intermediary hosts and vectors of malaria⌀, yellow fever⌀, filariasis⌀, dengue⌀, and a variety of other parasites and disease-causing agents.

MOTHER TERESA (1910–) An Indian Roman Catholic nun. Born Agnes Bojaxhiu in Yugoslavia of Albanian parents, she became an Indian citizen and founded the Missionaries of Charity, an order of men and women based in Calcutta, dedicated to helping abandoned children and the dying. The Order's first hospice was opened in 1950 and there are now over 400. Her work now spans five continents and her Order, which was recognized by the Vatican in 1965, now consists of 3,000 nuns, although Mother Teresa was forced to resign her leadership in 1990 through ill health. She was awarded the Nobel Peace Prize in 1979.

MUGABE, ROBERT GABRIEL (1925–) Zimbabwean statesman. He was imprisoned in Rhodesia for nationalist activities from 1964 to 1974. After his release, he and his supporters waged a guerrilla war against the Rhodesian authorities from neighbouring Mozambique, in alliance with Joshua Nkomo, leader of another nationalist movement. The war lasted from 1976 to 1980 when, following agreement between Rhodesia and Britain over independence, he became Prime Minister of the independent Zimbabwe. His presidency was originally plagued by differences with the opposition party under Joshua Nkomo and in 1985 he postponed the introduction of a multiparty state for five years and has campaigned strongly ever since for a one-party state.

MULTIFIBRE ARRANGEMENT (MFA) A complex system of bilateral quotas, established in 1974 to protect rich countries from Third World◇ competition in textiles by limiting imports. The MFA, which contradicts the principles of free trade◇, was extended for a further five years in 1986. In 1990 there were ninety-nine bilateral import-restraint agreements operating under the MFA. Calls to do away with it have been steadily growing, and Japan has proposed that the Arrangement should be scrapped in the interests of developing countries◇, with textiles being included under the General Agreement on Tariffs and Trade (GATT)◇.

World trade in textiles and clothing was worth $177 billion in 1988, with the developing nations exporting about $43 billion worth of goods. The World Bank◇ reported that if quotas were removed, the developing countries would stand to gain an estimated $3 billion through increased sales. Most of this would go to South Korea, China and Brazil, the three biggest exporters, with Hong Kong being the biggest loser. The World Bank also calculated that if all tariffs attached to textile imports were removed, it would result in a total $15 billion gain worldwide and the developing countries would receive $8 billion of this.

MULTILATERAL DEVELOPMENT BANKS (MDBS) The MDBS are major banks set up by international treaties that lend money almost exclusively to governments and government agencies to fund development projects. There are four – the World Bank◇, the Inter-American Development Bank◇, the Asian Development Bank◇ and the African Development Bank◇. Policy decisions within these banks are made by a board of governors representing the individual member nations. However, the number of voting shares of each government is dependent upon their financial contribution to the bank.

The MDBS are the largest public-development lenders in the world. Each bank operates a two-tier structure of lending. Funds are lent at close to commercial rates of interest through a programme referred to as a 'hard loan◇ window'. Loans are also made at little or no interest through a 'soft loan window'. At the World Bank, the International Bank for Reconstruction and Development (IBRD)◇ is the hard loan window and the International Development Association (IDA)◇ is the soft loan window.

The MDB's financing is provided by member countries in two forms. Paid-in capital is money that is actually transferred to the banks when the governments purchase capital stock. Callable capital is money that is pledged by the member governments, and for which they receive shares of stock, but it is not actually paid to the bank. Callable capital thus acts as a guarantee allowing the MDBS to borrow on the open market. It could, theoretically, be collected if required. In general, the paid-in capital of major donor governments is only 10 per cent of that outstanding as callable capital. As well as paid-in capital and pledges of callable capital from donor governments, money for the MDB's hard loan windows is obtained by selling securities to investors and raising money from the commercial market, whereas the funds for the soft loan windows are received exclusively through contributions from donor governments.

The MDBS are theoretically operated at a level which will make only enough profit to cover administrative costs and maintain high credit ratings. Indicative of the net flow of money from the developing world in interest payments, the World Bank has made a profit every year since 1947. The 1990 net income, some $1.1 billion, represented an 86 per cent increase over that realized at the start of the decade.

MULTILATERAL INVESTMENT GUARANTEE AGENCY (MIGA) An agency, formed in 1988 within the World Bank◇ system, designed to offer insurance for multilateral investments. The agency aims to promote private investment in less-developed countries by offering insurance against non-commercial risks, primarily war, expropriation and currency-exchange restrictions. It offers long-term political risk insurance, offering cover to investors against future takeovers by governments following civil war, revolution or changes in economic policy. Most other agencies offer only short-term cover, usually of three years' duration. MIGA is therefore designed to complement national insurance schemes such as Britain's Export Credit Guarantee Department's◇ scheme to protect overseas investment. It also attempts to create new business. MIGA's field of operation is to be restricted to countries that are members – eighty-three in 1990.

MULTIYEAR RESCHEDULING ARRANGEMENTS (MYRA)

Many heavily indebted developing countries⌀ are unable to meet their obligations under the terms of their loans and are forced to negotiate a new timetable for repayments, a process known as rescheduling. The so-called Paris Club⌀ is a multilateral forum in which officially guaranteed debts⌀ (those owed to governments) are rescheduled. The London Club deals with debts to commercial banks. The original idea was that only those debts falling due in the current year would be rescheduled, the new agreements giving debtors time to make good their economies so that future obligations could be met. In practice, an increasing number of countries found that one year was insufficient for this purpose. Rescheduling covering debts due in more than one year was found to be appropriate in some cases as it eased the administrative burden on both creditor and debtor and allowed the latter to plan for longer-term adjustment policies. The first commercial bank MYRA was granted to Mexico in 1984 and the first official MYRA to Ecuador the following year.

N

NARMADA VALLEY PROJECT

One of the largest and most controversial water-resource development projects undertaken anywhere in the world. Over the next fifty years or so, 30 major dams⌀, 135 medium-sized dams and 3,000 smaller dams are to be built on India's Narmada River and its tributaries. Of the total, 5 dams will be for hydroelectric power⌀ generation, 6 multipurpose and 19 for irrigation⌀. When completed, 50,000 sq km will be irrigated and 2,700 megawatts of electricity will be generated. It was originally envisaged that 11.5 million local villagers would benefit from the Project in addition to those living in major cities. The bulk of the outside financing is being supplied by the World Bank⌀.

Over 20 million people live in the Narmada Basin, and Hindus consider the river to be the holiest of all. The project will submerge 3,500 sq km of forest⌀ – 11 per cent of the forest in the Narmada valley – plus 200 sq km of cultivated land⌀ and some 400 sq km of grazing land. At least 40 per cent of the land to be irrigated is composed of soil⌀ that is susceptible to waterlogging and salinization⌀. It is estimated that a total of between 1 and 1.5 million people will be displaced by the dams. Many do not have title to land and thus will receive no compensation, contravening the World Bank's own guidelines on resettlement⌀ of tribal peoples. The World Bank temporarily suspended $450 million of finance for the first two dams – the Sardar Sarovar dam in Gujarat and the Narmada Sagar dam in Madhya Pradesh – pending a full review. In 1990, having already given nearly $20 million towards the purchase of hydroelectric equipment for the project, Japan also withdrew its support. Despite growing national and international pressure and warnings that the Project could cause considerable dangers to public health, including the spread of schistosomiasis⌀, a disease not usually found in India, the authorities have confirmed that it will go ahead.

NATURAL GAS

A naturally occurring mixture of gaseous hydrocarbons consisting mainly of methane⌀. It is obtained from underground reservoirs and is often associated with oil⌀ deposits. Like oil, it originates in the natural decomposition of animal matter. One major advantage of natural gas is its comparative cleanliness; burning gas produces only half the carbon dioxide⌀ produced by burning coal⌀. It is a relatively cheap and effective fuel and is generally expected to be the fastest-growing fuel source over the next decade, although in comparison to other fossil fuels⌀ such as coal, it is in short supply. Known reserves will be exhausted early next century if current consumption levels are maintained. Of the world's 86 trillion cubic metres (cu m) of recoverable reserves, 35 trillion cu m are in the Soviet Union, 11 trillion in Iran and 5.8

trillion in the United States. Natural gas also contains a variety of non-hydrocarbon impurities, the most important being hydrogen, which is extracted commercially.

Annual global production is in the order of 1.9 trillion cu m; the Soviet Union (770 billion cu m), the United States (472 billion cu m) and Canada (85 billion cu m) are the largest producers.

NATURAL RESOURCES   The human race is dependent for survival on resources such as food, fossil fuels◇, water◇, shelter and the climate. These are all 'natural resources' as they are provided by the natural environment. Some of these resources, particularly fuels such as coal◇, oil◇ and minerals◇, occur as the result of centuries of biological or physiological activity and exist in limited supplies which are not easily replenished. Once consumed, they will not be replaced within a time span that is useful for the human race. They are thus referred to as 'non-renewable resources'.

Other resources which replenish themselves comparatively quickly – such as water, timber, food crops and solar power◇ – can, if managed properly, provide a continuing yield to meet human needs. These are referred to as 'renewable resources'. Overexploitation or mismanagement of these resources can lead to their disappearance.

Natural resources such as fossil fuels and water are not distributed evenly around the world, and their use is also inequitable. Less than one third of the world's total population live in the industrialized developed countries, but they consume 70 per cent of the world's natural resources.

NATURAL RESOURCES LEGISLATION   In 1962 the United Nations◇ General Assembly passed a Resolution on Permanent Sovereignty over Natural Resources. The Resolution declared:

> The right of peoples and nations to permanent sovereignty over their natural wealth and resources must be exercised in the interest of their national development and of the well-being of the people of the state concerned.

The resolution was passed by 87 votes to 2, with 12 abstentions. Only France and South Africa voted against.

NEEM   The seed of the neem tree, *Azadirachta indica*, produces an insecticide◇ that affects at least forty major insect pests. The compound paralyses the insects' digestive muscles, causing weight loss, interrupted growth, and problems with moulting. The insecticide is fairly specific, harmless to birds and mammals and, unlike man-made chemicals, does not concentrate in the food chain. It also undergoes natural biodeterioration. The neem tree is common in Africa, India and some parts of Latin America, where Azadrachtin produced from the leaves is already used as an insect repellent, and in soaps and toothpastes. In rural parts of India small pencil-sized sticks are cut from thin branches of the neem tree; these, with one end crushed, are then used as toothbrushes without the need for toothpaste.

NEW INTERNATIONAL ECONOMIC ORDER (NIEO)   A term introduced in 1974 to refer to a far-reaching, comprehensive programme for radical restructuring of the world economy in the interests of developing countries◇ and global equality. The concept was evolved by the nations of the South◇, specifically members of the Group of 77◇ and the Non-Aligned Movement◇. Based on the ideas of national sovereignty and the right of states to choose their own methods of economic development, its main documents were the Declaration and Programme of Action concerned with the setting up of such a new order, passed by the sixth Special Session of the United Nations◇ General Assembly in 1974, and the Charter of Economic Rights◇ and Duties of States, adopted by the General Assembly later in the same year. The most significant component of the NIEO is the Integrated Programme on Commodities (IPC)◇ which, owing to lack of support from the wealthy industrial countries, has failed to have any significant impact.

NEWLY INDUSTRIALIZING COUNTRIES The World Bank◇ and the Organization for Economic Co-operation and Development (OECD)◇ are two global economic institutions that categorize developing countries◇ according to their economic performance. Both agencies use the category of Low-Income Country (LIC)◇. Both also use the term Middle-Income Country (MIC)◇, and the same subdivisions for the countries in this grouping. In the late 1980s, the term Newly Industrializing began to be applied to developing countries that are relatively far along the path to industrialization, such as Brazil, India and Singapore, irrespective of their LIC, MIC or other status.

NEWSPRINT   More than one-third of the world's commercial wood◇ harvest is turned into paper⊙, and as only a quarter of that paper is recycled, the demands of the paper industry are increasing. The newspaper industry remains one of the main consumers of wood-based paper. Paper used to make newspapers is known as newsprint, and a daily newspaper of 32 pages with a print run of 500,000 copies consumes around 75 tonnes of newsprint daily. Every day, at least 450 million people buy a newspaper, and the global demand for newsprint is 29 million tonnes per year – and rising. The average citizen in the industrialized countries uses about 18 kg of newsprint a year, as compared with 1.5 kg in the developing countries⊙.

Worldwide, at least 20 million tonnes of newsprint are manufactured from new material annually, with the United States consuming a third of the total. Newsprint production has a dramatic impact on the environment: trees must be felled and, broadly speaking, a sheet of paper consumes twice its weight in wood. Paper manufacture also consumes vast amounts of energy⊙ – in the United States 10 per cent of all industrial energy goes into making paper – and the large quantities of chemicals needed to treat and bleach wood-chip fibres are the source of a great deal of pollution⊙.

Alternative sources for newsprint production – such as the use of kenaf⊙ rather than wood, and mechanical rather than chemical means to separate wood fibres – are becoming more commonplace. Less energy is needed to turn kenaf into paper, far fewer chemicals are needed in the manufacturing process and fibre-whitening treatment, and kenaf is fast-growing and far more productive than wood. In one year, a hectare of land⊙ planted to kenaf can produce up to five times more paper than a hectare of trees. Tests in the United States have shown that the use of kenaf newsprint provides sharper print definition and good colour reproduction, giving newspapers a brighter, cleaner look. Furthermore, ink does not rub off the newsprint, less ink is required during printing and, more significantly, kenaf-based newspapers show no signs of the 'yellowing' with age that affects most conventionally produced newspapers.

NILE PERCH The Nile perch (*Lates niloticus*) is a fairly large and notably carnivorous freshwater fish⊙ native to Africa. In 1960 this fish-eating species was introduced into the mainly plant-eating community of fish species living in Lake Victoria in Africa. Its presence has dramatically upset the balanced ecosystem⊙ within the lake and adversely affected the majority of the human population⊙

living around its shores. The Nile perch quickly spread throughout much of the 68,600-sq-km lake, usually at the expense of the rich variety of plant-eating fish which had previously formed the basis of the fishing⊙ industry upon which most of the local population depended for their livelihood. By 1984, almost all the native fish of commercial importance had disappeared and fisheries around the lake had become dependent on the Nile perch and four other species, all of which were in decline. Future fishery productivity levels were forecast to stabilize at 80 per cent of what they were before the introduction of the Nile perch.

The local populace has no taste for the Nile perch, which consequently fetches only low prices in local markets. Unlike traditonally caught varieties, it cannot be sun-dried. Instead it has to be smoked, leading to increasing levels of deforestation⊙ as wood⊙ has to be cut for curing. Because of the greater size and strength of the Nile perch, fisherfolk require bigger and stronger nets. This often necessitates the purchase of new equipment, which has driven the poorer fishing families out of business and deprived many of their traditional means of livelihood.

NITROGEN CYCLE The process through which nitrogen, an essential element for plant growth and the formation of animal tissue, passes through the ecosystem⊙. Although the earth's atmosphere contains a great deal of nitrogen, it cannot be used by most organisms as it is in gaseous form. Nitrogen gas in the air is converted (or fixed) to ammonia⊙ naturally by the action of lightning and cosmic radiation. The ammonia can then be converted into nitrogen-containing compounds that can be used by plants. Certain soil⊙ bacteria can also carry out this fixing process, which is repeated artificially by humans during the industrial manufacture of fertilizers⊙. Bacteria convert atmospheric nitrogen to nitrites, then to nitrates, which can then be used by plants to make amino acids and proteins. Some species of nitrogen-fixing bacteria live in a mutualistic relationship in specialized structures on the roots of a specific group of plants – leguminous⊙ plants, including the peas, beans and certain trees. The presence of these plants greatly increases the nitrogen content of the soil and hence improves fertility.

Nitrogen in the form of inorganic compounds (nitrites and nitrates) in the soil is absorbed by plants and turned into organic compounds such as proteins. Some of the plant tissue containing the proteins is then eaten by herbivores; the nitrogen-containing compounds are released during the digestive process for use by the herbivorous animals themselves to form their own requisite amino acids and proteins. This process is repeated when herbivores are eaten by carnivores, and so on up the food chain. The nitrogen is returned to the soil, either through excreta or when an organism dies – the micro-organisms responsible for decomposing dead tissue return the nitrogen in its inorganic form.

NITROGEN OXIDES Nitrogen oxide, and other oxides of nitrogen (the mixture often being designated by the symbol NO_x), are commonly formed during the combustion of fossil fuels◇ and are vented from power stations and in exhaust fumes from vehicle engines. Nitrogen oxide acts as a catalyst in the formation of ozone◇ at ground level. This low-level, trophospheric ozone is thought to damage trees and vegetation, contribute to the formation of smog, and cause breathing difficulties, particularly among asthmatics.

In the stratosphere, oxides of nitrogen contribute to the destruction of the protective ozone layer. In the upper atmosphere one molecule of nitrogen can destroy ten ozone molecules, so that NO_x contribute 6 per cent of the greenhouse effect◇.

In 1988, two international treaties concerned with limiting emissions of nitrogen oxides emanating from anthropogenic sources were signed at a meeting in Bulgaria. Twelve countries signed a treaty which called for a 30 per cent reduction in emissions of these gases by 1998, a move endorsed by the European Community◇. The signatories were Austria, Belgium, Denmark, Finland, France, Federal Republic of Germany, Italy, Lichenstein, the Netherlands, Norway, Sweden and Switzerland. The second treaty, signed by twenty-four nations including the Soviet Union, the United Kingdom and the United States, calls for a freeze in nitrogen oxide emissions – at 1987 levels – by 1994. Nitrogen oxide emissions add around 12–15 million tonnes of nitrogen to the atmosphere each year.

In the Federal Republic of Germany, where around half of all nitrogen oxide pollution◇ arises from power plants and factories, a new blend of chemicals called NO_x-OUT is undergoing trials. The mixture reputedly removes nitrogen oxides from the exhaust gases of power plants before the NO_x can be voided, at a cost of one-fifth of current methods. The use of catalytic converters◇ on automobiles◇ also reduces nitrogen oxide emissions.

Recent research has found that bacteria living in the Indian Ocean produce significant amounts of nitrous oxide, and biological production of the gas in the oceans appears to be on a par with man-made emissions.

NOISE A form of pollution◇ that is often ignored. Noise disrupts everyday activity, disturbs sleep and prevents people from carrying out their job to the best of their ability. Excessive noise causes stress, damages hearing and precipitates other physiological and psychological problems, and many countries have produced legislation to specify maximum permissible noise levels. The worst noise pollution arises from transport, especially near roads, railway lines and airports, and it is growing in intensity and geographical spread. In 1985, the Organization for Economic Co-operation and Development (OECD)◇ reported that 130 million people – 16 per cent of the total population◇ in OECD countries – were being exposed to noise levels over 65 decibels (dBA).

One of the major sources of noise pollution is aircraft, with people near the world's increasing number of airports suffering the most. Where aircraft noise is concerned, 0.5 per cent of the population in the industrialized Western countries and Japan are exposed to noise levels above 65 dBA. In the United States the figure is higher – over 5 million people, or 2 per cent of the population. The noise caused by modern aircraft taking off and landing has become a major psychological and economic problem, growing in complexity and magnitude. Many airports around the world have been forced to ban night flights to help curb noise pollution. Residents around Heathrow Airport in London (UK) receive government subsidies to help soundproof their homes and, following a fourteen-year-long trial, 3,800 residents near Osaka Airport in Japan received $5 million compensation for disruption and health problems inflicted by excessive noise.

NOMADS Peoples who live in no fixed place but wander periodically with their goods and possessions, their movements governed by the seasonal availability of food, pasture, trade or employment. Nomadism is a traditional way of living that does not blend well into modern societies.

Transhumance is a form of pastoral nomadism in which livestock◇ are moved seasonally between mountain summer pastures and lower-lying winter pastures, or between northern and southern or wet-

and dry-season grazing areas. It has been practised by pastoralist tribes in Africa for centuries. For thousands of years, traditional hunter-gatherers, such as the Australian Aborigines⊙, used to exist in small family groups that spent anything from a few days to a few weeks in a particular vicinity before moving on. In Europe, until relatively recently, traders, tinkers, entertainers and those who provided a variety of services, such as gypsies, travelled widely and at whim seeking work.

Nomadism is now being stamped out in most part of the world. In Africa, the artificial boundary lines drawn up to form national frontiers took no heed of the traditional migrating patterns of many nomadic tribes, who now find themselves unable to cross international borders to carry on their traditional lifestyle. Increasing human population⊙ numbers and density are also restricting nomadism. Similarly, most authorities try to persuade nomadic people to settle in one spot so that they have access to modern medical facilities and other social services, and so that they can be taxed.

In certain parts of the world, notably in Africa, a nomadic lifestyle is the best means of existence in fragile environments. Well-digging programmes in the Sahel have caused pastoralists to stop ranging their herds, leading to subsequent overgrazing⊙ on land that is already marginal. Bans on cyclical movements to new areas of grassland and burning prior to grazing have increased the habitat for tsetse⊙ flies, and all the problems they create, and reduced the nutritional quality of dry-season grass. Moving of herds traditionally allowed pastoralists to exploit fodder sources when the rains failed, and the prevention of transhumance has proved counterproductive, for throughout much of Africa conditions are such that intensive cattle-ranching is simply not efficient and is environmentally damaging. Highly productive cattle breeds cannot be used where tsetse flies exist, and in other areas high densities of animals cannot be supported on the relatively poor soils.

NON-ALIGNED MOVEMENT
A theory originated by Jawaharlal Nehru, Prime Minister of India from 1947 to 1964, which, in general, opposed colonialism, neo-colonialism and imperialism. The theory was formally adopted in 1961 at a twenty-five-nation conference held at Belgrade, delegates mainly representing Asian and African governments, with India and Egypt playing leading roles. Countries within the group were those that did not wish to be considered part of any of the major international groupings existing at the time. The Movement's president is chosen from the leaders of the member nations, and in 1979, under Fidel Castro's⊙ leadership and with the backing of Ethiopia and Vietnam, the Non-Aligned Movement was shifted towards a 'natural' alliance with the Soviet Union, although this was against founding principles. Now, the term refers to those members of the United Nations⊙ who are not associated with either the Western European–North American grouping or the Soviet bloc. Exactly how the Movement will be affected in the long term by the breaking up of the Communist bloc remains to be seen.

The goverments of the Non-Aligned group press for completion of the decolonizing process, more generous terms of trade, development aid⊙ from the relatively rich North⊙, and, in principle, for Northern non-intervention in the South's⊙ affairs. A central tenet is the need for a New International Economic Order (NIEO)⊙. The Movement's membership has grown as more countries gained their independence, and it now incorporates 102 countries.

NON-GOVERNMENTAL ORGANIZATIONS
(NGOs) Sometimes referred to as voluntary agencies. Private organizations of a charitable, research or educational nature concerning themselves with problems of the global or local environment and development on a worldwide, regional or local scale, or the spread of knowledge about them. Some collect funds from the public to undertake development projects or disaster relief; others are active in research into development and environment issues and the greater and freer dissemination of information and findings. Globally, there are now nearly 5,000 international NGOs, representing millions of supporters, dealing with a variety of environment and development issues. Most are post-1945 movements formed out of growing concern for the environment, although the oldest, working on matters of peace and security, date back almost 200 years. The NGO movement is becoming increasingly involved in decision-making processes at both national and international level, now being allowed to contribute to the formulation of agenda at United Nations⊙ meetings. Many of the 'Green'⊙ political parties evolved from an NGO beginning.

NON-PROLIFERATION TREATY
The Treaty for the Non-Proliferation of Nuclear Weapons, initially signed by the Soviet Union, the

United Kingdom and the United States in 1968, came into force in 1970. Contracting parties commit themselves 'to pursue negotiations in good faith on effective measures relating to the cessation of the nuclear⟡ arms race at an early date and to nuclear disarmament'. Over 138 states have now signed the Treaty, including most of those countries that possess nuclear weapons. China and France are not signatories and several other nations with known or suspected nuclear capabilities, including India, Israel and South Africa, have also not signed. Barring the Intermediate Range Nuclear Forces Treaty agreed between the Soviet Union and the United States in 1987, the signatory parties have patently failed to meet their obligations. Moreover, despite the ending of the Cold War, prospects for global agreement and improvement in upholding the Treaty do not look particularly good. At the fourth review of the Treaty, in 1990, the eighty-four countries attending were unable to reach a consensus on the final end-of-conference document.

NORTH An unofficial and loosely defined term held to describe those countries above a dividing line, drawn and emphasized by the Brandt Commission⟡, which differentiates the better-off countries of the 'North' from the less-developed nations of the 'South'⟡. Basically, the North encompasses the wealthier, industrialized countries of Europe, Japan and North America, together with the two developed nations from the southern hemisphere, Australia and New Zealand.

NORTH ATLANTIC TREATY ORGANIZA-TION (NATO) A mutual defence alliance formed in 1949. The treaty, signed by Belgium, Canada, Denmark, France, Iceland, Italy, Luxembourg, the Netherlands, Norway, Portugal, the United Kingdom and the United States, provided for the collective defence of the member states against the perceived threat posed by the Soviet bloc. All member states are bound to protect and aid any member state against attack. NATO also seeks to encourage economic and social co-operation among signatory nations. Greece and Turkey joined in 1952, the Federal Republic of Germany in 1955 and Spain in 1982.

The chief body of NATO is the Council of Foreign Ministers, with representatives functioning in permanent session. Both the Supreme Allied Commanders in Europe and the Atlantic are Americans, although a British admiral customarily fills the position of Allied Commander, Channel. In 1960 a permanent multi-national Allied Mobile Force (AMF) was established to provide a rapid response should an invasion of any member nation occur.

NATO has been beset with problems since its inception, notably the hegemonial position of the United States and the presence in Europe of nuclear weapons under US control. The operational costs of the pact and what proportion individual nations should pay also proved to be matters of discord, as is the problem of weapons standardization.

NATO's military headquarters is called Supreme Headquarters Allied Powers Europe (SHAPE), and when France withdrew from the command organization – but not the alliance – in 1966, SHAPE moved from France to Belgium. Greece withdrew from NATO politically, but not militarily, from 1974 to 1981.

The role and effective future of NATO as an organization was thrown into doubt in 1990, after the virtual collapse of the opposing Warsaw Pact⟡, the reunification of Germany, the democratization process in Eastern Europe⟡, and the signing of the Conventional Forces in Europe (CFE) Treaty. Under the terms of this Treaty, not only would military forces and arms be reduced to near parity, but the sixteen NATO members and the six Warsaw Pact states formally confirmed that they no longer regarded each other as adversaries.

NUCLEAR ENERGY The energy⟡ produced through changes in atomic structure, via either nuclear fission or nuclear fusion. The energy liberated in fission occurs when a heavy atomic nucleus, such as uranium⟡, splits into two or more parts, the total mass of the parts being less than the mass of the original nucleus. The difference in mass is equivalent to the energy required to bind the original nucleus together. In a fusion reaction, two lighter nuclei, such as hydrogen and deuterium, collide and combine to form a stable nucleus, such as helium. As the nucleus formed is lighter than the sum of the two component nuclei, energy is given off. The fusion process is the basis of the hydrogen bomb.

NUCLEAR POWER In nuclear power stations, the energy⟡ produced by fission reactions is harnessed to produce heat at a controlled rate – usually to create steam to drive a turbine – in one of two ways. Both use natural uranium⟡, which contains less than 1 per cent of the fissionable U-235 isotope, the rest being U-238. If a nucleus of uranium-235 is struck by a neutron, a U-236 nucleus is formed. This immediately splits into two parts, liberating two or three more neutrons in the process. These neutrons will cause further splitting of nuclei and a chain reaction will build up.

In a thermal reactor, the chain reaction is prevented by use of a moderator which slows down the neutrons so that they are not absorbed by the U-238 nuclei. In a fast breeder reactor⌀, the core is surrounded by natural uranium. Neutrons escape into the surrounding blanket and collide with U-238, eventually forming the fissionable plutonium⌀ isotope Pu-239. More Pu-239 is formed than is used to enrich the fuel in the core. Fast-breeder reactors are fifty times more economical in uranium usage than thermal reactors, although they operate at extremely high temperatures and require liquid metal coolants.

The majority of commercial reactors are of the thermal type and the thermonuclear reactor is likely to be a major source of power until well into the twenty-first century; by 1990 these reactors already accounted for 5 per cent of global energy production.

The world's first commercial-scale nuclear reactor began to operate in Britain in 1956. By the end of the 1980s, 435 nuclear reactor plants around the world were producing 17 per cent of the total supply of electricity. In 1988, 114 further plants were under construction, 16 were on order and 96 were at the planning stages. During 1989, 10 new nuclear power plants were connected to national electricity grids. At the end of the year, 319,000 megawatts (MW) of electricity were being produced by nuclear power plants, 9,000 MW more than in 1988.

The history of the nuclear power industry has not been free of major accidents. The three most serious and well publicized have been those at Chernobyl⌀ in the Soviet Union in 1986 (caused by a leak from a non-pressurized boiling-water reactor, one of the largest in the Soviet Union), at Three Mile Island⌀ in the United States in 1979 (a pressurized-water reactor leaked radioactive matter due to a combination of mechanical and electrical failure as well as operator error), and in the United Kingdom in 1957 when fire destroyed the core of a reactor at Windscale (Sellafield⌀), releasing large quantities of radioactive fumes into the air.

The chequered history of the nuclear power industry, the aftermath of potentially catastrophic accidents, and the sheer costs involved – of both building and decommissioning⌀ plants – have begun to have a decisive effect, and the number of facilities planned and under construction at the beginning of the 1990s was falling. Between 1970 and 1987 149 nuclear plants were cancelled, adding to the 58 that had been retired between 1956 and 1987, having come to the end of their working lifetime.

All nuclear power produced has come from uranium. A tonne of natural uranium can produce more energy than 10,000 tonnes of coal⌀. Fears that wider and unregulated use of nuclear energy⌀ might bring a greater spread and variety of nuclear weapons stimulated the production of the Non-Proliferation Treaty⌀ (1970) and strengthened the lobby calling for nuclear power plants to be scrapped.

NUCLEAR SHIPPING The incidence of nuclear accidents involving naval vessels became of increasing concern during the 1980s. The Soviet Union's MIKE-class submarine which sank in the Norwegian Sea in 1989 carried two nuclear torpedoes and had twin nuclear reactors. The craft sank in waters up to 1,800 m deep, well beyond its operating depth. The reactors will be subject to extremely high pressure and attack by sea water⌀, and corrosion is also expected to release the plutonium⌀ from the nuclear warheads on the torpedoes. The total amount of radioactivity⌀ released by this vessel alone could be as high as 20 million curies, 40 per cent of that released by the nuclear accident at Chernobyl⌀.

By the end of the 1980s some 360 ocean going vessels were nuclear-powered. The number of naval nuclear reactors was estimated to be 544, in addition to some 15,550 nuclear weapons stationed aboard ships or submarines. By 1990, about fifty nuclear weapons and nine nuclear reactors had officially been lost at sea by the navies of the United States and the Soviet Union, including the loss of an H-bomb from the US aircraft carrier *Ticonderoga* off the Japanese coast in 1965.

NUCLEAR TEST BAN TREATY A treaty signed in 1963 banning nuclear testing on the ground, in the atmosphere, in space and underwater. The signatories were the Soviet Union, the United Kingdom and the United States, although many other countries agreed to adhere to its terms. It made no attempt to control or limit the stockpiling of nuclear weapons and was therefore rejected as ineffectual by both China and France.

Between 1945 and 1989, there were 1,819 recorded nuclear tests. Tests have been carried out on every continent, an average of one test every nine days. China, the Soviet Union and the United States carry out tests at isolated sites within their mainland territory. Britain makes use of the US testing site in Nevada, while France has two test sites in French Polynesia in the Pacific Ocean.

NUCLEAR WASTE   A term which embraces a vast range of materials which remain radioactive, and therefore dangerous to most forms of life, for hundreds, if not thousands, of years. The nuclear industry produces a large amount of radioactive waste◇, swollen by lightly contaminated overalls, gloves, and medical phials from hospitals and research institutions. In addition, tailings from uranium◇ mills and highly radioactive liquid produced during the reprocessing of nuclear fuel also has to be contended with. By 2000, the inventory of high-level wastes alone is projected to reach 150 billion curies.

The problem of what to do with radioactive nuclear waste has yet to be resolved. Before 1983, low-level radioactive waste in the United Kingdom and elsewhere was disposed of at sea. This practice was stopped after demonstrations by seamen and a moratorium was agreed under the London Dumping Convention◇ – although such waste can be dumped directly into the sea from land; this is done by both France and the United Kingdom. Low-level radioactive waste is not a major problem, as it is possible to bury it in such a way that it will not pose a risk higher than that from normal background radiation. At some stage, however, repositories must be designed that are capable of long-term containment of the levels of radioactivity produced by highly radioactive waste if future generations are to be protected. Uranium-235 has a half-life of 713 million years but decays far more quickly than Uranium-238, which has a half-life of over 4,000 million years.

Until the end of the 1980s, high-level-radiation wastes were generally stored near where they were produced, in stainless steel tanks, constantly cooled and monitored. Ultimately the intended goal is to solidify the waste in glass through the process of vitrification, for eventual burial in natural hard rock strata. Vitrification should reduce the volume of waste and make it easier to handle, but critics claim that the glass blocks may themselves disintegrate in the future before the radioactive waste is safe. Exploration for safe, deep underground sites in which to bury the wastes in Europe and North America have so far been relatively unsuccessful. Geological faults and the effect of possible earthquakes make it impossible to guarantee safety of any site for the length of time the waste must be kept.

Some nations with a nuclear power◇ industry have taken to sending contaminated fuel overseas for reprocessing and seek to dispose of their waste elsewhere. The Federal Republic of Germany cancelled the nation's major reprocessing complex and now sends fuel to the United Kingdom for reprocessing. In addition, Germany exports nuclear waste to China for burial in the Gobi Desert, providing nuclear technology and know-how in exchange.

O

OCEAN THERMAL ENERGY CONVERSION (OTEC)   A system, still at the development stage, for producing power by exploiting the temperature difference between the warm surface and cool bottom layers in tropical seas. The system can generate power as well as desalinate brine to produce fresh water◇. It is estimated that a system can be constructed which will produce 10 megawatts of electricity which, in turn, will produce 20 million litres of water◇, enough to supply the needs of 20,000 people.

Water from the top layers of the tropical ocean, already at 26°C, is converted to steam to run conventional turbines. When this steam is condensed, the process produces fresh water. The turbine heat-exchanger that condenses the steam is cooled by water at 6°C pumped up from around 1 km below sea level.

OGALLALA AQUIFER The largest source of fossil groundwater◇ in the United States, the Ogallala aquifer stretches from southern South Dakota to north-west Texas. Expansion of agricultural production in the 'Great Plains' of the USA has been dependent on irrigation◇, most of which was supported by overexploitation of the waters in the Ogallala. From 1944 to 1978 irrigated land◇ in the area rose almost fourfold, from 21,000 to 80,000 sq km. Approximately 20 per cent of all irrigated cropland in the USA is fed by water from the Ogallala. Since 1940 over 500 cu km have been withdrawn from the aquifer, far in excess of renewal. Depletion has reduced water◇ reserves in some states to half their normal level, with enough to permit pumping for another decade and no more. If current use trends continue, it is estimated that 20,000 sq km of land will have to be taken out of production by 2000. At least 6,000 sq km have already been taken out of production in the states of Colorado, Kansas and Nebraska. In all, the irrigated area in the six states that rely most heavily on the Ogallala has already shrunk, declining 15 per cent between 1978 and 1984.

OIL Crude petroleum, or rock oil, is the thick, dark-green mineral oil that occurs in permeable underground rock. It is derived from the remains of living organisms that died millions of years ago. Under the effects of heat and pressure the organic material changed into oil, which became trapped in underground reservoirs. The modern oil industry began in 1859 when oil was discovered in Pennsylvania. Petroleum cannot be used in its crude form and has to be refined by fractional distillation, in which the components of the oil are separated according to their boiling points. Blending of the final products can produce fuels such as petrol, kerosene and aviation spirit, as well as a range of chemicals for other uses.

In the 1960s coal◇ was replaced as the world's biggest source of industrial energy◇ by oil and the natural gas◇ that is found trapped with it underground. The price, conservation◇ and political importance of oil consequently began to have a major global impact. In 1961, the Organization of Petroleum Exporting Countries (OPEC)◇ was set up to protect producing countries, mainly in the Middle East, from exploitation. Its advent ended the era of cheap energy and precipitated a world energy crisis during the 1970s. Actions by OPEC caused oil prices to triple during 1973-4 and again between 1978 and 1980, During this time, the economies in the industrialized nations, which were heavily dependent on oil, entered a lengthy recession. Oil output fell by 10 per cent during the following decade as non-oil-producing countries tried to cut their energy demands or find alternate sources. In 1974, the International Energy Agency (IEA)◇ was established to protect consumers. Since then, various factors have combined to stabilize the situation, notably the discovery and flow of North Sea oil.

Global production of crude oil, only 275 million tonnes in 1938, had risen to 1·05 billion tonnes by 1960 and rocketed to 2·28 billion tonnes by 1970, before peaking at 3·1 billion tonnes shortly thereafter. Following the oil crisis, worldwide consumption declined during the latter 1970s and early 1980s, but began to rise again in 1986. By 1988, global production of crude oil was again over the 3-billion-tonne mark; the Soviet Union (624 million tonnes), the United States (409 million tonnes) and Saudi Arabia (255 million tonnes) were the largest individual producers.

Oil continues to maintain its position as the world's major source of fuel, supplying about 40 per cent of global energy. World reserves of oil are estimated to be sufficient to meet prevailing levels of demand for only a further forty years, with almost 60 per cent of these reserves located in the Middle East. The world's recoverable resources of crude oil in 1990 were estimated to be in the region of 94 billion tonnes; Saudi Arabia (23 billion tonnes) and Kuwait (10 billion tonnes) have the largest national reserves.

OIL PALM The oil palm, *Elaeis guineensis*, is native to the tropical lowland forest◇ areas of West Africa. It is now cultivated extensively on plantations in the tropical zones of Indonesia, Malaysia, South America, and throughout South-East Asia as the source of palm oil. Several varieties exist, each producing a distinctive fruit and nut; dura is the most common form. The trees are tall-growing (up to 15 m) and produce large, fleshy fruits which contain a white kernel with a hard black shell. Oil palm trees begin to fruit about four years after planting but the first bunches of fruits are usually small and of little economic value. Oil palms do not reach full maturity for eight to ten years. Harvesting is a continuous operation; the number of fruit bunches varies with the season.

Trees tend to produce their maximum yield between the ninth and twelfth year; this ranges from 1,500 to 4,500 kg of bunches per hectare (about 2,500 kg/ha is average).

Sterilization is the name given to the process of boiling the detached fruits and pounding and extracting the oil. It softens the fruit, kills disease-causing organisms and inhibits the enzymes which are normally present in ripe fruits. After boiling, the oil is separated by immersion in water.

Oil palms are of major economic importance. Palm oil is used in the manufacture of margarine and cooking fats, glycerine, oil paints and polishes. It is also used in the preparation of soups and stews and is a major source of Vitamin A. Apart from the oil which is extracted from the fruit, the kernel yields another type of oil which is also used in soap manufacture. One hectare of palms normally produces about 800 kg of palm oil and 100 kg of kernel oil. The trees have a number of other uses: the leaves are used to make brooms, fibres are used to make mats, and the bark is used to make baskets. Tree trunks can be used to make canoes or for building. The shells and fibre of the fruits are used as a fuel, and a highly nutritious palm wine can also be made. The residual material from both these processes can be used as a livestock⌀ fodder. In many plantations, cattle are raised under the tall palm trees as they maximize the productivity of the land⌀ and keep weeds and other shrubs at bay.

Around 4.5 million tonnes of palm oil are produced annually; Malaysia produces almost half the total. Nigeria and Indonesia are the world's other major producers.

OIL SPILLS The disastrous consequences of major marine oil spills first became evident in 1978 when the *Amoco Cadiz*⌀ ran aground off the French coast. Over 220,000 tonnes of crude oil⌀ were released into the sea, washing ashore and contaminating over 130 beaches. At least 30,000 seabirds died as a result, together with 230,000 tonnes of crabs, lobsters and fish⌀. Valuable oyster and seaweed beds were destroyed. Over 400,000 local inhabitants sued the owners of the tanker for compensation totalling $750 million.

An Argentinian supply ship which sank off the Western Antarctic⌀ peninsula in 1989 posed the greatest ecological threat in the continent's history. The *Bahia Paraiso* was loaded with over 1 million litres of diesel oil when it sank close to Litchfield Island. The island's unique flora and fauna faced the threat of extinction⌀ from the oil slick emanating from the stricken ship. Experts do not know how long the local wildlife will need to recover from the pollution⌀, for it takes up to a hundred times as long for diesel fuel to break down in sub-zero Antarctic conditions than is required in warmer climes. It is likely that the effects of the oil spill will still be evident in the year 2100.

The *Exxon Valdez*⌀ supertanker spilled about 240,000 barrels of oil when it ran aground in Prince William Sound in Alaska in 1989. Initial clean-up measures were woefully inadequate and it was ten hours before the first containment booms arrived. Failure to use booms meant that chemical dispersants had to be used. These alter the surface tension of the water⌀ so that the oil sinks below the surface. Subsurface oil, however, remains toxic and some of the dispersants also pose a danger to wildlife. Hydrocarbons derived from oil which are toxic to marine life appear in sediment in sub-Arctic waters for at least twelve months following an oil spill.

As the world's fleet of oil-carrying supertankers ages, so the incidence of oil spills is increasing. Most of the world's supertankers, or Very Large Crude Carriers (VLCCS), are fifteen to twenty years old and were believed to have a lifetime of twenty-five years at best, yet they are unlikely to be replaced. A VLCC costs at least $80 million to build, and to show a profit it would have to earn $40,000 per day. In 1990, VLCC charter fees had fallen to $12,000 per day. In efforts to avoid more disasters, international pressure is growing to force tankers to have double hulls – a measure first proposed over thirty years ago but dropped following lobbying by the oil companies.

ONCHOCERCIASIS More commonly known as river blindness, onchocerciasis is a form of filariasis⌀. It is a tropical disease which particularly affects much of West and Central Africa and parts of Latin America and Arabia. Onchocerciasis is a disease of the skin and underlying connective tissue caused by a parasitic worm, *Onchocerca volvulus*. Fibrous nodular tumours grow around the adult worm in the skin; these may take take several months to appear after infection. The skin becomes inflamed and itches, and if there are secondary bacterial infections the nodules degenerate into abscesses. The migration of larvae produced by the worms into the eye can cause total or partial blindness – hence the name river blindness. Where onchocerciasis is endemic, vast tracts of fertile and productive land⌀ along river banks are rendered redundant. Humans cannot cultivate the land for fear of contracting the disease, which is transmitted through the bite of blackflies, *Simulium*, which breed in the rivers.

The World Health Organization (WHO)⊃ report that 17 million people in 34 countries, mostly in tropical Africa, are suffering from the disease. Around 900 million people are at risk, and in Africa at least 320,000 have been blinded. In Latin America, 97,000 people are infected, of whom 1,400 have been blinded.

Onchocerciasis is difficult to diagnose in its early stages. If possible, nodules are removed as and when they appear. Drugs such as suramin and dimethyl-carbamazine have been used in treatment but have proved hazardous as the presence of large quantities of dead larvae or adults in the body can be problematic. A new drug, Ivermectin⊃, provides effective treatment. It has been available since 1987 and has been provided free of charge by the manufacturers to treat those infected with the parasite, but the problem of reinfection remains unsolved.

In West Africa onchocerciasis has been greatly reduced due to a widespread campaign to release biodegradable insecticides⊃ regularly into rivers to destroy larvae of the blackfly vector. Insecticidal bacteria are also being used in attempts to control the fly. A major Onchocerciasis Control Programme was launched in West Africa in 1974, its main strategy being the aerial spraying of breeding sites of the disease-carrying blackflies. As a result of the programme, around 4 million children born since the start of the project no longer risk going blind.

ORAL REHYDRATION SALTS (ORS)

Dehydration as a result of persistent diarrhoea has killed 150 million children since 1950. At least 750 million children suffer from diarrhoeal diseases annually, and at least 10,000 infants under the age of five die every day as a result of diarrhoea. Dehydration and malnutrition⊃ cause these deaths, 70 per cent of which could be prevented through the use of rehydration salts, a carefully balanced mixture of sugar⊃ and salt, which need to be mixed with water⊃ and given orally to affected children. Between 1983 and 1985 the use of ORS trebled, and the lives of 1 million children were being saved by the salts each year. The use of contaminated water for mixing the salts can have a counterproductive effect; boiling water before use reduces this risk. Researchers have also discovered that feeding infants water in which rice⊃ has been boiled has a similar effect to the use of rehydration salts.

In 1987, 60 per cent of the population of the developing world had access to packets of ORS, with three-quarters of the salts used coming from the developing countries⊃ themselves – manufacture of the salts had begun in fifty-eight nations. Each packet usually costs less than 10 cents (US), and normally two packets mixed with boiled water are sufficient to cure the dehydration caused by diarrhoeal diseases. Despite this success, by 1990 less than one-third of all the children in the world suffering bouts of diarrhoea were receiving ORS therapy, and in Africa fewer than one in four children were being given ORS treatment.

ORGANIZATION OF AFRICAN UNITY

(OAU)   Founded in 1963 at Addis Ababa, the OAU brought together the previously rival 'Casablanca' and 'Monrovia' groupings of African nations. Its aims are to promote unity among African countries, oppose colonialism, mediate on territorial disputes between member states, and discourage attempts to change frontiers that cut across tribal areas. The OAU also attempts to co-ordinate efforts to raise living standards throughout Africa via improvements in the economic, cultural and political spheres. It has helped to co-ordinate policies on the South African issue and has taken a leading role in promoting environmental improvement policies throughout the continent. The membership of fifty-one comprises all states on the African continent plus the Saharan Arab Democratic Republic, which is supported by Algeria and opposed by Morocco, and over which the OAU remains split. South Africa and Namibia are not members.

The Heads of State Summit Meeting which governs OAU activities is held annually, but a permanent Standing Committee is maintained in Ethiopia. As well as the permanent Secretariat, the OAU operates through a number of specialized commissions dealing with economic, social, technical, scientific, cultural and defence co-operation. The organization has been beset with disputes, and in 1982 divisions amongst member states prevented the Summit meeting from being held. The Joint African and Mauritanian Organization (*Organisation Commune Africaine et Mauritienne*) (OCAM), founded in 1962, works for African solidarity in Francophone African countries within the framework of the OAU.

ORGANIZATION OF AMERICAN STATES

(OAS)   Originally established in 1890 to encourage friendly relations between regional member states, the organization has undergone a number of changes of name. Between 1890 and 1910 it was known as the International Union of American Republics, and between 1910 and 1948 it adopted the title Pan-American Union. The OAS as it now is was formed in 1948 to foster mutual understanding and co-operation between all American republics and has a peacekeeping role, aiming to create collective security. It is based on the principle of the Monroe Doctrine of 1823, which formed the backbone of United States' foreign policy. This effectively warned European powers not to intervene in the Americas and, in return, the USA would refrain from intervention in Europe. In 1962, Cuba was expelled from the OAS because of its decision to house missiles from the Soviet Union. Members are the United States and thirty independent South and Central American countries and Caribbean states. The OAS is now largely devoted to social and economic development within Latin America.

ORGANIZATION OF CENTRAL AMERICAN STATES (ODECA)

The *Organizacion de Estados Centro Americanos* (ODECA) is an international organization formed in 1951. Its members include Costa Rica, El Salvador, Guatemala, Honduras, and Nicaragua. Based in Guatemala, its aim is to promote social, cultural and economic development in member states through joint, coordinated action. It was originally inaugurated to oversee the introduction of a 'common market' to improve trade and economic activity, but its aims were expanded in 1962 to encompass political, educational, defence and environmental matters.

ORGANIZATION FOR ECONOMIC CO-OPERATION AND DEVELOPMENT (OECD)

An intergovernmental organization of twenty-four industrialized countries, established in 1961 and based in Paris, which attempts to co-ordinate the economic policies of member states. It replaced the Organization for European Economic Co-operation, which was established in 1948 to promote economic recovery in Europe under the Marshall⊃ Plan, when Canada and the United States joined and the scope of the organization was extended to include development policies and overseas aid⊃.

The OECD aims to achieve the highest levels of sustainable economic growth and employment and a rising standard of living in member countries, to contribute to sound economic expansion in all member as well as non-member nations, and to further the expansion of world trade in accordance with international obligations. One specialist committee, the Development Assistance Committee (DAC)⊃, serves solely to improve the flow of resources to the less-developed countries in the Third World.

Member countries are Australia, Austria, Belgium, Canada, Denmark, Finland, France, Federal Republic of Germany, Greece, Iceland, Ireland, Italy, Japan, Luxembourg, the Netherlands, New Zealand, Norway, Portugal, Spain, Sweden, Switzerland, Turkey, the United Kingdom and the United States. Under the terms of a special agreement, the Yugoslav government is also represented at some OECD meetings.

ORGANIZATION FOR INTERNATIONAL ECONOMIC CO-OPERATION (OIEC)

An international body proposed as a successor to the Council for Mutual Economic Assistance (CMEA)⊃, which was dissolved in 1991. Member countries, ex-Warsaw Pact nations plus Vietnam, Cuba and Mongolia, will be joined by Germany as an observer. The new body is to help build up market-based links between companies and industrial sectors and promote trade and economic reforms across the Eastern European region. Its role is expected to be advisory rather than prescriptive, as was the CMEA.

ORGANIZATION OF PETROLEUM EXPORTING COUNTRIES (OPEC)

A body established in 1960 in Baghdad to co-ordinate the price and supply policies of the world's major oil-producing states. It was formed primarily as a way of creating a unified body which could represent the interests of the eleven chief oil-exporting nations (Abu Dhabi, Algeria, Indonesia, Iran, Iraq, Kuwait, Libya, Nigeria, Qatar, Saudi Arabia and Venezuela) in dealings with the major transnational oil⊃ companies. OPEC also aimed to strengthen and improve the position of Third World⊃ countries by forcing the developed world to provide them with technology and open up the markets in the developed world to the resultant products produced by the Third World countries.

Based in Vienna since 1965, OPEC is the only example of a cartel dealing with a primary product that has met with comparative success. Deciding in 1973 to double its share in the receipts from oil, it has continued to control the price.

Concerted OPEC action in raising crude oil prices sharply in the 1970s triggered worldwide recession while generating huge revenues for OPEC member states. Some of these sums were donated or recycled to poorer developing countries⌂. The global crisis precipitated during the 1970s, when oil prices more than tripled, caused the introduction of extensive energy-saving programmes throughout the industrial world and intensified the search for alternative sources of energy⌂, as well as the exploration for and exploitation of any oil deposits outside the jurisdiction of OPEC countries, no matter where or under what conditions they were to be found. In the 1980s, OPEC's dominant position was undermined by reductions in demand for oil in the industrialized countries and by rising production in non-OPEC nations, notably from North Sea oil controlled by Britain and Norway. These factors contributed to a sharp fall in the world price of oil and forced OPEC into limiting production by its members.

OSLO TREATY   Signed in 1972 by most of the countries bordering the North Sea, the Oslo Convention governs the disposal of wastes in the North Sea and north-east Atlantic. The Treaty came into effect in 1974. Like the London Dumping Convention⌂, the Treaty has a 'black' list of substances which can be disposed of only in trace quantities and a 'grey' list of substances that can be dumped only with full permission. On the black list are organohalogens, mercury⌂, cadmium⌂ (with all related compounds), and non-degradable plastics. Plans are currently being discussed to introduce regulations to reduce the build-up of nutrients, phosphorus and nitrogen⌂ in the sea. These are to blame for the increasing eutrophication⌂ conditions of the water⌂ and the frequent appearance of algal blooms. In order to accomplish this, nutrients will have to be removed from sewage effluent before it is discharged.

The United Kingdom is baulking at this, as the country's newly privatized water industry would be unable to meet the costs of complying with any such directive. Britain is the only nation that dumps sewage sludge into the sea – around 9 million tonnes each year, about 60 per cent of which goes into the North Sea. The alternatives are incineration⌂, landfilling and spreading sludge on farmland. The Oslo Convention countries also face major disagreement over the phasing out of the dumping of polychorinated biphenyls (PCBs)⌂. Denmark, Norway and Sweden want this accomplished by 1995, the Federal Republic of Germany by 1998, Belgium, Britain and the Netherlands by 2005, and France by 2010. A further dilemma over how to curtail the discharge by the UK and Norway of oil-

contaminated waste from offshore installations in the North Sea is also unresolved.

OTTAWA AGREEMENTS   A series of agreements concluded at the Imperial Economic Conference held in the Canadian capital in 1932. Under the new accords, preferential trade tariffs between Great Britain and the Dominions were established. The agreements marked the abandonment of the policy of free trade⌂ that had traditionally been adopted by Britain.

OVERGRAZING   The overstocking of land⌂ with grazing animals, usually small stock such as goats, sheep, and cattle. It is now a major threat to the world's grasslands and rangelands. Grazing lands and forage support most of the world's 3 billion head of domesticated grazing animals. The productivity of these lands is low. One hectare (ha) of fertile, well-managed pasture can support 3–5 animals; in arid lands 50–60 ha may be needed to sustain 1 animal. In Africa, Asia and the Americas, hills, arid and semi-arid land, or any land that is poor, with fragile soil⌂, is utilized to rear livestock⌂. However, disease and other environmental constraints severely restrict production. Africa has 300 million head of cattle, but they generally weigh less than half their counterparts in other regions of the world.

The pressure to maintain or raise livestock numbers mounts as the human population⌂ increases. Overgrazing is further exacerbated when temperate-forest livestock species are used in arid areas. Dryland ungulates, such as antelopes, are adapted to prevailing conditions and have small water⌂ requirements, whereas cattle, sheep and horses do not belong in arid areas; their water needs are too great. Yet between 1950 and 1976, the world beef herd doubled. Overgrazing in many traditional cattle-raising areas means that herds cannot now be increased to keep up with human population expansion and demands. Cattle and sheep graze on grasslands, whereas goats browse on shrubs and the lower parts of bushes and trees and eat virtually anything. Goats can be especially valuable as they provide milk and meat from plants that man cannot use and, like camels, they are extremely hardy and adapted to survive in drought⌂ conditions. But they are extremely destructive feeders, particularly when large numbers are kept on marginal land.

Overgrazing leads to the elimination of plant cover, especially in high-density areas such as those close to waterholes. Soil can be destroyed by faeces or irreversibly damaged by compaction, particularly where large densities of sheep are grazed. Soil is laid bare and prone to erosion by wind and rain and ultimately becomes desertified. In Australia, within 200 years of colonization, 2 million sq km of arid land, and over 800,000 sq km of semi-arid land, had been overgrazed by sheep and cattle, to the point of serious loss of fertility. Traditional pastoralists avoid these problems by ranging their herds extensively. Where pastoralists are encouraged to become more sedentary, problems quickly arise.

In Africa, overgrazing has become commonplace as traditional transhumance has been prevented. During the 1950s and 1960s aid⌁ agencies drilled deep-water wells in the Sahel to provide water⌁ for livestock in areas where pastoralists normally ranged their herds. When rains began to fail, nomads⌁ moved their herds to the wells. In parts of the Sahel, 6,000 head of cattle were surrounding wells and grazing land that could support only 600. Millions of cattle died – most from hunger, not thirst.

Overgrazing of rangelands is now a problem being faced throughout the semi-arid tropics. In the Sudano-Sahelian region of Africa, where livestock-rearing has been steadily intensified, 3.4 million sq km, or 90 per cent of the total, is desertified, much as a result of overgrazing. In North America 42 per cent of rangeland is desertified, as are 72 per cent of South American and Mexican rangelands. In the western United States cattlemen have used their political clout to overstock federal grazing lands and national forests⌁. Indigenous animals such as big-horn sheep and the desert tortoise are becoming endangered as a result. Overgrazing in India is also threatening indigenous wildlife.

OVERSEAS DEVELOPMENT INSTITUTE (ODI)

An independent organization set up in 1960 in the United Kingdom as a centre for research and information on economic and social development issues. It is financed by grants and donations from government and private sources in Britain and overseas. The Institute organizes meetings and seminars, publishes books and academic works, and produces a journal and briefing pamphlets. Fellowship schemes are also available which enable young economists to undertake two-year assignments in developing countries⌁.

OXFAM

The Oxford Committee for Famine Relief, established in Britain in 1942 to relieve famine and poverty worldwide and help finance long-term aid⌁ projects. It is an independent voluntary organization, with affiliated but autonomous national organizations in other countries, campaigning to involve as many people as possible in the cause of world development and raising funds for the relief of poverty and suffering among under-privileged communities anywhere in the world. OXFAM channels the funds raised through a variety of agencies working in the field. Funds are allocated for specific projects, predominantly those of a self-help development type. Specialist projects of less direct benefit to poor communities but of great importance to global development, such as the International Vegetable Germ Plasm Bank, also receive OXFAM support. By the end of the 1980s OXFAM was receiving an income of around $100 million, with 79 per cent of the total being spent on relief and long-term development projects overseas.

OZONE

A condensed, gaseous form of oxygen, pale blue, unstable and toxic. It is one of the main components of ground-level smog, which damages trees and vegetation and causes human ill health, especially respiratory problems. In the stratosphere, it provides a barrier against ultraviolet (UV) light and stops other harmful high-energy solar radiation from penetrating through to the earth's surface. Exposure to UV is closely linked with the incidence of melanoma, the commonest form of skin cancer⌁, and non-melanomatous skin cancer. During the 1980s global attention was drawn to the fact that chemicals being produced by industrial processes, notably chlorofluorocarbon (CFC)⌁ gases, were gradually destroying the protective ozone shield. Experts predict that cases of skin cancer may be expected to increase significantly over the next forty years. Incidence of non-melanomatous skin cancer will rise by up to 36 per cent, given a 6 per cent reduction in global ozone levels between 1970 and 2030. Every 1 per cent decrease is also predicted to make 100,000 people blind due to UV-induced eye cataracts. Aquatic organisms, especially marine phytoplankton and zooplankton and the larval stages of certain fish⌁, are extremely sensitive to even small increases in UV radiation.

Trophospheric Ozone – Ozone is a major constituent of photochemical smog at ground level, synthesized from the emissions from car exhausts in the presence of sunlight. In the northern hemisphere the lower-atmosphere concentration of ozone has been increasing by about 1 per cent annually. Measurements, most of which have been made in the northern hemisphere, suggest that as a result of this rise, less harmful UV is getting through to the ground – at least in northern countries. Ultraviolet light is scattered by air molecules and dust particles in the lower atmosphere, causing it to take zigzag paths. Short wavelengths are scattered more than longer wavelengths. In summer or at low latitudes, when the air is dry and dusty, a combination of scattering and extra ozone could explain why UV declines in the north. South of the equator, not only is there less ozone in the upper atmosphere but the air is generally cleaner, so reducing scattering and probably resulting in an increase of the UV reaching the ground.

Stratospheric Ozone – The ozone layer in the upper atmosphere is between 20 and 50 km above the earth. Ozone forms as the result of the dissociation by solar UV radiation of molecular oxygen into single atoms, some of which then combine with undissociated molecules to form ozone. It absorbs UV-radiation, so protecting the earth's surface from excessive amounts. Over the past ten to fifteen years, ozone in the upper atmosphere has declined by several per cent.

South of the equator the ozone layer depletion is most marked, especially over the polar area. The average annual depletion for all areas south of 60° South is already over 6 per cent. At all southern latitudes ozone in the stratosphere is being lost, even in regions not subject to polar weather. In the cold, still air of the Antarctic night clouds form, containing nitric acid and water ice. These stratospheric clouds drain nitrogen⌐ out of the air, leaving behind chlorine in the active form of its oxide. It is this which directly destroys ozone in the spring, destructive photochemical reactions beginning as sunlight returns. Outside the polar region, where there are few ice crystals, it is believed that the same process is occurring, with droplets of sulphuric acid, increasing through anthropogenic pollution⌐, taking the place of the crystals. In the Arctic, ozone is also being depleted, but less severely. This is partly due to the fact that the stratosphere above the North Pole is warmer than that over the South Pole, and partly because air above the Arctic mingles with other air flows and is not so isolated from the rest of the atmosphere during the winter months.

Many industrial pollutants with the potential to destroy ozone, including chlorine and nitrogen compounds, never reach the stratospheric ozone layer: they are trapped below this level and fall to earth as acid rain⌐. The exhausts of rockets and supersonic aircraft, however, which penetrate the ozone layer are particularly destructive. Where solid fuels are used, the damage is severe. A single flight of the United States space shuttle emits 187 tonnes of chlorine and seven tonnes of nitrogen compounds before it reaches a height of 50 km. One tonne of chlorine can destroy 100,000 molecules of ozone, and a molecule of nitrogen dioxide can destroy 10 ozone molecules. A single flight also produces 840 tonnes of other gases that remove ozone molecules from the atmosphere on a one-to-one basis. A single shuttle flight is capable of destroying up to 10 million tonnes of ozone, and as few as 300 flights could destroy the ozone layer entirely.

Several attempts have been made by the global community to introduce measures that would help protect the stratospheric ozone shield, most notably the Montreal Protocol⌐, the stumbling block being who should pay for their implementation. Having originally refused to contribute to an international fund suggested for this purpose, the United States, along with other developed countries, finally agreed in 1990 to contribute to a $230 million fund to be set up to pay for ozone-friendly technology for developing countries⌐ and Eastern Europe⌐, mainly to allow these countries to replace their CFC production with safe alternatives. The USA proposed to contribute $20 million annually – although it had made $1 billion in taxes levied on chemicals that have depleted the ozone layer.

P

PALAU *See* Belau

PALME, SVEN OLOF (1927–86) Swedish politician and international peace advocate. A Social Democrat, he was Prime Minister from 1969 to 1976, and again from 1982 for four years until his assassination in Stockholm. He was widely recognized as a leading campaigner for peace. From 1968 onward, his criticism of US policy during the Vietnam War⬦ and his support for US Army deserters who sought refuge in Sweden led to a period of strained relations between Sweden and the United States. Best known for his leadership of the Independent Commission on Disarmament and Security Issues, whose findings were published in *Common Security – A Programme for Disarmament* in 1982, he also acted as a United Nations⬦ special envoy to mediate in the war between Iran and Iraq.

PAN-AFRICAN INSTITUTE FOR DEVELOPMENT (PAID) An international organization, based in Switzerland established to promote and broaden co-operation and exchange between the Francophone and English-speaking countries in Africa. A major goal is the improvement of information dissemination and exchange between and within its own regional institutes based in Africa. The PAID organizes extension training, research, and consultative support programmes through these regional centres.

PANAMA CANAL A canal, over 82 km long, across the Isthmus of Panama which connects the Atlantic and Pacific Ocean. Construction was begun towards the end of the nineteenth century by the French Panama Canal Company, but excavations were halted in 1889 owing to a combination of the firm's bankruptcy and the effects of mosquito⬦-borne diseases such as malaria⬦ and yellow fever⬦, which killed off hundreds of construction workers. In 1903 the United States was granted construction rights by the newly independent Panamanian government and the Canal was officially opened in 1914, allowing ships to sail from the Atlantic to the Pacific Ocean without having to travel all the way round the coast of South America. Under the terms of the 1903 treaty the USA acquired sovereignty in perpetuity over the Panama Canal Zone, a region extending 5 km on either side of the Canal. In return, Panama received $10 million and an annuity. In 1978 both countries agreed that the Canal should revert to Panamanian sovereignty, together with the Canal Zone, by the year 2000. Panama assumed jurisdiction over the Zone in 1979.

The canal is of vital economic importance to Panama; tolls bring in $350 million annually, 8 per cent of the country's Gross National Product (GNP)⬦, yet deforestation⬦ along its length is threatening its existence. As soil erosion⬦ increases, sediment in the Canal builds up and dredging costs increase. The economic and strategic importance of the Panama Canal is now threatened by plans to build a similar interocean canal linking the Pacific and the Atlantic through Nicaragua. The Panama Canal is already too small, capable of accommodating only relatively small ships of up to 30,000 tonnes. A 1984 study estimated that the cost of enlarging the Panama Canal to allow passage for bigger ships would be in the region of $40 billion, while the cost of building a new canal through Nicaragua is likely to cost only a quarter of this figure.

PANOS INSTITUTE An independent, international organization based in London with offices in continental Europe and the Americas. PANOS undertakes research and information dissemination activities on all aspects of the environment and sustainable development⬦. Through a programme of regional partnerships, PANOS aims to strengthen the information capacities of the media and Non-Governmental Organizations (NGOS)⬦ in the Third World⬦. The Institute carries out a series of thematic information programmes, and publishes reports and analyses of development policies and actions at all levels, together with regular informative materials, trying where possible to use researchers and writers from the Third World. PANOS is funded by contributions from governments, international agencies concerned with the environment and development, and independent foundations.

PANTANAL   The Pantanal – 13 million hectares of marshland and seasonally inundated savannah in south-west Brazil – is the world's largest wetland⟳. The region is under a continuing and increasing threat. It is a major illegal wildlife trafficking centre, the caiman (*Caiman crocodilus*) being the major target, with an estimated one million animals being poached each year. The Pantanal is also a cocaine-smuggling centre and is suffering from widespread deforestation⟳, with increasing turbidity and siltation rates in all waters. Immigration, agricultural expansion, pesticide⟳ runoff, industrial development and the resultant pollution⟳ are all adding to the threat of irreversible damage to one of the world's most important wetland habitats. The Pantanal is home to 650 species of birds, 230 varieties of fish⟳, 80 mammal and 50 reptile species. Although the Brazilian government is planning to spend at least $2.5 million on 'Operation Pantanal' in an effort to protect the wetland, their plans have been jeopardized by television. A racy soap opera, 'Pantanal', set in the wetland area became a huge success, attracting hordes of tourists and inadvertently promoting an influx of immigrants.

PAPER   The bulk of the world's paper is produced from wood⟳, and more than one-third of the world's total commercial wood harvest is converted into paper. Only a quarter of that paper is currently being recycled. To make paper, trees are felled and the wood is pulped, either by the application of chemicals or mechanically by mashing and grinding. Pulp produced using only chemicals is frequently bleached to produce the white finish sought by consumers and also to improve resistance to yellowing with age. When chlorine gas is used as the bleaching agent, the waste produced is extremely polluting. The higher the grade of paper, the more energy⟳ and chemicals needed during the manufacturing process. Paper-production is an extremely energy-intensive operation; in the United States, paper manufacturing accounts for 10 per cent of all industrial energy⟳.

Globally, around 195 million tonnes of paper are now being produced each year – a 36 per cent increase over the amount produced in 1975. The United States, with 63 million tonnes, is the world's leading producer, followed by Japan (20 million tonnes) and Canada (15 million tonnes). As well as producing the bulk of the world's paper, the developed countries also consume a great deal more than the Third World⟳; the average person in the West uses 120 kg – fifteen times as much as the average consumption of a person in the developing world. Worldwide, total paper use has risen fivefold since 1950 and now stands at around 230 million tonnes. Yet despite the increasing trend towards recycling⟳ of paper goods, notably newspapers, vast quantities of paper are still thrown away each year. In the UK alone, 3.4 billion disposable nappies and 1.5 billion sanitary towels are jettisoned each year, while in the USA the average citizen is estimated to use as much as 311 kg of toilet paper each year.

PARAQUAT   The trade name for a yellow, water-soluble compound, dimethyl dipyridilium, widely used as a broad-spectrum weedkiller. Paraquat is a non-selective contact herbicide⟳ commonly used by farmers and gardeners. It adheres strongly to soil⟳ particles and so is not free to poison other plants. Consequently, new crops can be planted soon after spraying. Furthermore, it has the distinct advantage over many other weedkillers of being quickly degraded by soil organisms. Farmers in an increasing number of countries are using paraquat instead of ploughing to prepare soil for sowing. This is fairly economical, prevents soil damage and lessens soil erosion⟳. However, paraquat differs from most other herbicides in that it is highly toxic to mammals and humans. Less than a teaspoonful is fatal if swallowed, and the chemical is also absorbed through the skin. Where it is used intensively, paraquat can drain into rivers, poisoning fish⟳ and other wildlife.

Paraquat generally concentrates in the lungs and also causes kidney damage, usually with fatal results. Treatment for paraquat poisoning involves the use of activated charcoal or some other absorbing material, but it is effective only if it is carried out immediately after ingestion. No other treatment is available.

PARIS CLUB   An international forum of donor and recipient governments which commenced operations during the 1950s. It allows the rescheduling of those debts⟳ granted or guaranteed by official bilateral creditors to be fully negotiated. It has no fixed membership, or any firm institutional structure. During the 1980s, it was intimately involved in efforts to find viable solutions to the major debt crises being faced by many impoverished developing countries⌀.

PASSENGER PIGEON   The passenger pigeon, *Ectopistes migratorious*, serves as a perfect example of how rapidly humans can drive other wildlife to extinction⌀. These pigeons, considered to have been one of the most numerous birds on earth, were gregarious birds, native to North America, which flew in huge flocks that reportedly darkened the sky and created their own winds. In the early 1800s the species was so widespread that birds were hunted by the million for the commercial meat market. In 1813, the world-famous American ornithologist J.J. Audubon reported a flock that took three days to fly by. Another single flock was calculated to contain 2 billion birds. By the 1850s the pigeons were considerably less common, although in 1869 a single market in Michigan reportedly traded 12 million in a forty-day period. Intensive shooting during the latter part of the nineteenth century, coupled with the destruction of the bird's natural woodland breeding-ground and the fact that successful breeding seemed to be dependent on the presence of large flocks, combined to cause the birds' demise. The pigeons went from a position of billions to millions in fifty years, and from millions to extinction in thirty years. The last bird to be shot in the wild perished in 1908, and the last passenger pigeon on earth, a female called Martha, died in a zoo in Cincinnati in 1914.

PCB   *See* Polychlorinated Biphenyls

PEACE CORPS   A US government-backed agency founded in 1961 by President John F. Kennedy with the aim of providing skilled manpower, most significantly teachers and agriculturalists, to countries in the Third World⌀. The Corps is made up of volunteers who possess the requisite training and experience and must be US citizens over eighteen.

PEACE DIVIDEND   Following the thawing in relations between the United States and the Soviet Union during the late 1980s, and the signing of disarmament⌀ treaties such as the 1987 Elimination of Intermediate-Range Nuclear Forces Treaty (INF), there was widespread optimism that the vast amount of financial resources traditionally being poured into defence and armaments programmes could be released and devoted to social and environmental improvement projects instead.

Despite initial optimism, by the end of 1990 the likelihood of this happening before the turn of the century at the very earliest became remote.

Détente between East and West resulted in significant disarmament and a reduction in the military budgets of a number of nations. However, for political reasons, there appeared to be only a slim chance of a reduction in the amount of funds being allocated to military research and development – a total of around \$1 trillion in 1990. Indeed, some veteran peace campaigners were of the opinion that this figure would continue to rise owing to powerful vested interests. Academic and industrial pressures were adding to the political whims of nations such as the United States and Israel, which place great emphasis on excellence in all aspects of military technology as a means of improving national security.

In 1990, Iraq's invasion of Kuwait not only created the conditions for a possible third war involving the major powers of the world but also quickly destroyed any chances of a 'peace dividend'. Acting under a mandate from the United Nations⌀, many industrialized nations committed themselves to a huge build-up of military forces in the Gulf region. The costs of the deployment and maintenance of troops and weaponry in the area were high, as was the likelihood of environmental damage – caused either by war⌀ or by the mere presence of armed forces personnel in a fragile environment. The United States – in the attempt to ensure that Iraq withdrew from Kuwait, and mindful of a desire to maintain oil⌀ supplies from the region – made the greatest commitment in terms of troops, equipment and expenditure. By the start of 1991, US operations, codenamed Desert Shield, were already costing \$35 million per day. In addition, the 200,000 or more troops in the region were using up well over 10 million litres of water⌀ daily. Meals and drinking water, mostly provided in non-biodegradable plastic containers, posed a serious litter and sanitation problem, in addition to the pollution⌀ caused by their equipment.

The effect of disarmament on national economies, both East and West, will truly be realized only during the 1990s. In the former Soviet bloc countries, the democratization process begun in the late 1980s and the movement towards a free-market economy had served to emphasize the bankruptcy of many centrally planned economies. The historic Conventional Forces in Europe Treaty (CFE) in 1990 saw members of the North Atlantic Treaty Organization (NATO)⌀ and the Warsaw Pact⌀

agree to measures which called for the destruction of armaments within strictly defined geographic limits. But the Soviet Union's action in removing almost 80,000 tanks, artillery pieces and armoured vehicles eastwards, out of the zone covered by the Treaty, was thought to be partially caused by the desire to avoid the sheer cost that would be involved in destroying them to meet CFE obligations. Similarly, the changing emphasis on employment and the effect on the economy remain unknown factors. Traditionally, 11 per cent of the workforce in the United States and 9.7 per cent of workers in the Soviet Union are employed in some aspects of the armaments industry.

PEANUT *See* Groundnut

PEARSON, LESTER BOWLES (1897–1972)
Canadian statesman and diplomat. Following a posting as ambassador to the United States (1945–6), he became a delegate to the United Nations◇ and chairman of the North Atlantic Treaty Organization (NATO)◇ in 1951. He played a crucial role in helping to settle the Suez Crisis in 1956 and was awarded the Nobel Peace Prize in 1957, primarily for his role in helping to create the United Nations Emergency Force. He was leader of Canada's Liberal Party from 1958 and served as Prime Minister (1963–8); his government initiated a commission to examine means of establishing French–English equality in Canada. He was also appointed head of an independent Commission set up in 1968 to study the overseas aid◇ system and propose improved policy measures. The Commission, commonly known as the Pearson Commission, was the forerunner of the 1980s independent Commissions, the Brandt◇, Palme◇ and Brundtland◇ Commissions.

PEAT Partially decomposed, dark-brown or black plant and vegetable debris laid down over long periods of time in waterlogged conditions in temperate or cold climates. The remains of bog moss (*Sphagnum*) is a major constituent. Peat is the starting point for the formation of coal◇ and can itself be used as a fuel, although it has a relatively low carbon content. Peat generally lies close to the surface and can be cut, formed into bricks, dried and burnt. Alkaline peat, found in several areas such as the Fens in the United Kingdom, is used for horticultural purposes.

PEREZ de CUELLAR, JAVIER (1920–)
Peruvian diplomat. A delegate to the first United Nations◇ General Assembly in 1946–7, he subsequently held several ambassadorial posts before being appointed Secretary General of the United Nations in 1982.

PESTICIDES Chemical agents used to kill insects or other organisms harmful to crops, other cultivated plants, and to human health or industry. Many pesticides are broad-spectrum biocides toxic to humans, livestock◇ and wildlife as well as the target pest. Some, such as parathion, DDT◇ and dieldrin◇, have caused widespread poisoning in humans and livestock after accidental exposure, and their use in many industrialized countries has been prohibited or tightly restricted. Globally, up to 40 per cent of crops are lost to insects, other invertebrate and vertebrate pests, disease and weeds. Over 50,000 species of fungi cause 500 different diseases in plants; well in excess of 10,000 insect species are known pests; more than 1,500 species of nematode damage crops, and of the 30,000 species identified as weeds, 1,800 are responsible for major economic losses. Most pests tend to be of greater consequence in developing countries◇ where tropical climates prevail than in countries in temperate zones.

The control of plant pests and vectors of human and livestock disease for the past forty years has been mainly based on the extensive use of chemical pesticides, which have played a major role in the battle to maintain adequate food supplies. Around 90 per cent of pesticides are used for agricultural purposes, the remainder in the health protection field. Since the late 1940s pesticide use has increased elevenfold – at enormous cost, both environmental and economic. World sales of pesticides increased from $2.7 billion in 1970 to $11.6 billion in 1980, rising to $28 billion by 1985. In that year farmers around the world applied 3 million tonnes of pesticides, double the amount used in 1970. About 80 per cent of these deadly chemicals are used in developed countries. According to the United Nations Food and Agriculture Organization (FAO)◇, if agricultural output is to double between now and the year 2000 in order to meet demands for food, the consumption of pesticides in the developed countries will have to grow by 2–4 per cent annually. Moreover, the rate of use in developing countries will have to rise by 7–8 per cent, although many experts forecast that it is actually more likely to double during the next decade.

However, many questions are being raised as to the efficacy of pesticide use, the environmental consequences and the role of pesticides in agricultural production. In general, the amount of pesticide reaching a target pest is extremely low. Only around 0.1 per cent of pesticides applied to crops is considered actually to reach the pest, so over 99 per cent moves into the ecosystem◇ to contaminate land◇, water◇ and air. Ecosystems have no natural mechanism for breaking down novel man-made chemicals, which usually persist in the environment as a result.

A significant proportion of the pesticides used in the developing countries have been banned in the industrialized nations because they have been found to damage either human health or the environment. The most obvious example is DDT, which is still widely applied throughout the Third World◇. However, the bulk of pesticide production is controlled by companies within the developed world, and a mere fifteen industrial producers in five countries account for 90 per cent of global pesticide production.

Despite the massive and extensive use of man-made pesticides for over forty years, which has caused widespread environmental damage, world-wide crop losses to insects have almost doubled from 7 to 13 per cent. The FAO report that over 1,600 insect species have developed resistance to major man-made chemical pesticides since the 1940s, primarily because of long-term, non-selective use of these novel chemicals. Furthermore, rats have begun to develop resistance to the latest generation of pesticides, meaning that there may soon be no effective rodenticides to combat one of the human race's greatest and most destructive pests. Overall, pests are reducing crop production by roughly the same amount as they were when synthetic chemical pesticides came into use. The battle against insect vectors of diseases has also not fared well, with the most significant affliction, malaria◇, showing a worldwide resurgence.

PESTICIDE COCKTAILS Following the withdrawal from widespread use in industrialized countries of pesticides◇ such as DDT◇, which were considered too hazardous, a 'safer' range of chemicals was introduced. These newer pesticides generally had to undergo strict testing to prove that they had relatively little adverse effect on the health of either humans or wildlife. An example of the 'safer' pesticides is malathion, which testing showed to be relatively harmless. In 1989, however, scientists reported that when it is used in fields or areas where another fairly non-toxic chemical, the fungicide◇ prochloraz, is also being used, malathion can become lethal. Other fungicides have been shown to exert the same effect, increasing the toxicity of malathion to dangerous levels. Although testing and licensing of individual chemicals is being being monitored and improved in many nations, there are absolutely no requirements to test any interactive effects of pesticide cocktails in the environment.

PESTICIDE POISONING Many of the original synthetic pesticides◇ proved to be so persistent in the environment that millions of people throughout the world continue to ingest minute amounts of toxic pesticides such as DDT◇ and dieldrin◇, even in areas where application was halted several years ago. Many of these chemicals are fat-soluble and become concentrated in the human body – often in mother's milk, through which toxins are passed on to newborn children.

In addition to slow accumulation in body tissue, direct consumption of foodstuffs treated with pesticides can prove fatal. Over 500 people died in Iraq in 1971–2 through eating bread made from wheat◇ which had been treated with a toxic fungicide◇.

Despite almost forty years of experience in using pesticides, surprisingly little is known about their effect on human health. There are little or no reliable data on poisonings and death tolls. It is widely reported that nearly half of all pesticide poisonings and 90 per cent of pesticide-related deaths occur in developing countries◇, although these countries account for only 20 per cent of world pesticide use. Misleading advertising, inadequate labelling, widespread illiteracy◇, lack of regulation and policing, and indequate medical care all contribute to the high toll.

The United Nations◇ estimates that 40,000 people in developing countries are killed each year as a result of pesticide poisoning and a further 2 million suffer some form of injury. By comparison, the World Health Organization (WHO)◇ estimates that pesticides poison 1 million people each year. The number of cases of cancer◇ that will arise as a result of exposure to pesticides will not be known for decades, but the National Academy in the United States predicts that a million will occur over the next seventy years.

The Food and Agriculture Organization (FAO)◇ Code for Safe Pesticide Use in the Third World, adopted in 1985, continues to be widely ignored.

The Code details the responsibilities of manufacturers and governments in helping to prevent poisonings. It specifies the warnings that manufacturers should print on their labels and what level and form of advertising is acceptable. In 1987 there were proposals that the FAO Code should be strengthened by the inclusion of a 'prior informed consent' (PIC)⬦ principle. Although most countries were in favour, the eight major pesticide-manufacturing countries – Belgium, Canada, Federal Republic of Germany, France, Japan, Switzerland, the United Kingdom and the United States – opposed the move. In 1989 the FAO finally agreed to toughen its Code and introduced a 'Red Alert List' of more than fifty pesticides and chemicals that have been banned or restricted in five or more nations. Products on the list are subject to the PIC principle. PIC forces exporters to inform importers of the reasons why the pesticides have been restricted or banned in the exporting nation and provide detailed information on the hazardous properties of each pesticide. Critics argue that the new list excludes many of the pesticides which cause widespread poisoning in the developing world.

PESTICIDE RESISTANCE Many pest species, especially those with a comparatively rapid life cycle, exhibit the ability to develop resistance to chemicals used to try and control them. The process is a dynamic, multifaceted phenomenon dependent on biochemical, physiological, genetic and ecological factors.

Resistance has been observed since 1911 but has occurred at a greatly accelerated pace since the introduction and widespread use of synthetic pesticides⬦ in the late 1940s. Scientists have so far discovered insects, mites, ticks, fungi and rodents that have all developed some degree of resistance to chemical and biological control agents. In general, resistance has appeared mainly in pests of major economic importance which have been subjected to prolonged pesticide applications over wide areas.

Between 1970 and 1980 the number of arthropod pest species (mainly insects) showing resistance almost doubled, and by 1990 the United Nations Food and Agriculture Organization (FAO)⬦ was reporting that 1,600 insect species had developed resistance to the most commonly used insecticides⬦. Many became resistant to a range of different chemical pesticides. During the 1980s, rats and other rodent pests started to exhibit resistance to the so-called 'second-generation' rodenticides, having developed resistance to the 'first-generation' compounds, such as warfarin⬦, less than five years after they were introduced in the 1950s. By the late 1980s house flies had begun to show resistance to *Bacillus thuringiensis*, a bacterium developed as one of the first of the new range of biological pesticides.

PET FOOD The Food and Agriculture Organization (FAO)⬦ report *World Agriculture Toward 2000* concluded that most people in the world should be eating more in 2000 than they eat today, but the total number of hungry people is actually going to increase, with the number of undernourished people rising by over 100 million. Most of these will be Africans and Asians. In all probability, few Americans or Europeans will be adding to the list of the hungry. What is certain is that livestock⬦ and pets in the industrialized world will not be going short of food.

In global terms, livestock farmers in the United States are major consumers of all feedstuffs utilizing land⬦ and food that could have been used to feed hungry human mouths. By the time an average beef cow goes to market in America, it will have eaten 2,655 kg of highly nutritious feed, according to US Department of Agriculture figures. This equates to 636 kg of feed for every 45 kg of beef produced. A laying hen will eat 28 kg of feed for every 100 eggs laid. To keep their animals well fed, US farmers pay in excess of $20 billion every year. More significantly, in view of the vast numbers of starving people around the world, European and North American pet-owners also pay out huge sums of money to keep their pets well fed. The Washington-based Pet Food Institute reports that US pet-owners spend well over $5 billion dollars a year. In excess of $3 billion goes to feed the nation's 60 million dogs and another $2 billion or more goes on food for the country's 50 million cats.

PHOSPHATES Naturally occurring nutrients which are essential for healthy plant growth but can cause serious environmental pollution⬦ in large quantities. Phosphate rock occurs as sedimentary deposits of phosphatic limestone, as guano, which accumulates from bird droppings, and as an igneous mineral, apatite. At least 75 per cent of the phosphate rock quarried annually is used as fertilizer⬦. World consumption of phosphate fertilizer continues to rise in line with the use of other agricultural chemicals: from 21 to 31 million tonnes between 1970 and 1983. Around 130 million tonnes of phosphate are mined each year; the

United States (51 million tonnes), the Soviet Union (25 million tonnes) and Morocco (19 million tonnes) are the main producers.

Misuse of phosphate fertilizers is leading to wide-spread degradation of watercourses. The main effect caused by excessive phosphate build-up is eutrophication⬦, which occurs when phosphates and other nutrients such as nitrates enter and accumulate in lakes, slow-moving rivers or any other body of water⬦. There are two main sources of phosphate pollution, both due to human activities and problems are increasing all over the world as a result. Runoff from agricultural land⬦ is a major source of pollution, since phosphates constitute an important component of fertilizers which are applied in amounts far greater than can be taken up by target crops. Phosphates from human sanitation systems also enter waterways in the form of treated and untreated sewage effluents. Sewage causes a far greater problem when it contains high concentrations of water-softeners, chemicals with a high phosphate content commonly used in synthetic detergents.

PHYSICAL QUALITY OF LIFE INDEX (PQLI)
A comparative measurement of a nation's general well-being, devised by the United States Overseas Development Council, based on an analysis of the average of three of an individual country's social indicators: life-expectancy⬦ (at age one), infant mortality rate⬦ and adult illiteracy rate. Each component is measured on a scale from 0 to 100. If life-expectancy at age one is thirty-eight years, the country scores 0. If it is seventy-seven, it scores 100. The infant mortality scale ranges from 229 deaths before age one (per 1,000 live births) down to 7 per 1,000. Literacy rates for those over fifteen years of age are taken as direct scores.

PLAGUE
An acute disease of rats and wild rodents caused by the bacterium *Pasturella pestis*, which can be transmitted to humans by the bite of rat fleas. It often occurs in epidemics of varying magnitude, mostly in areas where living standards and sanitation are poor. Bubonic plague, the most common form, has an incubation period of two to six days. Headache, fever, weakness, aching limbs and delirium develop, followed by painful swellings of the lymph nodes (called buboes). In some case the buboes burst after a week, releasing pus, then heal. In others bleeding under the skin, producing black patches, can lead to ulcers which can be fatal – hence the name Black Death⬦ given to a plague epidemic which ravaged Europe during the Middle Ages. In badly infected patients, bacteria can enter either the bloodstream, to cause septicaemic plague, or the lungs, to cause pneumonic plague. If untreated, these diseases are nearly always fatal. Treatment with modern antibiotics is effective in controlling the disease. Control of rats and their fleas is a much more advisable preventive measure.

PLATINUM
A precious silver-white metal, found in nature in limited supply in igneous rock and nickel ore. It is malleable, ductile, has a high melting point and is resistant to oxidation and attack by acids. It can absorb large quantities of hydrogen and is widely used as a catalyst, particularly in the manufacture of ammonia⬦, sulphuric and nitric acids. It is also used in petroleum refining, in thermocouple wire and in jewellery. Around 110,000 kg are produced each year; South Africa (80,000 kg), the Soviet Union (20,000 kg) and Canada (5,000 kg) are the biggest producers.

PLUTONIUM
Perhaps the most significant of all 'man made' elements. First produced in 1941, it is derived from uranium⬦ for nuclear purposes. By 1945 only a few kilograms had been made, following research which cost an estimated $2 billion. The plutonium manufactured proved enough for testing programmes and the manufacture of at least two atomic bombs used during the Second World War. Plutonium is made in nuclear reactors by bombarding uranium-238 (U-238) with neutrons. The U-238 transforms to plutonium-239 by absorbing a neutron into its nucleus. Plutonium can then be extracted from the reactor fuel by a series of chemical reactions referred to as 'reprocessing'⬦. It can be used as fuel either in conventional or thermal nuclear reactors or in fast-breeder reactors⬦.

Countries that reprocess spent nuclear reactor fuel have large and ever-increasing stocks of plutonium and are facing problems of what to do with it. The United Kingdom's nuclear programme, for example, has so far produced 47 tonnes of plutonium, of which 30 tonnes are stockpiled. There is no official tally of the amount of plutonium in the world, although nuclear reactors around the globe are thought to be producing over 70 tonnes annually, and may well be producing twice this amount by the year 2000.

POLIO
Polio is an infectious viral disease which affects the central nervous system and can lead to

paralysis. It occurs in different forms, mostly attack-ing children. Infection with mild forms self-cures relatively easily and can provide lasting immunity.

The causative virus is excreted in the faeces of infected people; consequently the disease is most common in areas where sanitation is poor. Epidemics can and do occur in regions of good hygiene amongst individuals who have not acquired immunity during infancy, either naturally or through vaccination programmes.

Symptoms commence seven to ten days after infection and in the majority of cases paralysis does not occur. The milder, less dangerous form is characterized by muscle stiffness and the common symptoms of influenza or a stomach upset. In severe cases these may be followed by weakness and even-tual paralysis of the muscles, although the paralytic form is now uncommon.

There is no specific treatment apart from simple measures to relieve symptoms. However, immuniz-ation using either oral or injected vaccine proves highly effective; vaccinated children develop lasting immunity.

By the year 2000 the World Health Organization (WHO)⊃ will attempt to eradicate polio through the combined efforts of its member states and several major Non-Governmental Organizations (NGOS)⌒, co-ordinating a global immunization programme targeted on the world's children. Currently, around 208,000 children around the world, mostly in devel-oping countries⌒, are stricken with the disease. Around 10 per cent of those infected with the para-lytic form die. An estimated 10 million people in the world today have some degree of lameness as a result of infection with the polio virus, predominantly during childhood.

POLLUTION The addition to the natural environment of substances that, through either their composition or the amount released, cannot be rendered harmless by normal biological processes. A certain background level of pollution occurs through natural processes, such as volcanic erup-tions and forest fires, as well as anthropogenic sources. Normal weathering of rocks causes 5,000 tonnes of mercury⌒ to be added each year to the world's oceans, about the same that arises from human activities, but it is the scale and prolonged intensity of man-made pollution that pose the greatest threat.

Human agencies significantly increase the levels of carbon dioxide⌒ and other pollutant gases well beyond those that occur naturally. Many modern

industrial processes have produced effluents which have led to the pollution of virtually every aspect of the biosphere⌒, land⌒, rivers, seas and the atmos-phere⌒. Pollution levels have worsened because many of the man-made, toxic pollutants (such as pesticides⌒ and fertilizers⌒) are persistent and can-not be broken down by normal biodegradation sys-tems. Pesticides such as DDT⌒ accumulate in the environment and in the bodies of animals until they reach toxic levels. Industrial development has led to complicated, long-term problems such as how to dispose of radioactive waste⌒ and other hazardous wastes⌒, increasing amounts of heavy metals, atmospheric pollution by greenhouse gases⌒, and the disposal of human sewage and refuse. In many cases technical solutions to these problems have been discovered, but they are not put into operation because they are regarded as being too costly.

POLLUTION TRADING An idea proposed during the discussions on global climate changes in the late 1980s which would allow nations to meet stricter pollution⌒-emissions standards. Some factories in a country would be allowed to emit pollutants above accepted emissions standards, so long as others in the same country fell well below them and the total national emission of pollutants was equal to or below that which would have occurred if all factories had conformed to emission control levels.

POLONOROESTE PROJECT The Polono-roeste Project in Brazil was described as one of the biggest land reform efforts ever attempted when it was initiated in 1982. The scheme involved the clearing of 400,000 sq km of land⌒, substantially covered by rainforests⌒, in Rondonia in north-western Brazil to make way for dams⌒, cattle-ranching, mining and cash-cropping⌒. If current rates of deforestation⌒ continue, an area in excess of 240,000 sq km will be denuded by the mid-1990s. The full cost of the Project was estimated to be $1.6 billion, of which the World Bank⌒ pro-vided a series of loans totalling $435 million. Of this sum, $256 million was spent on producing a single paved road, Highway 364, built to provide access to the Project area. In 1985 alone, 200,000 migrants arrived in the state of Rondonia. Most of the people travelling to the resettlement⌒ site are poor land-less⌒ peasants, although some have been given title to land through the Federal Land Agency. All have found poor tropical soil⌒ virtually unfit for agricul-ture⌒, and as many as 80 per cent of settlers have already sold or abandoned their land and left the area. By 1990 some 8,000 sq km of new ranchland in Amazonia⌀ had already been abandoned.

POLYCHLORINATED BIPHENYLS (PCB)

A series of toxic artificial chemicals, first synthesized in 1881 and used on a commercial scale since the 1930s. Their low flammability, high level of heat-resistance and low electrical conductivity made them particularly valuable components in a variety of products. PCB's became widely used in the production of fluorescent light bulbs, adhesives, hydraulic fluid and, most importantly, electrical transformers and capacitators. They are also used as plasticizers, improving the flame-retardance of plastics and increasing resistance to chemical attack.

Part of the problem with PCBS is their stability. They are extremely stable compounds and can be safely destroyed only through careful incineration⌀ at temperatures above 1,200°C. Incomplete combustion can lead to the formation of other, more toxic compounds such as the dioxins⌀. PCBS themselves can be released through the incomplete incineration of waste plastics.

The initial discovery of the toxic effects of PCBS was made and documented in 1936, but virtually no attention was paid to this until the 1960s. It was not until 1976 that the European Community⌀ banned the use of PCBS, although their use was still permitted in sealed equipment. The problem of what to do with PCBS in old and damaged equipment has not yet been successfully addressed. The scale of the problem is becoming more and more apparent as research has found that deep-ocean killer whales and polar bears from remote parts of the Arctic have dangerously high levels of PCBS in their body fat.

POPULATION Estimates of the number of people in the world vary substantially, but it is generally accepted that the global population doubled between 1950 and 1987, and now exceeds 5 billion. Another billion will be added by 1998, with 90 per cent of the growth expected to occur in the developing countries⌀. The rapid pace of population increase is of growing concern: it took over a hundred years for the human population to double from 1.25 to 2.5 billion, but only thirty-seven years for the next doubling. Medium-level United Nations⌀ projections forecast that the human population will reach slightly over 6 billion by the year 2000, rising to 8 billion by 2025 and finally stabilizing at 10 billion towards the end of the next century.

Since the first human beings appeared on the earth, at least another 46 billion people have inhabited the planet at some time, meaning that approximately 11 per cent of this total are alive today and many countries, particularly resource-poor nations with high population-increase rates, are faced with an increasing dilemma: how to satisfy the basic needs⌀ of their people.

Simply feeding the global population adequately has been widely recognized as a major problem for forty years or more, and the problem of hunger⌀ has plagued human communities for centuries. The world has over 8 billion hectares of land⌀ suitable for agricultural production of some description. This equates to about 1.6 hectares per person, but the distribution pattern of this land does not match that of the human population. In many parts of the world, food is being produced in insufficient quantities; in other regions surpluses are being produced, but in an unsustainable fashion. Over the last forty years the amount of cropland *per capita* in use has actually fallen by a third, while fertilizer⌀ use has multiplied by a factor of five, enabling crop production to register an increase, thus maintaining or slightly increasing global food supplies. As more and more land becomes unproductive through overuse, the human population continues to increase. Every year until 2000, the world's population will be swelled by almost 90 million people.

Yet in general, only when the population is stable or falling can nations, especially poor ones, make any progress in feeding themselves and satisfying their basic needs. In Europe, the problem of population increase was solved partly by the plague⌀ which wiped out a quarter of the population in the Middle Ages, and by mass emigration. Between 1846 and 1930, 50 million Europeans left the continent for the United States, Latin America, Australia and parts of Africa.

History has shown that only when a nation attains a high standard of living can a stable population-replacement level be reached. For the future, the underlying question that will have to be resolved is what standard of living for a population of 8 to 10 billion will be sustainable? Arthur Westing, a prominent ecologist, calculates that the current world population will have to be at least halved if an affluent standard of living (the consumer-driven example prevailing in the most-developed nations) is to be achieved and become commonplace around the world. In contrast, around 70 per cent of the world's present population could be supported if a standard of living equivalent to that seen in nations exhibiting the world average *per capita* Gross National Product (GNP)⌂ were to be generally adopted – countries with near-global average *per capita* GNP include Gabon, Greece and Malta. However, even if worldwide agreement over which standard of living should be the global goal were reached, severe problems of land reform⌂, redistribution of wealth, access to natural resources⌂, and the maintenance of freedom and equality will have to be overcome if the goal is to be secured in the long term.

POST-HARVEST LOSSES Between 1945 and the early 1970s the problem of losses of grains and perishable crops was overlooked as efforts focused on improving crop production to meet food needs. In 1975 the United Nations⌂ General Assembly recognized the growing concern among the international community and passed a resolution that stated:

> the further reduction of post-harvest food losses in developing countries should be undertaken as a matter of priority with a view to reaching at least a 50 per cent reduction by 1985. All countries and competent international organizations are to co-operate financially and technically in the effort to meet this objective.

The resolution sparked a decade-long effort orchestrated by the Food and Agriculture Organization (FAO)⌂, which proved far from successful. In 1984, almost at the end of the decade of action, approximately 180 million tonnes of grain, representing 10 per cent of the world's harvest, were lost or spoilt owing to inadequate storage facilities or through faulty handling, drying and processing. This grain was worth an estimated $18 billion but, more importantly, could have fed about 818 million people.

The scale of post-harvest losses remains massive, despite the UN-directed programme. Yet as less land⌂ becomes available for agricultural expansion and yield increases begin to stagnate, saving and using more of what is actually produced may be the only way to ensure a greater availability of food in the coming decades.

Despite extensive research into post-harvest losses, there are no firm data on how much food is lost. Losses of cereal grains range between 5 and 30 per cent of the harvest, and between 15 and 60 per cent of fruit, vegetables, roots and tuber crops are lost, depending on factors such as climate, storage

conditions and type of pests encountered. In addition, around 20 per cent of fish⌂ catches spoil or become unfit for human consumption.

The estimated annual post-harvest losses in developing countries⌂ alone is conservatively estimated to be 10 per cent of all durables (cereals and legumes⌂) and about 20 per cent of all perishables (root crops, vegetables and fruits). The total annual cost of these losses was calculated to be in excess of $10 billion as long ago as 1978. Food losses increase the need for and levels of food imports. The situation is worst in developing countries in the tropics, where prevailing conditions of heat and high humidity are ideal for the growth of micro-organisms. In many of these countries, farmers lack the facilities for drying and storing their crops safely. Food which is not stored properly can easily be consumed or spoilt by a variety of insects, rodents, birds and micro-organisms. A lack or scarcity of equipment and appropriate technology forces a dependence on primitive processing, transporting and marketing systems. Shortages of formally educated technicians and advisers to oversee improvements in storage, handling, drying and processing exacerbate the problem, yet agricultural colleges rarely offer courses in post-harvest technology. Poor roads, inadequate markets and transport difficulties make the problem worse.

The FAO calculates that 30 per cent of all stored food in the Third World⌂ is lost to pests and moulds. Unsatisfactory food-storage methods are often blamed for this loss, although analysis of many traditionally used storage methods found that they were highly efficient. In the Third World, the adoption of modern, temperate-region monocropping systems has often proved counterproductive. The high moisture content of high-yield cereals, providing perfect breeding conditions for moulds and bacteria, has contributed to the problem, as have changes in cropping patterns caused by introduction of more industrialized farming systems⌂.

POTATO The potato, *Solanum tuberosum*, first known to have been cultivated in the Bolivian and Peruvian Andes in about AD 200, was introduced into Europe in the late sixteenth century. It quickly became an important constituent in the diet of many nations, especially amongst poor communities, and is now a valuable staple food in many countries throughout the world, especially in Europe. More than 150 varieties are grown, producing yields of around 7 tonnes per hectare (ha). The best varieties, with good management, can yield up to 40 tonnes/

ha. Strains have also been developed which are resistant to potato blight, the disease which led to the disastrous Irish Potato famine which started in 1846.

Potatoes are good energy foods and are highly nutritious, containing more protein than many cereals or other root and tuber crops. They will grow at high altitudes where maize⌀ does not flourish, commonly producing twice as much protein and 25 per cent more starch. Under good management, potato yields can be stored for several months – only two months in tropical conditions, but far longer if they are stored in cool, dark conditions. Light spoils potatoes and turns them green, making them poisonous. Potato cultivation is expanding in many developing countries⌀ but overall, only 15 per cent of global production occurs in Asia, Africa and Latin America. Worldwide, 266 million tonnes are produced each year; the Soviet Union (63 million tonnes), Poland (34 million tonnes) and China (30 million tonnes) harvest the most.

PRECAUTIONARY PRINCIPLE A term coined at the Bergen Conference⌀ in Norway in 1990, attended by thirty-four nations from Canada, Eastern and Western Europe, and the United States, held as a follow-up meeting on the findings of the World Commission on Environment and Development (WCED)⌀. The governments represented agreed that the global community must take action to stave off potential large-scale environmental disasters, such as global warming⌀, without waiting for scientific proof about their cause or extent. Furthermore, the wealthier nations must help poorer countries to protect their environments for the good of the global community.

When the meeting was held, global warming was an unproven hypothesis based on imperfect computer models, with inconclusive evidence both for and against. On this basis several of the donor governments, including the United States and United Kingdom, had previously declined to commit themselves to the provision of funding and technology until scientific proof had been established, although in 1990 the UK revised its position and decided to contribute funds to research into the phenomenon.

PREFERENTIAL TRADE AREA FOR EASTERN AND SOUTHERN AFRICAN STATES (PTA) A regional agreement, dating from 1984, which sets out a wide-ranging, comprehensive programme of co-operation between signatory states, its ultimate goal being the establishment of a full economic community. A reduction in tariffs and completion of an agreed list of goods from member states that would qualify for specialized trade considerations were prominent among early initiatives. Reductions of up to 70 per cent on capital goods and 10 per cent on luxury goods were agreed. It is intended gradually to extend the list of preferred goods until a common market is established in 1992. Accord was also reached for the movement towards harmonization of Customs regulations and trading standards. There were fourteen original member countries: Burundi, Comoros, Djibouti, Ethiopia, Kenya, Lesotho, Malawi, Mauritius, Rwanda, Somalia, Swaziland, Uganda, Zambia and Zimbabwe. Tanzania has since joined, and membership is also open to Angola, Botswana, Madagascar, Mozambique and the Seychelles.

PRIMARY HEALTH CARE (PHC) In 1979 an international conference on Primary Health Care (PHC) was attended by a combination of 134 member states of the United Nations⌀ and 67 major Non-Governmental Organizations (NGOs)⌀ and intergovernmental agencies. The conference decided unanimously that PHC was the correct approach to achieve the goal of Health For All⌀ by the year 2000.

PHC is a concept which calls for the provision of a basic level of health care, attention and advice to all people throughout the world. Through community-based strategies, programmes will involve health monitoring and education⌀ as well as treatment. Above all, PHC is intended to ensure that basic medical help and advice reach every individual, especially those in greatest need.

Most developed countries are approaching full coverage, but major problems of social, economic, ecological, political and biomedical importance still have to be overcome if the goal is to be reached on a global scale.

Illiteracy and poverty rates correlate directly with individual and community health levels. Improving literacy rates raises health awareness and general welfare, yet the numbers of illiterate women in developing countries⌀ is growing larger rather than decreasing, and the literacy gap between men and women is also widening. Poverty, a leading factor in the creation of poor health, is also on the increase and 1 billion people, 25 per cent of the world's population⌀, are living in absolute poverty. Although many of these live in the developing world, millions of people in industrialized countries also live in poverty. It is believed that up to half a million children in the United States, most living below the poverty line, have no direct access to a general practitioner.

Where economic crises are common or nations are poor, governments are unable to provide the populace with an adequate health-care system. The provision of PHC therefore often requires a social revolution in health matters. Consequently, the achievement of global equity in health care is a long way off. Inequities are highlighted by comparisons of life-expectancy⌐, which varies between fifty years in some countries and over seventy where health-care systems are efficient, of a high standard and widely available.

PRIMARY PRODUCT   Any agricultural product, from farm, forest⌐ or fishery⌐, or any mineral⌐, in natural or unprocessed form, or which has been subjected to only such minor initial processing as is required to prepare it for entry into international trade. Primary products are, in general, produced by developing countries⌐ for sale to the industrialized nations where most of the processing occurs; these nations have a major role in setting the prices for these products on the world's markets.

Many Third World⌐ countries have become dependent on one or two primary products for the bulk of their export earnings. A United Nations Conference on Trade and Development (UNCTAD)⌐ survey of 84 developing countries found that 43 were dependent on primary commodities for at least 90 per cent of their total export revenue. More than 80 per cent of these countries depended on a single primary commodity for at least 75 per cent of their export earnings.

As most primary products are of an agricultural nature, vagaries of climate and environmental perturbations wreak economic as well as environmental havoc, causing total crop loss and loss of export earnings. Paradoxically, increased production is not necessarily beneficial, as it also forces world prices down. During the 1980s, all the common primary commodities showed a significant drop in price on the world market compared to the average trading prices seen over the preceding two decades. Tea⌐, jute⌐ and rubber⌐ fell by 60 per cent or more between 1960 and 1981, while bananas⌐, oil palm⌐ and sisal showed a price decline in real terms of around 40 per cent.

Price discrepancies also arise according to the origin of the primary product, reflecting the industrialized countries' power in fixing world prices. The price for agricultural products exported predominantly by the industrialized countries during 1988 rose by 12 per cent, whilst those emanating from developing countries declined by nearly 3 per cent over the same period. Wheat⌐ and corn⌐ showed the greatest increases. This will eventually cause an inflation in the food import bill of developing countries of some $5 billion, whilst, as a result of decreasing commodity prices, their export earnings fall.

PRIOR INFORMED CONSENT   A principle used to regulate the movement of toxic chemicals⌐, nuclear waste⌐, or other hazardous substances. Before any exporting country can dispatch a shipment, it must have the importing country's consent, usually in writing. Before this is received, the exporting country must first have provided the importing country with full and detailed information on the content and nature of the intended export, allowing the recipient nation to make a full assessment of the risks involved in importation.

PROJECT TIGER   An international conservation⌐ project to save the Bengal tiger from extinction⌐. When it was launched in India in 1972, the number of Bengal tigers had fallen to around 1,800 owing to a combination of hunting⌐ and habitat loss⌐. In 1900 there had been an estimated 40,000 animals. Fifteen years after the introduction of the Project, which saw large tracts of land⌐ set aside as reserves in which the tigers could live and breed unmolested, surveys found that there were 4,000 tigers, with some estimates rising as high as 6,000.

The success of Project Tiger in increasing tiger numbers owes much to Indira Gandhi⌐ who, at its initiation, led the Indian government's campaign to establish the fifteen reserves where grazing, hunting and logging were either prohibited or strictly controlled. In these reserves, forest⌐ decline has been halted or reversed and tigers have flourished along with several other species of wildlife that were also thought to be endangered. International agencies such as the Worldwide Fund for Nature (WWF)⌐ committed $1 million to the Project, with the Indian government allocating $12 million.

The Project highlights one of the major problems of conservation – should the conservation of endangered species⌐ of wildlife take precedence where human communities suffer as a result?

Over fifty people are killed every year in and around the special reserves set up to allow the tigers to recover. All the villages within the original core area of the reserves were forced to relocate – the villagers were supposedly provided with alternative land⌒, construction materials and all resettlement⌒ costs as compensation. Although entry to the core areas of the tiger reserves is prohibited, many of the original villagers continue go there in search of fish⌒, honey, sugar⌒ cane or other products and food. For many this is their only source of livelihood, one which has been exploited for several generations.

The Bengal tiger has been saved from extinction, but at a significant cost financially and in terms of the lives of the displaced villagers. In future projects of this nature those displaced must be provided with better access to fuel, fodder and services if they are to respect the reserve boundaries.

PROTECTED AREAS Creating protection against human exploitation has long been the best method of conserving natural areas and species of wildlife. Since the establishment of the first national parks during the late 1800s, some 4,500 major areas have been designated for nature conservation⌒. By 1988 at least fifteen nations had designated over 10 per cent of their land⌒ as protected areas, although the viability of the protection system and enforcement of restrictions varies markedly. In total, sites around the world designated as protected in some form cover an area of 4.54 million sq km. In addition to national parks, nature reserves, natural monuments and other areas governed by national legislation, the total includes a substantial amount of land protected under international agreements. Scattered around the world, some 404 wetland⌒ sites of international importance, encompassing 2.86 million sq km, plus 269 biosphere reserves⌒, covering 1.43 million sq km, and 78 sites of special natural interest designated under the World Heritage Convention⌒, receive some degree of protection under internationally agreed regulations.

PUGWASH CONFERENCE A series of specialized international conferences on science and world affairs, the first held in the village of Pugwash in Nova Scotia, Canada, in 1957, at the suggestion of a group of eminent scientists, including Albert Einstein and Bertrand Russell. Delegates discuss problems concerned with scientific and technological research and application, disarmament and social responsibility, and the role of the scientist in matters of global and social development.

PYRETHRUM Any of a variety of Old World plants of the genus Chrysanthemum, commonly grown for ornamental purposes in temperate regions and commercially for the production of biological pesticides⌒ in Africa. The name pyrethrum is commonly used to describe an annual herb, *Chrysthanthemum cinerariaefolium*, which grows well in highland regions and is now grown primarily for the production of biological pesticides. Pyrethrum (or pyrethrin) is the potent insecticide⌒ derived from the flower heads of the plant.

C. cinerariaefolium grows best on well-drained, fertile soils at altitude, above 2,000 metres, but annual rainfall needs to average at least 1,000 mm per year and be evenly distributed throughout the growing season. The plants reach peak yields about three years after planting; subsequent yields rapidly begin to decrease. Flowers are picked by hand, and yields of around 300 kg per hectare are average.

Pyrethrum, the chemical with insecticidal properties extracted from the dried flowers of the herb, became the first widely used biopesticide. It has several advantages over the synthetic pesticides: it is relatively safe to use near food or humans, it has a repellent as well as deadly effect on insects, it breaks down rapidly in the environment, and insects do not seem to be able to develop resistance to it. As a result, it is now commonly used in household aerosols and as a powder in grain-storage warehouses.

Q

QAT Qat (or khat) is a privet-like white-flowered evergreen shrub, *Catha edulis*, native to parts of Africa and Arabia, which is extensively cultivated in many Arab countries, most notably in Yemen. The leaves, which are chewed or prepared as a drink, have narcotic properties similar to those of coca◠, and in Yemen they are chewed daily by virtually all men and an increasing number of women as a stimulant. In North Yemen, the nation's economy is based on agriculture◠. Small-scale farmers use every available piece of land◠, many of them farming high on terraced mountainsides. On the lower mountain slopes the famous mocha coffee◠ used to be cultivated, but production has fallen dramatically over the past few years because the conditions best for growing coffee are also best for growing qat. The United States Embassy estimates that qat production in North Yemen is worth $1 billion annually, a hundred times more than the value of all the nation's exports. It is much more profitable to grow qat than coffee, as the former will produce four or five crops per year and fetch a much higher price. Qat leaves are usually chewed fresh, so the market has, in the past, been based solely on national demand. With the advent of rapid international travel, the demand for qat overseas is now being realized.

Doctors in the United Kingdom have expressed growing concern over the availability of qat, which they claim may cause violent behaviour and paranoid psychosis. Many immigrants from Muslim countries – where alcohol consumption is banned and the chewing of qat leaves has become a social habit – took full advantage of the availability of the leaves. In the UK during 1989, qat leaves, mostly imported from Ethiopia and Kenya, sold for $4 per bundle, with communities of Somalian refugees◠ being the most avid purchasers. The death of a refugee and a child from a fire started by an adult reputedly under the influence of the amphetamine-like substance from the qat leaves, sparked a movement to curb the availability and use of qat in the UK. Governments elsewhere are considering the introduction of legislation to restrict the habit, and qat-chewing has already been officially banned in Somalia.

QUELEA Quelea are small birds, members of the weaverbird family. There are three species, all living in Africa south of the Sahara. These birds are recognized to be one of the world's most serious agricultural pests. There are more red-billed quelea (*Quelea quelea*) in the world than any other type of bird – several billions. They form huge colonies, breeding in tropical Africa and migrating south in such enormous numbers that they have been compared to swarms of locusts◠. They cause extensive damage to traditionally grown small grain crops of sorghum, millet◠ and teff, as well as to field crops of wheat◠, barley and rice◠. The birds are not nomadic; they remain in an area as long as seed is available, often causing localized famines as a result. All sorts of measures have been introduced in the attempt to control them, including poisoning, introducing diseases and dynamiting their colonies, all to no avail, and a decade-long campaign of research coordinated by the Food and Agriculture Organization (FAO)◠ proved fruitless.

QUOTA Each member country of the International Monetary Fund (IMF)◠ has a quota, equivalent to its subscription to the Fund. This is payable as one-quarter in a freely usable currency and three-quarters in national currency. The quota forms the basis upon which entitlements to draw on the IMF's facilities are allocated, and its size is related to the country's national income and foreign trade. Quotas undergo regular review and increase.

R

RABIES   A viral disease, endemic in a hundred countries, which can have a high mortality rate in humans if effective treatment is not given during the early stages. The virus which causes the disease can also induce immunity, and treatment usually involves a series of immunizations. The disease-causing virus is usually introduced into humans through the bite of an infected animal, most commonly a dog or a fox. In ninety countries dogs remain the main vector, accounting for 98 per cent of human deaths. The disease develops in a painful manner which eventually ends in muscular spasms of the throat brought on either by drinking or by the sight of water, referred to as hydrophobia (aversion to water). Where treatment is available or preventive vaccines are administered, rabies can be cured or avoided. If allowed to progress, it is incurable and fatal.

The disease has varying impact due to the discrepancies in medical treatment between the developed and the developing world. Expressed as deaths per 1,000 cases, the figure for Africa is 82.7, while in Europe it is 0.3.

The best and most economical way to tackle the problem is to eradicate stray dogs. Several countries, notably small island states, have carried out massive anti-rabies campaigns and, having eradicated the disease, have managed to maintain a disease-free status by strictly quarantining all animals entering the country. Such programmes have eradicated or substantially reduced rabies in China, Cyprus, Hong Kong and Spain.

In larger, landlocked regions, lasting control is much more difficult. Millions of dollars have been spent to prevent dogs spreading rabies in Western Europe, but other animals have become vectors and rabies is now spreading, advancing at 25 km per year. Britain, which is rabies-free and maintains strict quarantine regulations, has expressed concern over the possible transmission of the disease when the island is connected to mainland Europe by the Channel Tunnel. Europe reports 20,000 cases of human infection annually among people bitten by rabid animals.

In Europe, wild foxes now act as a reservoir of the disease, and in North America skunks perform a similar role. A vaccine which provides effective immunity has been developed, but plans to try and eradicate rabies-carriers in Europe and South America have encountered fierce opposition, primarily because the eradication programme will rely on the release of either attenuated or genetically engineered live vaccine, the long-term effects and dangers of which are not fully realized. Between 1984 and 1987, 50,000 sq km of the Federal Republic of Germany were declared rabies-free following provision of vaccine bait to foxes. No cases of vaccine-induced rabies were reported.

In Latin America over 200,000 people are bitten each year by rabid dogs but other animals, notably vampire bats, also spread the disease. However, throughout the region bovine rabies is a far greater problem and is estimated to cost Latin American cattle-producing nations more than $50 million annually. In 1988 tests of a genetically engineered vaccine were carried out in Argentina under the auspices of the Pan-American Health Organization, the regional branch of the World Health Organization (WHO)⊃, although the Argentinian government and local population were not informed of the nature or extent of the trial.

RADIOACTIVITY   The atoms of certain elements are unstable and may easily lose part of their nuclear material. These elements are liable to transform into other similar elements by spontaneously ejecting, or 'radiating', various particles or energy⊃ waves from their atomic nucleii. These particles or energy waves can be dangerous because they damage or destroy living cells.

The period of time needed for half the number of atoms in a given mass of one of these unstable elements to transform is known as the 'half-life'. The new element formed as a result of the decomposition may or may not itself be radioactive and thus unstable. The rate of radioactive decay, or half life, of any substance is highly specific. Uranium-238, the commonest isotope of Uranium⊃ found in nature, has a half-life of 4.5 billion years. Other isotopes produced by nuclear reactors have much shorter half-lives, measured in days or even hours. Iodine-131, for example, has a half-life of eight days.

Depending on the element and its isotopes, radioactive decay takes different forms, alpha, beta or gamma, depending on which particles are lost from the nucleus. All are potentially hazardous.

Alpha radiation is caused through the loss of an alpha particle. These particles have a relatively great weight and size and can be stopped by skin or a sheet of paper. They do not penetrate human tissue or other matter to any depth, and cause most biological damage when they are ingested.

Beta radiation arises from the expulsion of beta particles which are around 2,000 times smaller than alpha particles. They are thus considerably more penetrative and can travel through at least 1 cm of living tissue, passing through many cells before thay are stopped. They also cause most damage when ingested rather than when skin is irradiated.

Gamma radiation, in electromagnetic⬦ wave rather than particle form, can easily penetrate skin, is relatively far more dangerous and requires a thick layer of a dense material such as lead⬦ or concrete to provide an effective barrier.

RADIOACTIVE WASTE The waste products from nuclear reactors, uranium⬦-processing plants, industrial facilities producing radioisotopes for the scientific community, and all materials from hospitals or research institutions that are or have become radioactive. The disposal of these wastes, which include gases, solids and liquids, constitutes a major problem for present-day authorities, as well as those of years to come. A final solution may never be found, for the wastes can remain toxic for centuries and, in some cases, for millions of years.

Solid waste is usually in the form of irradiated fuel elements and other equipment from nuclear power⬦ stations. Waste of low activity was traditionally packaged for disposal at sea. High-activity waste may be combustible (with valuable plutonium⬦ being recovered from the incineration⬦ process), or it can be buried, this latter option being controversial. Waste disposal commonly involves storage in underground tanks, or the waste can be placed in drums weighted with concrete and dumped at sea – although international bans on the disposal of solid radioactive waste in the world's oceans have been in place for some time.

Liquid waste poses a more daunting problem due to the greater possibility of leakage. Storage of liquid wastes in salt mines, granite rock formations, in clay basins or under the seabed have all been proposed as solutions to the dilemma. All these plans have met with great controversy because no container has yet been developed which can be guaranteed to withstand decay or natural geological disturbances such as volcanic action, and stay intact for the 10,000 years or so that will have to elapse before the waste becomes safe. Current technology can produce only containers that would corrode after 1,000 years and probably dissolve within the next 1,000. The most hopeful prospects for disposal of liquid wastes lie in vitrification into solid glass cylinders which could then be placed in titanium⬦-cobalt alloy containers and deposited in underground repositories. Thought has also been given to the possibilities of sending these containers to other planets, the dangers inherent in this process being made all too apparent by a series of explosions of space launch vehicles and rockets, including one which destroyed one of the United States space shuttles during the 1980s.

Radioactive waste in gaseous form is, at present, of fairly minimal concern as it is produced only in relatively small quantities. This waste is usually released into the atmosphere in small quantities which are deemed to be safe, although no long-term environmental impact data exist to support or oppose this assumption.

RAINFOREST Global concern over deforestation⬦ focuses on tropical moist closed forests⬦, usually described as rainforests. Tropical moist forests contain a wealth of timber and other products, as well as having important environmental functions. They are also repositories of biological diversity⬦. A 2.5-sq-km area of tropical rainforest can contain 750 species of trees, 400 species of birds, 125 mammals, 1,500 varieties of flowering plants and thousands of invertebrate species. But rainforests also exist in temperate regions – Australia, Chile, Japan, New Zealand, Norway and the United States. They have the greatest biodiversity⬦ in temperate ecosystems⬦ and host the tallest trees in the world, the redwoods of California and the giant eucalyptuses⬦ of Tasmania. Temperate rainforests are also being decimated on the same scale as tropical forests. Douglas fir rainforest ecosystems have been eliminated in western Canada, and some 85 per cent has been cut down in Oregon and Washington in the United States. Coastal and Andean forests in the Bio-Bio and Maule regions of Chile have been destroyed, and only 28 per cent of original cover is left. In Australia, less than 10 per cent of the tall eucalyptus forests that grew in Tasmania 200 years ago remain.

Rainforests, temperate and tropical, are being cleared so that poor families can grow subsistence crops and for timber – most timber from all rainforests is exported to Japan. Yet comparative studies of tropical rainforests in Peru have indicated that it makes economic sense to conserve forests rather than destroy them. Based on the assumption of sustainable timber and annual collection of fruits and latex in perpetuity, tree resources in a hectare of virgin Peruvian rainforest were estimated to be worth $6,820. Using the land⌾ for cattle ranching would bring a return of $2,960, whereas managing the area as a timber plantation would realize only $2,055. In view of economic, environmental and social pressures surrounding the plight of the rainforests, the Colombian government has created a precedent by taking the innovative step of handing over 20 million hectares of Amazonian rainforest to the local Indian population, giving them the right and the means to live according to their traditions and, in so doing, protect the forest.

RAJASTHAN CANAL The world's largest irrigation⌾ system project, launched in India in 1958. The Canal was planned to be 500 km long, with 6,500 km of distributaries. Designed as a solution to the water⌾ problems of the drought⌾-stricken Rajasthan state, the Canal has been plagued with lack of funds, shortage of materials, political indecision and widespread corruption. Logistics problems, including the daily deployment and co-ordination of 25,000 workers and thousands of donkeys and camel carts, also contributed to construction delays. Originally due for completion by 1968, the Canal is unlikely to be operational by the end of the century. Initial construction costs of $39 million had soared to $600 million by the mid-1980s. A waterlogging problem also threatened to render the whole project ineffective. In 1978, engineers calculated that waterlogging caused by irrigation would not be a problem for at least 200 years. However, a few years later a thick, hard pan of subsurface gypsum⌾ was discovered under part of the land to be irrigated. Of the total 7,000 sq km area, 8 per cent is already waterlogged, and if the current rise in water table is maintained, as much as a quarter of the entire area will be similarly affected.

Lack of careful consideration of the social consequences of the project has also become apparent. In the region through which 80 per cent of the Canal will pass, people are being resettled on the newly developed farmland. Over 75,000 families have been allocated land⌾, 98 per cent of these people from other regions. Much of the land was traditionally used by nomadic pastoralists, who now have nowhere to graze their herds of cattle.

RAMPHAL, SIR SHRIDATH SURENDRA-NATH (1928–) Guyanese statesman, also known as 'Sonny'. He was Minister of Foreign Affairs and Justice in Guyana (1972–5) and Secretary-General of the Commonwealth⌾ from 1975 until his retirement in 1990. He was an influential member of the Brandt Commission⌾ and the Independent Commission on International Humanitarian Issues⌾.

RAMSAR CONVENTION Common name for the Convention on Wetlands of International Importance Especially as Waterfowl Habitat, signed in Ramsar, Iran. Finalized in 1971, the Convention is one of the world's oldest international conservation⌾ treaties: it came into force in 1975. It is unique in that it protects a specific type of ecosystem⌾ – wetlands⌾ – on a global basis. The signatories to the Treaty – fifty-eight in 1990 – agreed to designate at least one wetland which they will either protect or replace with one of equal worth if the listed site is destroyed. In total, over 500 wetland sites, covering in excess of 30 million hectares, have been nominated.

Wetlands are extremely important because they tend to be highly productive, useful sources of food, fuel, and, above all, are crucial to over 60 per cent of the world's fish⌾ catch, serving as major breeding-grounds and nursery areas. Many wetlands also play a role in purifying water⌾ or protecting inland areas from harsh weather. They are threatened by a variety of human activities, including drainage for agriculture⌾, dam⌾ and canal construction, peat⌾ cutting, forestry, and pollution⌾ attributable to industry or the overuse of fertilizers⌾ and pesticides⌾.

The work of the Ramsar Convention has traditionally been shared by the Worldwide Fund for Nature (WWF)⌾, the World Conservation Union⌾ (IUCN) and the International Wildfowl Research Bureau, but repeated calls have been made for a permanent, funded Secretariat to be established. In the late 1980s, it was decided that an annual budget of $440,000 should be made available, collected from the then fifty-two signatory states on an 'ability to pay' basis, through which it was hoped administration could be upgraded and new parties would be encouraged to join the Convention. According to conservationists, at least twenty of the wetland sites nominated under Ramsar are endangered, with several sites in Greece under special threat as they are likely to be drained as a result of development projects funded by the European Community⌂. Other sites in Africa, Germany, Jordan, Pakistan, South Africa, the Soviet Union and Uruguay were singled out as being under particular threat.

Proposals put forward in 1990 for the creation of a special fund to be devoted to the conservation of wetlands in developing countries⌂ met with limited support from the donor community. Despite an effort by environmental groups and some industrialized nations to establish an initial fund of $150,000, only 10 per cent of the suggested sum was raised. Australia, Canada, Ireland and the United Kingdom all opposed the fund.

REAGAN, RONALD (1911–) American politician. Initially a Hollywood film actor of limited success and fame, he entered the national political arena as Governor of California from 1967–1974. He fought two unsuccessful campaigns for the Republican presidential nomination before succeeding to the presidency, serving from 1981–1989. Throughout his tenure he regularly increased defence spending and adopted a hard line towards the Soviet Union. He also implemented a strong and active policy in Central America, especially Nicaragua. His administration sanctioned intervention in Lebanon and Grenada. Approval for the US invasion of Grenada and for covert Central Intelligence Agency (CIA) operations in Nicaragua was given despite opposition from Congress. An aggressive foreign policy was highlighted by his sanctioning of a destructive long-range bombing raid on Libya. His Strategic Defence Initiative⌂ (popularly known as Star Wars), which involved a military use of outer space, proved increasingly controversial as well as being one of the most expensive research and development programmes ever proposed.

He survived an assassination attempt in 1981 and retired from the presidency following two terms of office, the longest allowable under the US Constitution.

RECYCLING The reclamation of potentially useful material from household, agricultural and industrial waste. Recycling basically involves the manufacture of a complex product, either by natural or man-made agency, followed by the breakdown of the final or used product into those constituents that can be reused, or conversion of the product into a form that has a secondary useful purpose. This general pathway is a normal process in ecosystems⌂, and farmers throughout the world have for centuries recycled organic materials in forms such as compost, or to produce low-cost animal feed or fertilizer⌂. Recycling of commercially manufactured goods has only relatively recently been adopted as an economic necessity to reduce pollution⌂ and save energy⌂ and unnecessary expenditure on scarce raw materials, whilst slowing down the rate at which non-renewable resources are depleted and minimizing environmental degradation.

Several countries around the world now have active recycling programmes, official and unofficial, for consumers and for industry – with varying degrees of effectiveness.

In Europe, glass recycling grew from 1.33 million tonnes in 1979 to about 2.7 million tonnes by 1984. Sixteen per cent of bottles are recycled in the United Kingdom, compared to 50 per cent in Denmark, the Netherlands and Sweden.

Recycling of aluminium⌂ cans in the United States grew from 24,000 tonnes in 1972 to 510,000 in 1982, and 55 per cent of cans in the country are now recycled. This compares to 65 per cent in Canada, 42 per cent in Japan and a 13 per cent average in Europe.

In the United States the amount of paper⌂ collected for recycling doubled every year between 1975 and 1980, reducing the pressure on forests⌂ in the process. One tonne of recycled newsprint⌂ saves a tonne of wood⌂, equivalent to almost a dozen trees.

Through co-ordinated schemes some developing nations, such as Mexico, have become world leaders in the recycling of certain materials discarded by consumers. In others, such as Brazil and Egypt, recycling is well advanced, but only because thousands of poor inhabitants derive their livelihood from recycling materials dumped on public waste-disposal sites.

In the developed world, following a combination of economic and social pressures, industry has also been forced to develop recycling programmes. In the USA industrial solvents are being recycled in a business that is expected to become worth $1 billion by the year 2000. In the former German Democratic Republic an estimated 30 million tonnes of industrial wastes were being recycled each year by the early 1980s, providing 12 per cent of the raw materials needed for industry. In Hungary, 29 per cent of all industrial wastes were being recycled by 1985. In certain European nations waste exchange networks have been established, with over 150 waste products listed in the exchange scheme.

As concern for the environment has become more widespread, especially in the industrialized countries, the value of recycling has become accepted by most members of society. People from all levels of the community commonly deliver paper, glass and other substances to recycling centres. Unfortunately, consumer tastes in these countries have not changed to the same extent and an aversion to buying recycled products has only partially been overcome. In the USA collection of old paper rose by 34 per cent between 1983 and 1985, but reuse rose by only 5 per cent during the same period. Over 1 million tonnes of old paper is in store at a price of $5 per tonne compared to an original $40 tag. In the Federal Republic of Germany, some reports suggest that only 3 per cent of glass, paper, plastic and tin collected for recycling is in fact recycled.

As more attention is paid to recycling processes, anomalies as to its true worth have begun to emerge. It requires 20–25 times more energy to make aluminium by smelting bauxite⌒ than it does to melt and produce 'new' aluminium from scrap. The energy required to recycle copper⌒ is only one-tenth that used to produce the copper originally. Savings when steel is made from scrap amount to 47 per cent, while recycling newspapers saves 23 per cent of the energy emobodied in the product. But the recycling – breaking up, remelting and reforming – of glass is not all that efficient, producing only an 8 per cent saving over the energy needed to make new glass. Glass is better reused, not recycled.

RED CROSS – INTERNATIONAL An organization founded under the 1864 Geneva Convention⌒ to provide care for all casualties of war. Inspired by Henri Dunant, a Swiss philanthropist, it is based in Switzerland, with national branches in many countries throughout the world. The Red Cross emblem, recognized in most countries in the world, represents the Swiss national flag with the colours reversed. In Muslim nations, the crescent has been adopted as the emblem of the society. Volunteers form a large part of the organization's staff, and through their efforts in various national agencies the Red Cross has extended its field of operations to encompass a variety of problems associated with war or armed conflict. The broadened scope of activities now involves dealing with issues concerning refugees⌒, the disabled, and

the alleviation of human suffering and misery caused by epidemics, floods⌒, earthquakes and other natural or man-made disasters. The International Red Cross has twice received the Nobel Peace Prize (1917 and 1944) in recognition of its efforts.

REFORESTATION Name given to programmes which aim to replace forests⌒ that have been removed or destroyed. By the late 1980s, 14.2 million hectares of land⌒ each year were being reforested, the bulk of it in China and the Soviet Union. Reforested plantations in the tropics cover a relatively small area. In Africa reforestation and afforestation programmes offset a mere 10 per cent of the loss of trees due to logging operations.

Reforestation projects can be for industrial or non-industrial purposes, but the majority have tended to be little more than commercial plantations, including many of those orchestrated under so-called social forestry⌒ initiatives, many of which have proved counterproductive. In the Third World⌒, these were intended to provide villagers with fuelwood⌒ and other resources but tended instead to provide local industry with trees for pulp. The choice of fast-growing trees such as eucalyptus⌒ and conifers, commonly adopted in reforestation programmes, tends to have an economic rather than ecological basis. This has led to degradation of soils⌒, loss of nutrients and lowering of water tables⌒ in areas already under pressure as a result of previous forest depletion.

Successful reforestation projects can be achieved, and analysis of these has shown that to be really successful they must be either started and run by the local population or based on the active participation of many sectors of the local community who will themselves benefit from the programme's success. South Korea is a model example. In 1970 the country was barren, denuded of trees and plagued by soil erosion⌒. Hillsides were eroded and land had lost the capacity to retain water. By 1977 almost 650,000 hectares (ha) had been planted with fast-growing trees. The key to success was the federally linked Village Forestry Associations and the villagers' participation. Every hectare planted with fuelwood trees was matched by half a hectare planted with chestnuts. The production of mushrooms, fibre, bark, resins and other non-wood products was encouraged, and by 1978 these products were generating an income of $100 million for the 2 million families involved in the programme.

The replacing of forests with new trees and replanting areas where forests have been destroyed is proposed as one way to combat the release of carbon dioxide⌀ into the atmosphere and so reduce the greenhouse effect⌀. Like all vegetation, trees absorb carbon dioxide from the atmosphere through the photosynthetic process. The Dutch government has taken a global lead in this respect and intends to plant tropical trees over 250,000 ha – an area the size of Luxembourg – at a total cost of $460 million. Around 10,000 ha of forests are to be replanted each year to offset the carbon dioxide pollution⌀ expected to be released from two new power stations being built near Amsterdam and Rotterdam. The Dutch are to plant the trees in South America on land in Bolivia, Colombia, Ecuador and Peru where tropical forests have been razed. The cost of planting a tree in these countries is less than 10 per cent of the cost in the Netherlands.

In 1990, researchers discovered that replanting with new trees might not significantly reduce atmospheric carbon dioxide. Replacing old trees with new ones or converting old-growth forests to managed timber production may, in fact, be counterproductive. This process is estimated to have contributed 2 per cent of the total carbon released by land-use⌀ changes over the last century. It will take 200 years for young trees to absorb an amount of carbon dioxide equivalent to that which is being produced through burning wood⌀ and natural bio-deterioration.

REFUGEES   The number of refugees around the world, fleeing for political or economic reasons or to avoid war and oppression, was estimated in the mid-1980s to be between 12 and 13 million, a total that was increasing annually. Many nations have now resorted to various legislation or screening procedures to limit refugee influx.

There are at least twenty major focal points of refugee flight in the world, four of which record movements of over 1 million refugees. In Afghanistan some 5.9 million people fled to neighbouring Iran or Pakistan to avoid the civil war. In Southern Africa, 1 million Mozambicans have fled to Malawi, South Africa and Zimbabwe to escape civil conflict. In Ethiopia, around 1.1 million people have sought refuge in Sudan and Somalia to escape the wars raging in Tigray and Eritrea provinces. In South-East Asia at least 1.2 million people have fled from Vietnam since 1975, seeking asylum in any country that will take them. Civil strife in Liberia and Iraq's invasion of Kuwait in 1990 created two new foci of refugee flight. In some instances these refugees are

already being sent back, even from countries or dependencies which have traditionally offered asylum.

Over the past few decades a new factor, environmental degradation, has been identified as a major component contributing to the increases in refugee numbers. People fleeing from environmental degradation, estimated to be around 10 million worldwide, now make up the newest but largest class of refugees. In the Sahelian drought⌀ during the early 1970s nearly 1 million 'environmental refugees' were forced to leave Burkina Faso and a further 500,000 left Mali. The depletion and degradation of agricultural land⌀ is currently displacing more people than any other factor. An estimated 50 million are being forced to live off land that is rapidly deteriorating in biological productivity. They are becoming unable to meet their basic food and fuel needs and will soon be adding to the environmental refugee total. Poisoning of land by toxic wastes, pollution⌀ and natural disasters caused by human activities are increasing the magnitude of the problem. If, as a result of pollution caused by human activities, global warming⌀ causes the estimated one metre rise in sea level some predict by the middle of next century, a further 50 million new environmental refugees will be created.

REGIONAL SEAS   The world's regional seas, notably the Baltic, Caribbean, Mediterranean and the North Sea, are all badly polluted. These seas are essentially either enclosed or semi-enclosed, support fairly dense human populations and exhibit slow rates of water⌀ renewal – the Mediterranean, for example, is thought to take 80–100 years to renew its waters. As a result, the regional seas do not have the same cleansing capacities as oceans, but they are continuing to receive increasing amounts of pollution⌀ from a variety of sources, including industrial and municipal discharges, direct dumping from ships, oil⌀ pollution and agricultural runoff.

The Regional Seas Programme of the United Nations Environment Programme (UNEP)⌀ draws up Action Plans intended to reduce pollution damage in all the world's regional seas and, where possible, improve water quality. UNEP's ten Regional Sea programmes are supported by 120 countries. However, due to the nature and scale of the problem, these programmes have not met with major success.

In the Baltic, 100,000 sq km, half the deep water, remains badly affected by pollution. The Mediterranean is worse hit as the pollutants introduced by all the bordering nations tend to accumulate. In 1976, a UNEP-led Mediterranean Action Plan⌒, designed to clean up the sea, was initiated. Although it was endorsed by seventeen regional nations, all of which have undertaken water-improvement projects, the Mediterranean remains one of most polluted seas in world. In all, billions of tonnes of waste continue to enter the sea each year from land-based sources – 430,000 million tonnes were dumped in 1977 alone. Clean-up costs⌒ are rising steadily with each passing year; in 1983 costs were estimated to be in the order of $10–15 billion and at the beginning of the 1990s international agencies urgently made $1.5 billion available as the first step towards cleaning up the Mediterranean Basin by the year 2000.

In the Caribbean, pollution from pesticide⌒ runoff, industrial discharge and sewage effluent, rising in line with the increasing local population and the growing tourist⌒ industry, is worsening. In global terms, it is the most damaged sea as a result of pesticides, and also suffers from the fact that it is one of the busiest oil-tanker traffic routes in the world.

REMOTE SENSING Remote sensing is a term used to describe the rapidly developing process of obtaining and interpreting images of the earth, and is now commonly applied to the use of data from satellites orbiting the planet. In the 1950s, Australian scientists developed a method of mapping patterns of land⌒, soil⌒ and vegetation using data from aerial photographs. In 1972 the United States launched the first earth-observation satellite, LANDSAT 1. This had a resolution of 30 metres, referring to the amount of detail – or land area – that could be seen. Initially, satellite imagery proved to have two main disadvantages compared to aerial photography: the ground resolution was worse and the images were only two-dimensional, whereas aerial photographs could be examined stereoscopically. As the science of remote sensing developed, stereoscopic examination techniques were also evolved. In 1986, the French launched a high-resolution (15 m) satellite called SPOT which has a stereoscopic capability. Satellites capable of resolving very small images, down to 5 cm, have since been launched but these are used only for military purposes.

The type and quality of data transmitted from the satellites and the frequency with which it is received depend on the altitude of the satellite and the type of sensor being used.

For assessing and monitoring conditions at ground level, satellites usually describe a low-level orbit (700–900 km) in a north–south direction. They return to their original paths every few weeks, allowing information on any area or subject to be updated at regular intervals.

High-altitude satellites orbiting at 36,000 km above the earth's surface are used to record and analyse meteorological data. They travel at more or less the same speed as the earth rotates and so remain fixed over a given point on the planet's surface and are called 'geostationary'.

The ground resolution that can be seen from satellites is to some extent governed by the choice of how frequently observations are to be made. The greatest detail can be seen from cameras from which the film must be returned for processing. These allow objects or features of only a few metres wide to be recognized.

Scenes can be obtained on a more regular, recurrent basis from low-level orbiting satellites known as 'scanners'. These have a narrow field of view providing 10–80 m ground resolution every 12–26 days – but only when there is no cloud cover. The same satellites can also be used to examine a wider field of view, 1–4 km resolution – much more frequently: up to once every twelve hours. Geostationary satellites can provide imagery much more frequently – every thirty minutes – but with a resolution of only 5–10 km.

Satellite images provide a large pictorial representation of the earth's surface. The images are produced in four bands from different parts of the electromagnetic spectrum: one from the blue-green, one from the red and two from the near infra-red which can be used to measure vegetative changes. Images can be collected from different seasons and years with comparative ease. Changes in tree cover, crop growth and farming patterns can be continually monitored and the effect of pests or conditions under which outbreaks of pests such as locusts⌒ are likely can be determined. Early warning systems for pest outbreaks, drought⌒ or crop losses are perhaps one of the most important facets of remote sensing. Satellite data also help in locating mineral⌒ deposits and monitoring dams⌒, rainfall patterns and floods⌒. The high-resolution material collected from SPOT has also allowed the destruction of tropical forests⌒ to be monitored with some degree of accuracy. In 1990 the Soviet Union made photographs from a satellite with a resolution of 5 m commercially available. These allow changes in silt loads in rivers to be assessed, road conditions to be monitored and agricultural diseases to be identified. The Soviets also made material available from Synthetic Aperture Radar (SAR). Although it has a resolution of only 20 m, SAR has an advantage over optical images in that it can be used day or night in any weather. In arid areas radar can penetrate up to a few metres below the surface, revealing subterranean river channels and rock strata.

Brazil, China and Indonesia are all planning to launch their own remote-sensing satellites, following the lead taken by India which, in 1988, launched its own system (IRS-1) which rivals SPOT in the quality of its images.

Images from satellites can also be used for military and espionage purposes. Commercial operators have refused to make material available at times of conflict, such as the Iran–Iraq war and after Iraq had invaded Kuwait in 1990. Many developing countries⌒ are expressing growing concern at the use to which satellite imagery is being put and the matter of control over remote-sensing operations.

RENEWABLE ENERGY The sun provides the earth with energy⌒ in the form of electromagnetic radiation⌒. Without it, the surface temperatures of the earth would plummet and plants and animals would not be able to survive. Only a minute amount of the sun's total energy is intercepted by the earth. Of this, 30 per cent is reflected back into space: the remainder is absorbed by the atmosphere⌒, land⌒, oceans and living organisms.

The world's population⌒ has become dependent on 'unrenewable' sources of energy – coal⌒, petroleum and natural gas⌒ – the production of which takes centuries. Most 'renewable' forms of energy harness solar energy, either directly or indirectly. Solar panels trap the sun's heat directly and photovoltaic (solar) cells turn the sun's energy directly into electricity. Living vegetative material, or biomass⌒, stores solar energy.

Energy from the sun is used by vegetation to manufacture simple sugars, using carbon dioxide⌒ in the air or water. These simple compounds are then converted into more complex organic molecules. When fuelwood⌒ is burnt for heat and light, the sun's energy is effectively released. The sun also evaporates water⌒ from oceans, lakes and rivers. Cooling leads to condensation. The sun effectively lifts the water; this is exploited by hydroelectric power⌒ installations. As the water falls back to the sea it is used to drive turbines. The energy contained in wind and waves is also derived from the sun. As the earth's surface heats up, it causes differences in atmospheric pressures which force the air to move. The wind propels the water in the world's oceans to form waves. The rise and fall of ocean tides is also affected by the sun, although it is primarily due to gravitational pull from the moon. Tides are highest when the earth, moon and sun are aligned.

The incoming solar energy absorbed by the earth in one year is equivalent to twenty times the energy stored in all the world's reserves of recoverable fossil fuels⌒. The earth's core is also heated by the decay of naturally occurring radioactive elements such as uranium⌒ and thorium. This heat can be extracted either by drilling into natural aquifers of hot water or by forcing water under pressure through hot rock. Renewable sources such as biomass and hydropower already provide 21 per cent of the total energy consumed worldwide, but if only 0.005 per cent of solar energy reaching the earth could be harnessed through the combined use of biomass fuels, wind turbines, solar collectors, hydropower, wave-energy⌒ converters and other methods, it would supply more energy in a year than is obtained from burning coal, oil⌒ and gas combined.

By the year 2000 the contribution of all renewable sources to global energy production is expected to exceed that of oil and its derivatives.

REPROCESSING Nuclear reactors use up about one-third of their fuel each year. From 1986, almost 77,000 tonnes of this spent fuel were officially inventoried by reporting nations, with a further nine countries known to have operational nuclear power⌒ plants adding an unknown quantity to this total. All irradiated fuel has to be removed and is usually sent to a plant where unused uranium⌒ and plutonium⌒ are chemically separated from the other radioactive waste⌒ products. The waste has to be disposed of, as it remains potentially harmful for thousands, if not millions, of years. The reprocessing of irradiated fuel increases the volume of global hazardous radioactive waste more than a hundredfold.

Storage of both waste and spent fuel is reaching crisis point. It was originally believed that spent fuel could be stored underwater for five to ten years, but corrosion in fuel elements has been observed after one year in some instances. This ultimately slows down reprocessing and may eventually force reactors to shut down until there is spare capacity to receive their spent fuel.

Reprocessing spent nuclear fuel to recover plutonium is fraught with difficulties and potential risks to the environment and human health – not least of which is that the end product represents the cheapest and easiest means to make atomic weapons. Reprocessing plants also generally tend to release some radioactivity⌒ into the atmosphere as the spent fuel is separated into useful material and useless radioactive decay products. Reprocessing is also uneconomic. New uranium is far cheaper than recycled fuel and reprocessing is expensive: costs in Britain doubled between 1978 and 1983. Reprocessing plants around the world are also notoriously incompetent at matching theoretical plutonium yields with actual ones, and recovered plutonium can often not be accounted for.

RESETTLEMENT PROGRAMMES   Major infrastructure development projects – roadbuilding, urban renewal, large dams⌒, and large-scale agricultural or industrial schemes – can sometimes proceed only if people are moved to make way for them. In the Third World⌒, programmes of this nature can require the removal of millions. The Indonesia/Transmigration⌒ colonization programme is by far the largest human resettlement project in the world. By 1984, 3.6 million people had been resettled and the government plans to relocate a further 65 million inhabitants over the next twenty years. People are being moved from densely populated islands to less densely populated ones, but only after tropical forests⌒ have been cleared to accommodate them. The programme has received substantial backing from the World Bank⌒ and other development banks.

Between 1979 and 1983, the World Bank funded projects that forced the resettlement of at least 500,000 people, despite the fact that the social consequences of such programmes are often ignored. Where tribal peoples are involved, forcible removal from their traditional lands can lead to an almost complete destruction of their culture. These people are frequently denied any compensation for their forced removal, as they usually have no official title to the land⌒. Where land is offered in compensation, it is often of poor quality. Where alternative housing is also offered, it too is generally inferior. In 1980 the World Bank laid down a set of guidelines for resettlement projects. However, a 1986 review found that these guidelines were not being followed. In almost one-third of the projects evaluated during the review, the authorities involved had not even identified the areas where the displaced people in question were to be resettled.

RESPIRATORY INFECTIONS   About 4 million children, mostly in developing countries⌒, die each year from acute respiratory infections, primarily pneumonia. Normally, every child develops four to eight episodes of respiratory infections each year, most of them minor infections of the upper respiratory tract. The incidence of more acute, life-threatening lower respiratory tract infections is far higher amongst children in developing nations due to a variety of contributory factors: poor nutrition, low birth weight, indoor air pollution⌒, inadequate sanitation and lack of medical facilities.

Most pneumonias in children cannot be prevented by vaccines, but they can be cured by standard, inexpensive antibiotic treatment. Although pneumonia can be fatal, effective cure can cost as little as $4 per head. Other respiratory diseases such as pertussis (whooping cough) and the lung complications brought on by measles⌒, can be prevented by the administration of effective vaccines at a cost of less than $2 per child. Globally, over half of all visits to health-service facilities are related to acute respiratory infections in children. This tremendous strain on health units and personnel, especially those in the Third World⌒, could easily be alleviated by a relatively small, well-planned and cost-effective investment.

RHINE   The largest and allegedly dirtiest of Europe's rivers, rising in Switzerland and travelling 1,320 km to the North Sea. The governments of the five nations which border the Rhine are participating in a concerted programme aimed at cleaning up the waterway by the year 2000, attempting to make the river's water⌒ fit enough for humans to drink. In the late 1980s, the river remained one of the most polluted in Europe. After the worst polluters of the river have been identified, it is hoped that the level of pollutants discharged will be halved by 1995. This target may be difficult to reach as previous clean-up programmes have met with little success: the acidity of the river rose 37 per cent between 1976 and 1986 when water-improvement projects were supposedly being carried out.

In 1982 a report from a Dutch environmental group identified forty-five companies which were polluting the river and found that despite repeated clean-up campaigns, the Rhine remained as polluted in the 1980s as it had been in the 1970s. Nevertheless, some 20 million Europeans still draw their drinking water from the river. The Netherlands are particularly dependent on the Rhine, which provides 80 cubic kilometres of fresh water, representing an eightfold increase over the amount that would otherwise be available in the country. At its mouth, the Rhine dumps some 10 million cubic metres of silt. This has to be removed, but it is so contaminated that it cannot be disposed of at sea and the Dutch authorities responsible for dredging the river mouth have to keep it isolated in a special dump.

The scale of the pollution⌀ problem in the Rhine was brought to public attention once again by a massive fire at a chemical plant in Basel in 1986. The fire resulted in thousands of tonnes of toxic chemicals⌀ being washed into the Rhine, rendering a 100–150-km stretch of the river effectively lifeless. Sandoz, the company which owned the chemical plant, is to pay $7.5 million to the French government in compensation for the accident, although only $2.8 million of this total will be spent on restoring the river's ecosystem⌀.

RHINOCEROS   Rhinceroses are large, thick-skinned pachyderm animals from Africa and southern Asia with a horn or two horns made from fused hair on the nose. Since 1970 more than 60,000 have been killed. Most have been slaughtered for their horns, which are used to make ornamental dagger handles in North Yemen or for use in traditional-medicine potions throughout much of Asia. Counting animals from all five species, there are fewer than 11,500 left worldwide.

The most dramatic drop in population has been that of the black rhino, *Diceros bicornis*, which is native to Africa. Over 95 per cent of the total population has been killed in the past twenty years: numbers have dropped from an estimated 65,000 in 1970 to fewer than 3,000. Within the last decade, the black rhino has disappeared from Angola, Chad, Ethiopia, Mozambique, Rwanda, Somalia, Sudan, Uganda and Zaïre. Zimbabwe is the only country where the animals are maintaining their numbers, and the wild population in Zimbabwe's Zambezi valley is one of the last remaining.

The other African species is the white rhino, *Ceratotherium simum*, which has a northern and a southern subspecies. The southern populations were virtually exterminated, but a selective breeding and protection programme has managed to raise numbers to about 4,000. The northern subspecies is on the verge of extinction⌀, with fewer than twenty animals surviving in Zaïre and a few in southern Sudan.

In Eastern Asia practically the whole rhino – horn skin and blood – is used in the production of tradi-tional medicines to treat fevers, stomach disorders, and various other complaints. Extracts from the horn are also valued as an aphrodisiac. Throughout their range, the three Asian species of rhinoceros have long been subject to heavy poaching and only relict populations now exist. In north-east India and Nepal an estimated 1,500 great one-horned rhino, *Rhinoceros unicornis*, still survive. The smaller Javan one-horned rhino, *Rhinoceros sondaicus*, is found only in a single reserve in Java, and only fifty animals are left. The total population of the Sumatran rhino, *Dicerorhinus sumatrensis*, is believed to number a few hundred, living in the forested hills of Malaysia and Sudan.

The effects of habitat loss and hunting⌀ have been exacerbated because the animals have a long and complicated breeding cycle that is not fully understood. It is likely that females can give birth every two years but can breed only for a very limited period, meaning that they can produce at best three to four offspring each. The rhinoceros is now one of the world's most endangered large mammals, and the Worldwide Fund for Nature (WWF) has channelled over $3 million into a campaign to try and save all five species.

RICE   An annual cereal grass. The genus *Oryza* consists of twenty-four species out of which two, *O. sativa* (Asian rice) and *O. glaberrima* (African rice) are extensively cultivated. *O. sativa* is the more important. Probably native to India, it is now cultivated in tropical, subtropical and warm-temperate regions throughout the world. It is one of the world's most important food crops and is the main cereal crop and staple food of half of the world's population⌀ – most notably for those living in monsoon⌀ South-East Asia, where 90 per cent of the world's rice is produced and consumed.

Primarily as a result of the Green Revolution⌀ there are now thousands of varieties of rice, mostly developed through plant-breeding techniques. Although most varieties are grown in standing water⌀, 'upland' types can be grown on dry land⌀ where the rainfall is heavy. Virtually no rice grows where precipitation is under 100 cm annually. Most varieties tolerate high temperatures and humidity and produce good yields under continual cultivation systems.

Rice is predominantly grown on irrigated land, seedlings being transplanted into flooded paddy fields. Fields can be drained where necessary to allow mechanical harvesting, although in the developing countries⌀ harvesting is most commonly done by hand. Milling can remove the outer husk of the grain, yielding brown rice, or both the husk and the underlying bran layer, yielding the Vitamin-B-deficient white rice.

The provision of irrigation⌂, use of high-yield varieties⌂ and application of large quantities of fertilizers⌂ and pesticides⌂ under the Green Revolution saw rice yields in developing countries rise as a whole by 1.9 per cent annually between 1961 and 1980. In nations such as Indonesia and the Philippines, which are best suited to rice production, yields increased by over 3 per cent a year. However, the performance of rice in Asia is far below that achieved in temperate industrialized regions, mainly because of pests and disease. Over 500 diseases of rice grown in the tropics have been reported, against 54 reported in rice grown in temperate zones.

Global production is around 480 million tonnes annually, and steadily increasing. China (171 million tonnes), India (101 million tonnes) and Indonesia (41 million tonnes) are the world's largest producers.

RIGHT LIVELIHOOD AWARD   Also known as the 'Alternative Nobel Prize', the Right Livelihood Award is given each year 'to reward people with practical, exemplary and replicable solutions' to development problems. In addition to the main award, other cash grants are distributed to worthy groups and organizations. The awards were established in 1979 by Jakob von Uexkull, a Dutch-Swedish businessman whose offer to start an Environment and Third World award was rejected by the Nobel authorities. The Right Livelihood Award is presented each year in the Swedish Parliament the day before the Nobel ceremony, following an annual invitation from Swedish parliamentary members of all political persuasions. The initial $825,000 fund for the awards was raised through the sale of von Uexkull's stamp business. The original funds have now run out and the prize is currently financed by contributions from trusts and individuals around the world, although money from commercial sources is not accepted.

RINDERPEST   A contagious viral disease, also known as cattle plague, affecting cattle and certain wild animals throughout much of Africa and Asia. Symptoms occur three to nine days after infection and include loss of appetite, fever, mouth ulcers, dysentery and emaciation. Amongst livestock⌂, at least 90 per cent of acute cases are fatal: in chronic cases mortality rates are much lower, but in all cases productivity is seriously reduced. Compulsory slaughter of all infected herds or vaccination programmes are usually employed to control outbreaks. The Pan-African Rinderpest Campaign, launched in 1986, was one of the world's most ambitious anti-rinderpest efforts, aiming to eradicate the disease and revitalize the livestock industry in thirty-four countries across the continent, using well over 50 million doses of vaccine.

RIVER BLINDNESS   *See* Onchocerciasis.

RONGELAP   One of the Marshall Islands⌂ in the northern Pacific Ocean, lying 140 km from the United States's nuclear testing sites at Bikini and Enewetak. Between 1946 and 1958, fallout from the nuclear weapons testing carried out by the United States on nearby atolls fell over the island, mainly in the form of pulverized coral⌂ dust. In 1954, within five hours of the largest test, code-named Bravo, fallout began to settle on Rongelap, causing skin rashes, sickness and loss of hair. A day after the test explosion the USA evacuated the islanders, allowing them back three years later when Rongelap was proclaimed safe to inhabit.

Over the ensuing years islanders suffered from abnormally high levels of cancer⌂, leukaemia, still-births and birth defects. In 1979 the US authorities admitted that parts of the island were still dangerously radioactive⌂, but persistently refused the islanders' requests to evacuate them. The Rongelap people finally approached the Greenpeace⌂ organization to help in evacuating the island. In 1985 the Greenpeace flagship vessel, *Rainbow Warrior*, ferried the 308 islanders and their possessions to the island of Mejato, 190 km away. The ship then sailed on to New Zealand, *en route* to protest at French nuclear testing in the South Pacific, where it was bombed and sunk by agents of the French government.

RUBBER   A synthetic or natural tough, elastic organic polymer. Natural rubber is made from latex, a milky fluid collected from rubber trees. The rubber tree, *Hevea brasiliensis*, is native to Brazil but is now most widely cultivated throughout South-East Asia. Originally, seeds were shipped from Brazil to the United Kingdom in 1876, and then by the British to colonies in Asia, where large tracts of land⌂ have been given over to plantations. Rubber trees yield latex from about the sixth year of growth. When the bark is cut, the fluid exudes and can easily be collected using unsophisticated techniques: collection containers strapped to the tree. Rubber is coagulated from latex through processing with acids, then pressed into sheets and air-dried.

A variety of synthetic rubbers are also made from petrochemicals. Styrene-butadiene rubber, the commonest of the manufactured rubbers, is used to make automobile tyres⬦. It is often mixed with natural rubber to improve its resilience. Despite the billions of vehicle tyres and other rubber goods produced over the years, and the combined efforts of scientists and research institutions around the world, there is currently no environmentally sound and effective means of disposing of synthetic rubber compounds.

Of the global total of 4.8 million tonnes of natural rubber produced annually, most comes from the South-East Asian region. Malaysia (1.6 million tonnes), Indonesia (1.1 million tonnes) and Thailand (970,000 tonnes) are the leading producers. Some 10 million tonnes of synthetic rubber are also produced each year: the Soviet Union (2.5 million tonnes), the United States (2.3 million tonnes) and Japan (1.3 million tonnes) produce the bulk of this total.

RU486 A pill that has been tested and widely recognized as a safe and effective means of terminating pregnancy during its early stages, also referred to as the 'abortion pill'. Chemicals in the pill work by blocking the action of the hormone progesterone.

This effectively prevents any fertilized egg from attaching itself to the uterine wall and – usually within forty-eight hours – produces an effect similar to that of miscarriage. Dr Etienne-Emile Baulieu, who developed the pill, claims that it is a far safer method of abortion than any surgical operation, particularly in countries where medical facilities are poor. Globally, around 200,000 women die each year as a direct result of illegal or unsafe abortion operations; another 600,000 receive some form of mental or physical injury. The majority of deaths occur in the Third World⬦.

Following discovery and testing of the pill, the parent transnational⬦ pharmaceutical company, Hoechst, based in the Federal Republic of Germany, blocked distribution, fearing a boycott of its other products from consumers opposed to abortion. In France, strict government action resulted in the pill being made available, and by mid-1990 at least 40,000 abortions had been carried out in France using RU486. Side affects can include cramps, nausea and diarrhoea, as in normal menstruation. Rarely, severe bleeding occurs, as in spontaneous miscarriages, and blood transfusions may be required. In France, two women were reported to have developed heart problems during treatment with the abortion pill. Dr Baulieu, who is also an adviser to the World Health Organization (WHO)⬦, reports that the pill will probably be manufactured or acquired illegally in many nations if legitimate production and use are not sanctioned.

S

SADAT, ANWAR (1918–1981) Egyptian statesman. Twice Vice-President, he finally succeeded Gamal Abdel Nasser as President of Egypt in 1970. During his term of office he moved Egypt away from the Soviet camp and towards increasing United States influence. He handled the Egyptian campaign in the 1973 war against Israel and later, in 1977, visited Israel to try and reconcile the two countries. He shared the Nobel Peace Prize with Israel's Menachem Begin⬦ in 1978 for their efforts in formulating the Camp David Agreements⬦ and subsequent peace treaty. He was assassinated by Islamic⬦ fundamentalists.

SALINIZATION The process through which salt concentrations in the soil⬦ build up is called salinization. In areas of regular rainfall, salts are flushed out of the soil into underlying groundwater⬦ or carried away to the sea. However, where rainfall is low and evaporation rates are high – as in the arid tropics – soils tend to have a high salt content: salts can comprise up to 12 per cent of the soil profile in the worst areas. Salts inhibit root activity and limit plant growth. They build up in the top few centimetres of soil where roots are most active and, in severe cases, form a salt crust on the land⌖ surface.

Perennial irrigation⌖, when land is used year after year without fallow periods, is now a major cause of salinization. Land must be well drained or water-table levels rise, bringing the salt in the soil to the surface, where it is concentrated by evaporation. The United Nations Environment Programme (UNEP)⌖ reports that between 30 and 80 per cent of the 227 million hectares of irrigated land in the world suffers to some degree from salinization. Mainly as a result of mismanagement, tens of thousands of hectares are now too salty to support plant life and 22 million hectares are thought to be waterlogged, allowing salt concentrations to build up. Many estimates now suggest that as much land is now having to be taken out of production as a result of salinization and waterlogging as is being brought into production through irrigation.

Between 1950 and 1970 the global total of land under irrigation doubled. Just under 20 per cent of the world's cropland is now irrigated. However, salinity is a serious problem on 20–30 million hectares, almost 10 per cent of the world's irrigated land. Moreover, the salinization process is intensifying at a rate of 1–1.5 million hectares annually. Salinization is becoming an acute problem in China, India and other parts of Asia where the Green Revolution⌖ was based on the use of irrigated land and high-yield varieties⌖. Northern Africa, Australia and the United States are also witnessing increasing levels of salinity. In the Colorado Basin in the USA, annual salt build-up of 10 million tonnes causes crop yield losses of over $113 million – a sum that is projected to increase to almost $270 million a year by 2010.

Globally, around 20 million hectares of land are thought to be naturally saline, mostly in wetlands⌖ and coastal areas in arid regions. Latest figures suggest that about 30 per cent of the world's potentially arable land is affected by salt. Although salinized soils can be rehabilitated by draining away excess water⌖ and flushing out the salts with copious amounts of fresh water, both are extremely expensive, and therefore uncommon, operations.

SALVINIA An aquatic fern native to South America, *Salvinia molesta* has, thanks to transmission by man, become a significant problem in waterways in many parts of Africa, Asia and the South Pacific Islands. The freshwater fern has a phenomenal growth rate and can double its area within two days. It covers bodies of water⌖ with dense, free-floating weed mats which prevent or impede travel by boat, ruin fisheries⌖, prevent access to drinking water, choke irrigation⌖ schemes and provide a perfect habitat for a variety of parasites and insects injurious to human health.

Less than two years after its accidental introduction into the River Sepik in Papua New Guinea in 1980, *Salvinia* had virtually covered all the lakes in the lower half of the river's floodplain as well as many areas upstream. In 1981 the Papua New Guinea government, together with the United Nations Development Programme (UNDP) and the Food and Agriculture Organization (FAO)⌖, undertook an aerial herbicide-spraying programme to combat the weed but found that repeated spraying, at a cost of $50 per hectare, was needed merely to keep it in check. Biological-control⌖ scientists then discovered a weevil, *Cyrtobagous salviniae* which attacks the fern. Adult weevils eat the buds and slow down the plant's growth rate. Weevil larvae also burrow into the floating body of the fern, eventually causing it to rot and sink. By 1985, over 850,000 adult *Cyrtobagous*, together with millions of eggs, larvae and pupae, had been released into the Sepik floodplain. The weevil effectively destroyed around 1.5 million tonnes of salvinia, clearing 170 sq km of water surface and freeing more than sixty lakes from infestation. In these lakes the weevil persisted in small numbers on marginal vegetation, ensuring that the fern remains strictly under control with no threat of reinvasion.

SANDOZ ACCIDENT In 1986 a fire broke out in a warehouse at the Sandoz chemical factory in Basel, Switzerland. As a result, over 30 tonnes of pesticides⌖, fungicides⌖, dyes and other chemicals entered the River Rhine⌖. In just two hours the fire released more pollutants into the river than it normally receives in a year. Fortunately, nearby storage tanks containing the nerve gas phosgene were undamaged.

Over thirty different agricultural chemicals of varying toxicity were officially stored in the warehouse. Cleansing crews later found high concentrations of atrazine, a chemical which was not listed as being stored at the Sandoz site. Only then was it discovered that another chemical firm, Ciba-Geigy, had suffered an accident the day before the Sandoz fire, releasing the chemical into the river. The Sandoz fire focused public and scientific attention on the Rhine, and within a month twelve major pollution⌂ incidents had been reported. Experts calculate that it will be at least ten years before the worst-affected stretches of the river return to any degree of normality – if ever. Half a million fish⌂ were killed outright, mostly as a result of mercury⌂ pollution – 200 kg of mercury were released by the Sandoz incident. The water⌂ supplies of many towns and villages along the river downstream from the warehouse site were cut off for public health safety reasons. They have since been restored but dangerous, persistent chemicals may eventually permeate through the riverbed and into groundwater⌂, threatening future supplies.

SASAKAWA, RYOICHI (1899–) Japanese benefactor, believed to be the world's leading individual donor to environment and development projects, having given away $3 million annually for the past thirty years. Apart from governments he is the largest single donor to the United Nations⌂, having contributed in excess of $40 million. Charitable organizations associated with Sasakawa – notably the Japanese Shipbuilding Industry Foundation (JSIF) – have donated over $12 billion to international charities.

Having made a fortune as a commodity broker before the Second World War, Sasakawa later introduced motorboat racing into Japan and grew even wealthier on the proceeds. Motorboat racing, Japan's biggest spectator and betting sport, averages annual profits of $100 million and 3.3 per cent of all revenue generated passes to the JSIF, founded and still run by Sasakawa. The World Health Organization (WHO)'s⌂ successful smallpox⌂-eradication programme benefited substantially from Sasakawa's contributions and, amongst many other supportive measures, he helped to underwrite the Palme⌂ Commission on Disarmament, despite publicized comments that he preferred not to give support to organizations that are influenced by 'left-wingers or Communists'. For this reason – and because of his associations with the Moonies, the late President Somoza of Nicaragua and the late President Marcos⌂ of the Philippines – a number of institutions, most notably the Tokyo-based United Nations University⌂, have refused his offers of funding. Philanthropic Sasakawa Foundations have been set up in the United Kingdom, the United States and several other countries.

SAUDI FUND FOR DEVELOPMENT (SFD) Established in 1974. The fund's objective is to participate in financing projects to improve living standards in developing countries⌂. It mainly provides loans for capital projects but has also provided money for feasibility studies, mostly those concerning infrastructure⌂-improvement projects. Although most loans have gone to Arab and other Islamic countries, many other African, Asian and Latin American nations have also received assistance.

SAVE THE CHILDREN FUND (SCF) An independent voluntary organization established in London in 1919 to care for needy children irrespective of race, religion or nationality, particularly those in the developing world⌂. It operates in over fifty countries, with well over 1,000 field workers, including doctors, nurses, welfare workers and administrators. It provides relief in times of disaster and is especially involved in long-term projects in the field of maternal and child health care.

SCHISTOSOMIASIS Also known as bilharzia⌂: a debilitating parasitic disease thought to affect over 200 million people in tropical countries. At least 600 million people are at risk of contracting schistosomiasis, especially those who perform daily activities which bring them into contact with untreated water⌂ through swimming, fishing⌂, irrigated farming, washing and bathing in streams or ponds. The disease is contracted by exposure to water containing a certain species of water snail which acts as a host to the first larval stage of flukes of the genus *Schistosoma*, the disease-causing agent. When these larvae leave the snail in their second stage of development they are able to pass through human skin, become sexually mature, and produce vast quantities of eggs which then pass into the human intestine or bladder. The release of the characteristically spiked eggs causes anaemia⌂, inflammation, and the formation of scar tissue. Diarrhoea, dysentery, enlargement of the spleen and cirrhosis of the liver may occur. The human host may die, but before then numerous eggs will have passed out of the body in urine or faeces to continue the cycle.

The disease contributes to malnutrition⌂, especially in children; in adults, apathy and tiredness contribute to lost productivity; estimates of work days lost per case range from 600 to 1,000. Seventy-six developing countries⌂ report cases regularly, and at least 200,000 people die from the disease each year. Praziquantel, a drug developed to control schistosomiasis appeared in the 1970s. For less than $1 a patient can be cured, but the problem lies in the high rate of reinfection. Molluscides added to water to control the snail intermediate hosts are effective but more expensive. In the last decade, cases have been reduced by 10 per cent and the development of a vaccine, as yet in its early stages, may significantly improve this figure.

SCHUMACHER, E.F. (1911–77) Economist and philosopher. Born in Germany, he attended university at Oxford in the UK, where he developed his deep suspicion of Western growth-centred economics and his commitment to the need to build caring, socialist economies. Following a lecturing post in the United States, he worked in business, farming and journalism before re-entering academia as an economist in the mid-1940s. Calling on his varied background and mindful of his distrust of Western economics, he was the originator of the concept of Intermediate Technology for developing countries◇, and founder and chairman of the Intermediate Technology Development Group◇ in the UK. His thoughts and philosophy for economic development, particularly with regard to the situation in the developing countries, was encapsulated in his best-selling reappraisal of Western economic policies and attitudes, *Small is Beautiful*, and he became a regular adviser to governments throughout the world on problems of rural development.

SCREWWORM The screwworm fly, *Cochliomyia hominivorax*, is a parasite of warm-blooded animals, particularly livestock◇. Female flies lay several batches of eggs – 400 every three days, in any skin abrasions where the larvae develop. Under a wound the size of a pinprick a mass of maggots the size of a tennis ball can develop, eating the living flesh of the animal. Any cattle infected are quickly and severely weakened and, unless treated, will die within a few days. Human deaths have also been reported. Inspection and treatment costs around $1–2 per animal. The worm is a native of Central and South America, where larval infestations of animals are one of the major causes of economic loss. It is regarded as the most important insect pest of livestock in the western hemisphere, causing annual losses of $100 million from the 1950s to the 1970s.

During the late 1980s the worm was inadvertently introduced into North Africa, first appearing in 1988. The fly established itself in Libya, infesting more than 200 people and 2,000 domestic animals,

and threatened to spread to Algeria, Egypt, Morocco, Sudan, and Tunisia. The rich and diverse wildlife of Africa was put at risk by the infestation. The costs of controlling the screwworm on the continent, using sterile-male release techniques pioneered and developed in North America, are put at $20 million a year. In comparison, if the infestation spreads, it could affect 70 million head of livestock, and annual losses of $250 million have been forecast. In 1990 President Bush relaxed a ban on US citizens travelling to Libya to allow scientists to provide guidance on how best to tackle the worm. An $85 million eradication programme began in late 1990 – after numerous delays – using sterilized male flies: 100 million flies per week were released over a twenty-week period.

SEA-LEVEL RISE Around 15,000 years ago, the sea was 120 metres lower than it is today. Only in the past 5,000 years has it been at its present level.

Human activities are now playing a significant role in determining the level of the sea. Scientists have calculated that by 2000 about two-thirds of the world's total flow of water◇ from land◇ to sea will be controlled by dams◇. Over the past thirty years, reduction in river discharges to the oceans has actually led to a drop in sea level of at least 2 cm. However, overall, over the past century sea level has risen by approximately 15 cm, within a range of uncertainty spanning 10–20 cm. Most of this rise has been due to the average ambient temperature of the world rising by 0.3°–0.6°C.

As the world warms and cools, so the level of the seas change. Sea water expands when warmed, and polar caps and inland glaciers melt accordingly as temperatures rise. Global experts gathered at the Villach Conference◇ in 1985 to examine the problem of global warming◇ and climate change concluded that, based on observations of changes during this century, continued global warming of between 1.5° and 4.5°C would result in a sea-level rise of 20–140 cm. A rise of 1 metre would drastically affect over 300 million people in low-lying coastal areas around the world. Island states such as the Maldives and Kiribati could disappear underwater. A rise of only 50 cm would displace 16 per cent of the population of Egypt. In the Netherlands, two-thirds of the country already lies below sea level, and in the southern United States, the Gulf of Mexico would creep 53 km inland. The cost of protecting people and investments from encroaching waters, as calculated by the International Panel on Climate Change (IPCC), will be over $20 billion per year.

Computer models on climate change, global warming⌂ and sea-level rise are mostly guesswork at present and are steadily being improved. By mid-1990, it was being predicted that a warming of 1.5°–4.5°C would create a sea-level rise of between 14 and 24 cm by 2030, caused by thermal expansion of the oceans and by diminishing alpine glaciers. There is also likely to be a net gain of ice in Antarctica⌂, leading to a slight drop in sea level, but this will be offset by a small positive contribution from melting ice in Greenland.

Predictions of a 6 cm rise per decade over the next century were accompanied by a degree of error of plus-or-minus 3–10 cm. In 1990 the IPCC, drawing on the knowledge of 300 expert scientists from around the world, refined their calculations and reported that if no cuts in emissions of greenhouse gases⌂ were implemented, sea levels would rise 20 cm by 2030 and exceed current levels by 65 cm in the year 2100, although there was no evidence to show that levels during the 1980s were rising any faster than in previous decades. The IPCC also concluded that if no corrective measures were taken, sea-level rise would accelerate at three to six times the prevailing rate. There is a consensus of opinion that natural ecosystems⌂, such as wetlands⌂ and coral⌂ reefs, could adapt and cope with a sea-level rise rate no more than 2 cm per decade.

SEAWEED FARMING In many coastal regions of the world large marine algae, or seaweeds, are cultivated to produce nutritious human food, animal feed or for industrial purposes. Seaweed mariculture is commonly practised in coastal waters off California, Florida and Hawaii, in the South Pacific Islands, throughout the Caribbean, around China, Japan and the Philippines, and off the coasts of European countries. Cultivation of the edible brown seaweeds, *Laminaria* and *Undaria*, is well developed in temperate waters off China and Japan, where seaweed farming is commonplace.

Young plants are raised in glasshouses before being transplanted to long lines tethered in coastal waters. Mature plants grow to a length of 6 m and yield about 5 tonnes (wet weight) per line. In parts of the world where *Laminaria* and the similar *Porphyra* form a regular part of the diet, the incidence of some forms of cancer⌂ is markedly reduced. *Porphyra* is also reported to reduce blood cholesterol levels. Calcium alginate fibres produced from *Laminaria* can be used to produce fibres and cloth. These make excellent wound dressings – the mesh remains highly absorbent and moist, so promoting healing. The calcium content also helps blood-clotting. Plasters and dressing made from the fibres are now being used in several hospitals.

Several red seaweeds are also important sources of food and industrial products. There is a large and growing demand for carrageenan, a fine white powder made from dried red seaweeds. It is used as a stabilizer and thickener in products ranging from ice cream, canned foods, fruit drinks, beer, cough syrups and pet foods⌂, through to cosmetics, shoe polish and paints. Red seaweeds are cultivated in various parts of the world. The name carrageenan comes from Carragheen in Ireland, where the edible red seaweed *Chondhus crispus* grows in abundance.

In the Philippines and the South Pacific the most commonly grown red marine alga is *Eucheuma*, which is relatively fast-growing: it doubles its weight every two weeks and can be harvested after eleven weeks of growth. Cultivated plants can be harvested at three-monthly intervals, giving annual yields of 10–15 tonnes (dry weight) per hectare. Two other red algae, *Gelidium* and *Gracilaria*, are cultivated on a global scale. These are used to produce another colloidal substance, agar. This is used as an agent for gelling, stabilizing and emulsifying. Of greater significance, it is used as the medium for all microbiological cultures carried out in hospitals, laboratories and schools.

SECTOR LOAN Loans provided by Multilateral Development Banks (MDBS)⌂ that are not used for specific projects, such as a road or hydroelectric power⌂ scheme, but are meant to bring about the restructuring of an entire single component, or 'sector', of the recipient country's economy. The loans may apply to sectors such as agriculture⌂, energy⌂ or telecommunications. These loans tend to be much larger than those provided for projects, and the conditions attached to them tend to be much more stringent. In this way, the MDBS can exert a large degree of control in guiding the actions of the government receiving the loan. Sector loans, coupled with Structural Adjustment loans⌂, make up what is referred to as Programme Lending.

SEDIMENTATION Soil⌂ washed or blown into watercourses can either settle or be carried downstream, depending on the speed of water⌂ flow, eventually reaching the sea. When the soil particles – or silt – settle, they form a sediment; this process is called sedimentation. As much as 24 billion tonnes of soil are estimated to be carried from the land⌒ into the world's oceans by rivers each year. Only about 9 billion tonnes occur as a result of natural processes; the rest – some 15 billion tonnes – result from human activities. Whenever forests⌒ or other factors which prevent soil erosion⌒ are diminished or removed, erosion and silt loads increase. Silt prevented from moving downstream, usually where a dam⌒ or other barrier restricts water flow, builds up on the riverbed, eventually causing flooding. It can also foul dams and prevent hydroelectric power⌒ units from working properly.

In river deltas and other low-lying areas where embankments have been built to prevent flooding, the riverbed can quickly rise and the threat of breaches and flooding is magnified. A single breach of the right bank of the Brahmaputra in Bangladesh during the 1988 floods⌒ inundated 1,000 sq km of land. The sedimentation problem is particularly acute in China, where deforestation⌒ and massive agricultural operations have led to vast amounts of soil being eroded. The Sanmexia Dam was commissioned in 1964 and closed the same year due to excessive sedimentation. The Laoying Dam suffered a worse fate: the reservoir silted up before the dam became operational. Where the nation's Yellow River crosses the plains, and embankments have been constructed to prevent flooding, the riverbed has risen five metres above ground levels. A breach in the embankment could have catastrophic consequences.

In Africa, the impact of the agricultural activities of a rapidly growing population⌒ on fragile land is also beginning to have a major impact on sedimentation levels. In 1987 the Food and Agriculture Organization (FAO)⌒ concluded that sedimentation levels in many African rivers are increasing roughly 1.5 times as fast as the population⌒ in the catchment areas.

SEED BANKS Also known as gene banks or germ plasm banks, seed banks are repositories of seeds or vegetative tissue, mostly obtained from a wide range of primitive strains and wild crop varieties. The material is kept under conditions of low temperature and low humidity as a resource to help maintain genetic diversity. Although similar banks exist for the storage of animal tissue, seed banks have not only been in existence longer, they have brought the commercial and political problems involved in such ventures, especially those connected to the advent of biotechnology⌒, to global attention.

In 1984, the International Undertaking on Plant Genetic Resources was adopted by the member states of the Food and Agriculture Organization (FAO)⌒, with only the United States voicing a protest. The agreement proposed that all germ plasm, including seeds improved by biotechnological means, should be freely available to all. Within two years commercial considerations and national interests had virtually rendered the agreement obsolete. Brazil, Ethiopia, India and Taiwan had all prohibited the export of germ plasm from specific crops and the United States had refused to allow genetic material held in a gene bank to be sent to Nicaragua.

The issue of manipulation and control of germ plasm threatens to be one of the major problems of the 1990s. The International Board for Plant Genetic Resources (IBPGR)⌒ is responsible for running the world's gene banks of vegetative material. By 1985 sixty such banks had been established, around a third of them in the developing world. Developed countries hold over 90 per cent of known genetic plant-material stocks whilst the Third World⌒ maintains only 33 per cent, despite the fact that the Third World actually donated over 90 per cent of all the material held in gene banks around the world. The IBPGR is part of the Consultative Group on International Agricultural Research (CGIAR)⌒, which is controlled by the industrialized nations. In 1990 a decision was taken to move the IBPGR to Denmark, although this was later revoked and the Board remains in Rome, under the patronage of the Italian government and close to the FAO, where developing countries⌒ have a much greater say in policy decisions. An FAO Commission on Plant Genetic Resources was set up to oversee the work of the IBPGR and support global equality with respect to access to and use of genetic material.

The reasons for the conflict over seeds and germ plasm are clear. Genetic resources are raw materials that have been donated by the poor countries of the world to the rich. Once there – and housed in sophisticated facilities – the material became, to some extent, controlled by the patent laws or political whims of the countries in question. By 1989, the global seed business was worth $13 billion a year. Over the preceding decade, however, what had previously been an extremely broad-based plant-breeding sector had declined and become dominated by a small number of transnational⌒ petrochemical and pharmaceutical companies.

In the industrialized world, legislation designed to protect plant-breeders – so-called plant-breeders' rights – gave patent-like control to these companies over any new varieties they might develop, including those produced from gene-bank material originally supplied by developing countries. An extension of the patent laws to the Third World would give the companies control over crop varieties in countries which had donated the original material. The use of biotechnology significantly decreases the length of time needed to breed new varieties and strains, the use of which could soon become unavoidable if the same companies ceased to offer traditional varieties for sale.

The Third World countries, through the FAO, pressed for the establishment of a system of 'farmers rights', through which the farmers in the Third World who have developed strains over years of cultivation can also be rewarded. One way to facilitate this would be the creation of a World Gene Fund which could support the conservation⌀ and development of seed resources within the Third World. At the FAO's 1989 conference the International Undertaking was reaffirmed and a resolution passed which endorsed the concept of farmers' rights. The International Fund for Plant Genetic Resources was also established, and it is envisaged that government contributions will help to finance the Fund, together with a tax of up to 3 per cent levied on all seeds sold throughout the world.

SELLAFIELD   One of the world's largest nuclear power⌀ generation and fuel reprocessing⌀ complexes. The area of the site – originally known as Windscale – in Cumbria in the United Kingdom has been a source of controversy ever since the first nuclear reactor began operating there in 1950. This event marked the beginning of the world's commercial nuclear power programmes. Two years later, the world's first nuclear fuel reprocessing plant was constructed at the site.

In 1957 the original reactors were closed down following a disastrous accident when the nuclear core caught fire. At least 11 tonnes of uranium⌀ and the graphite moderators began to burn. Vast quantities of radioactivity⌀ were released to pollute the neighbouring countryside, including 20,000 curies of iodine-131. Although full details were not released at the time, subsequent investigations suggested that as many as thirty-three people may have died as a result of the accident. Since commencement of operations, over 270 serious accidents have been reported at Sellafield.

In addition to radioactivity released through accidents, liquid radioactive waste⌀ is deliberately discharged into the Irish Sea. Strontium, caesium

and plutonium⌀ are released, the first two substances being carried away by sea currents into the North Sea, the plutonium mainly settling onto the seabed. Over the past forty years or more a total of half a tonne of plutonium may have been discharged. The effect of increased radiation on the local population⌀ has been a matter of controversy since the complex began operations, but it is widely accepted that, at the very least, there is a higher-than-normal incidence of childhood cancer⌀ in communities around the Sellafield site.

By the beginning of the 1990s, the reprocessing activities at Sellafield had become a major concern. The original reprocessing plant had been superseded by another in 1964, designed to handle irradiated fuel from the UK's commercial and military reactors. In 1976, permission was sought to build a much more advanced reprocessing facility, the Thermal Oxide Reprocessing Plant (THORP), to deal with spent fuel from reactors in Britain as well as from overseas. Construction of the THORP plant began in 1984, with completion set for 1992. Several independent studies in 1990 predicted that the THORP facility would not be economically viable: one analyst estimated that the plant would sustain losses of more than $5 billion during its first ten years of operation. Environmentalists had also begun to express profound concern at what was described as a government decision to make money from becoming the world's dustbin for spent nuclear fuel. By 1991, contracts had been secured from European, Japanese and domestic sources to ensure that THORP would be operating at some degree of capacity past the year 2000. A contract signed with Germany, worth $350 million, was due to become effective from 2002 onward and account for half the plant's capacity during the subsequent ten years. Even though 40 per cent of Germany's power is derived from nuclear plants, the country does not have its own reprocessing facilities and has been sending significant quantities of spent fuel to the UK since the late 1980s.

Although usable nuclear fuel can be obtained through reprocessing – the viable fuel is returned to the source of the spent fuel – large quantities of radioactive waste are produced. At present, these remain in temporary storage at or near the reprocessing plants, for no safe way of disposing of them has yet been developed. The UK government has investigated options to create nuclear waste⌀ dumps in underground rock formations at Sellafield and at another nuclear complex at Dounreay in Scotland. Test drillings at Sellafield were abandoned before completion, those at Dounreay progressed further. A target date of the year 2005 was set for completion of a safe, permanent storage facility. This would necessitate commencement of construction in 1996 and submission for planning permission by 1993. In 1990, the International Atomic Energy Agency (IAEA)⌂ reviewed the UK government plans at both sites and concluded that the geological information to satisfy safety standards was lacking. The IAEA concluded that leakage of radioactive gas, geological faults and the corrosive effect of salt water on nuclear waste containers were among the problems that must be investigated and solved before the UK could go ahead with plans for a nuclear repository, adding that it would be difficult to find answers to these problems before the deadline for planning permission.

SEVENDARA Sevendara, (*Vetiveria zizanioides*), also known as *khus-khus*, is a densely tufted, hardy perennial plant with prolific fibrous roots. Native to India and Sri Lanka, it has been used there for centuries to protect clothes, as the sweet-smelling, aromatic roots are believed to repel insects. The oil extracted from the roots is also used to manufacture perfume and in Ayurvedic (herbal) medicine. Sevendara has many other uses and is now being employed as a fast-growing plant to prevent soil erosion⌂ and water runoff. It is already being used fairly widely in Brazil, Fiji and parts of Africa for both soil-conservation⌂ and commercial purposes. In India, where about 12 billion tonnes of topsoil are currently being lost each year, the plant is being touted as a possible solution to the nation's soil-erosion problem.

Sevendara grows well under virtually any conditions and kind of soil⌂, and can withstand both drought⌂ and floods⌂. Although it is a perennial which has to be propagated vegetatively because its seeds are sterile, it requires very little attention once planted. Its root system is very strong and deep-growing; the coarseness of the grass makes it unpalatable to livestock⌂ and resists pest and disease attacks. In addition, the grass is virtually fire-proof. More importantly for impoverished areas, the plant has considerable commercial potential and can be produced with little initial input or labour. After cleaning and drying, sevendara is used for making mats, fans, screens, awnings, pillows and bags. The viscous essential oil – vetiver oil – distilled from the roots is used extensively in the perfumery industry in Western countries. The roots are usually harvested fifteen to twenty-four months after planting, but can be harvested after less than a year if required. Global production of vetiver oil is between 100 and 150 tonnes. In the past, Java used to be the world's principal source but Haiti, Réunion and India are now the main producers, with subsidiary production in Angola, Brazil, the Congo and Guatemala.

SEVESO A town in northern Italy which, in 1976, was the scene of a poisonous gas leak after a vat used in the manufacture of a bacteriocide (hexachlorophene) exploded. A cloud of poisonous gas, mostly dioxin⌂, escaped from the production plant and contaminated 18 sq km of surrounding land. Almost 1,000 people had to be evacuated immediately, and crops and animals later had to be destroyed. Following the release and settling of the gas cloud, the soil⌂ surrounding the factory became so polluted that the top 20 cm was scraped up and taken away for disposal in a plastic-lined pit. A comprehensive environmental clean-up⌂ campaign was undertaken before residents were allowed back in 1977. A year after the accident, birth defects amongst those affected had risen by 40 per cent. The factory was finally dismantled two years later and two tonnes of dioxin-contaminated waste had to be removed and disposed of. Costs of damage arising from the accident were estimated to be $150 million.

SHELTER BELTS The wind is capable of eroding all types of land⌂, especially in arid areas. Shelter belts of trees, hedges or man-made materials can be used to protect land, crops, buildings or roads from the effects of wind erosion. Wind erosion blows away fine soil⌂ particles while heavier ones are rolled along the ground, to end up against any obstacles in their path. The best protection against the wind is a cover crop, but this may need shelter in order to establish itself. Shelter belts of trees, which can also provide a source of fuelwood⌂, fodder and building materials, are ideal. Shelter belts generally offer protection from the wind over an area the width of the belt and a downwind distance at least twenty times the height of the trees. The trees must not be planted too close together – wind must be able to penetrate them, otherwise destructive eddies will be set up. The same principles have been applied in desert areas where merely shaping sand dunes and encouraging them to grow can protect roadways. Sculpting dunes on the windward sides of roads in the shape of an aircraft wing accelerates the flow of wind over the road, and sand is blown further away rather than falling on the road.

SHIFTING CULTIVATION *See* Slash-and-Burn Agriculture.

SICKLE-CELL ANAEMIA A condition, first identified in 1910, arising from the production of an abnormal form of haemoglobin (the oxygen-carrying pigment of red blood cells). The disease is hereditary but not contagious, affecting only people of African, Asian and Mediterranean descent, including ethnic minorities in developed countries. When the blood is deprived of oxygen, the abnormal haemoglobin crystallizes and distorts the red cells into a sickle shape. These misshapen cells are removed from the blood in the spleen, eventually leading to anaemia⌂. The cells of sufferers live for between 30 and 60 days compared with the normal 120. The deformed cells cannot flow properly and have a restricted oxygen carrying capacity. Misshapen cells cluster together and form blockages which cause intense pain, heart and lung seizures, and blindness.

About 20,000 people die each year from the disease, and many others succumb to secondary infections. Severely affected children do not survive to adulthood. About 200,000 babies are born with the condition each year: of those who survive childhood, few live beyond the age of forty. Those less severely affected do survive and 'carry' the disease: they are capable of developing it but never do so. These individuals tend to have a built-in resistance to *Plasmodium falciparum*, the organism which causes an acute form of malaria⌂, which may help explain why a harmful disease such as sickle-cell anaemia, which is caused by a single gene, is retained in the population.

There is no satisfactory treatment; what treatment is available is long and painful. However, deaths can be reduced by between 15 and 90 per cent if the disease is diagnosed before the age of four months.

SLASH-AND-BURN AGRICULTURE This form of agriculture (also known as shifting cultivation) involves clearing land⌂, farming the cleared area for a few years and then moving on to another patch of land to repeat the process once the original site's soil⌂ has become depleted. The original plot of land is returned to and reused after a suitable fallow period. Some 250 million farming families

around the world, mainly in developing countries⌂, practise slash-and-burn agriculture, an ecologically sound method of farming common in the tropics. It is the only sustainable way of farming in rain-forests⌂ because of the soil's low fertility.

The system is still widely practised in tropical forests⌂ and savannah. Small areas of forest or bush are cleared and burnt, releasing most available nutrients which, in tropical forests, are locked up in the vegetation, allowing them to be returned to the soil in a form that is quickly and easily used by other plants. After two or three years the soil is exhausted and yields decline. Farmers then move on to repeat the process elsewhere. Land lies fallow, traditionally for up to fifteen years, during which time the natural vegetation regenerates and the plot of land can then be cleared again to repeat the cycle. Cleared and farmed soils are covered at all times and the natural layers of the rainforest are usually repeated in crop-production systems which, as a wide variety of indigenous crops are grown, often provides a year-round harvest. As a result of increasing human numbers and social pressures limiting access to land, fallow periods have declined or disappeared altogether. When this happens, slash-and-burn becomes a distinctively destructive means of farming, giving the land no time to recover its fertility.

In tropical areas the regeneration of forests on cleared land depends on the availability of seeds, the proliferation of tree stumps, the availability of suitable habitats for young plants and the supply of nutrients, all of which are minimized by shifting cultivators. Where tropical forests are cleared for cattle-ranching, the creation of pasture drastically reduces the levels of nutrients, relatively large areas are cleared – meaning that seeds have to travel greater distances to re-colonize – and mechanized farming systems⌂ mean that fewer tree stumps or suitable habitats for young plants remain. Consequently, of the two forms, slash-and-burn agriculture is far less damaging to rainforests than cattle-ranching.

SMALLPOX The only disease of major global significance to be eradicated by co-ordinated human action. Smallpox is a highly infectious viral disease, transmitted by direct contact, marked by a skin rash that leaves permanent pitted scars. Symptoms begin eight to eighteen days after infection; the initial fever, prostration, severe headache and backache are followed by the appearance of the rash. Secondary infections often prove fatal. There is no specific treatment, but an attack, particularly in infancy, usually confers immunity. The development of a vaccine and widespread co-ordinated immunization programmes have led to the eradication of the disease.

In 1967 smallpox was endemic in thirty-one countries. In that year alone, between 10 and 15 million people were stricken – some 2 million died, and millions of survivors were left either blind or disfigured.

The last known case of smallpox was detected in Somalia in 1979, and the disease was believed to have been eradicated from the globe following two years in which no cases were reported, apart from a single outbreak in the United Kingdom which was caused by a laboratory-kept virus infecting scientific personnel. It is believed that samples of the disease-causing organism are being kept in laboratories in South Africa, the Soviet Union and the United States.

SOCIAL FORESTRY   Social, or community, forestry is the name given to reforestation◇ programmes that are meant to ease and promote rural development. Social forestry supposedly performs three distinct functions: it increases village self-sufficiency in fuelwood and other forest products, thus reducing the pressure on existing forests◇, and curbs migration from rural areas; it lowers unemployment levels; and it improves the environment. Successful projects have been implemented, mostly those which concentrated on getting local populations to plan and execute their own projects on a self-help basis. Many original programmes failed, for growing trees, unlike farming, requires relatively little year-round labour once the initial planting has been completed, so limiting employment opportunities and providing little incentive to encourage villagers to support reforestation projects. Local food supplies were sometimes compromised because much of the land used◇ for social forestry was previously used to grow food crops for local consumption and many of the original social forestry projects ended up producing material for commercial operations, not for the benefit of rural communities.

SOCIETY FOR INTERNATIONAL DEVELOPMENT (SID)   The SID was established in 1975 and is the world's largest, most geographically extensive, not-for-profit membership organization devoted to international development. Its main aims are to promote international dialogue, understanding and co-operation to help stimulate the social and economic development of all peoples throughout the world; to carry out research and produce educational materials to help achieve this end; and to foster a sense of community amongst those committed to the cause of development. Since its foundation, the Society has become a worldwide forum for the exchange of information and ideas on all aspects of economic, political and social development.

The SID now boasts over 10,000 members, drawn from all walks of life, and over 200 institutional members, located in 132 countries around the world. A significant portion of the Society's work is undertaken through the 87 SID chapters which provide specific programmes geared to the interests of the communities and local members they serve. Every three years the members elect a twenty-four-strong Governing Council which then chooses the President and other officers. The Society's International Secretariat is in Italy, although the SID is not affiliated to any government; its finances are raised through membership fees and contributions from a wide range of development organizations, foundations and institutions. The SID supports the North–South Round Table, a panel of 100 world experts on development matters who meet twice a year.

SOIL   Soil is formed through natural and time-consuming processes such as the weathering of bedrock, the type of soil differing in accordance with the rock from which it came. There are five main types of soil: clay, sand, loam, peat◇ and calcareous (chalky) soil derived from limestone rocks. All soils contain a range of organic matter in varying states of decay, together with minerals◇ and millions of living organisms. The depth of soil varies according to the prevailing climate, contour of the land◇ and rainfall. Soil washed off mountains builds up in valleys creating a deep soil profile, producing extremely fertile land.

In total, 8 billion hectares (ha) of land are capable of being agriculturally productive to some degree, but it is the top few centimetres of precious topsoil that support the growth of food and cash crops◇ upon which the world depends. In temperate zones the top 23 cm of good soil contains enough plant food to support over 100 crops, although some of the nutrients are in a form that plants cannot naturally use. Soil formation occurs at a rate of about 1 mm every hundred years, or around 150 kg per hectare. Once removed, it is not replaced in a hurry.

The 1989 average of 0.28 hectares of cropland per person is rapidly declining as the human population increases. Since 1945 the *per capita* amount of cropland actively cultivated and in production in the world actually fell by well over one-third as topsoil became either lost or exhausted through overuse. Unless more land is opened up for cultivation, *per capita* cropland will decrease further – to 0.17 ha by the year 2000.

Fortunately, crop production has not shown a similar decline – world food production rose by 3.2 per cent in 1989. Increasing production has been made possible largely through artificial replenishment of the soil's plant-supporting capacity – fertilizer⌂ use has increased fivefold since the mid-1940s. During this entire period many of the world's farmers, notably those in the industrialized nations, have been mining the nutrients from the soil and growing food in an unsustainable manner. The continuing use of fertilizers tends to degrade and exhaust topsoil, making it more prone to erosion. Inorganic fertilizer use is also no long-term solution: conventional organic farms have been shown to lose up to four times less soil each year than those using inorganic fertilizers. Yet fallow periods, crop rotations and intercropping⌂, all of which allow the soil to regenerate its tilth, have been abandoned in pursuit of profits and intensive cultivation of a single crop.

In many developing countries⌂, the soil supporting the nation's agricultural sector is the mainstay of the economy and supports the well-being of the people. The pursuit of an industrial agriculture⌂ and the production of cash crops⌂ result in peasants being ejected from arable land. Pressure merely to eke out a living has forced these landless⌂ peasants throughout the Third World⌂ similarly to mine their soils. Pastoralist herdsmen are being forced to give up their traditional nomadic⌂ lifestyle and become sedentary farmers, placing extra stress on already impoverished soils.

Human action and abuse are leading to the loss of vast amounts of irreplaceable soil at rates far in excess of replacement levels. On steep slopes in Ethiopia, losses of 296 tonnes/ha have been recorded. In India, it is estimated that the annual harvest depletes the soil of 18.5 million tonnes of nutrients, but only 10 million tonnes – made up of organic waste and fertilizers – are replaced. Yet the country must somehow double its food output on this decaying land if it is to feed the 1 billion people expected to inhabit it by the year 2000.

SOIL EROSION Soil⌂ – or, more importantly, the surface layer of productive topsoil – can be eroded from land⌂ by moving wind or water⌂.

Wind erosion – Large areas of the world are affected by wind erosion, one of the key causes of desertification⌂. It occurs when soil is left bare of vegetation and is most severe in arid and semi-arid lands exposed to overgrazing⌂ by livestock⌂. Over 22 per cent of Africa north of the equator and 35 per cent of land in the Near East are susceptible to erosion by wind. Topsoil is stripped away and land, buildings, machinery and fences can be fully or partially buried or damaged. In extreme conditions, 150 tonnes of soil can be blown from one hectare of land in an hour. Africa and Asia between them may lose almost 1 billion tonnes of soil annually through wind erosion. The dustbowls⌂ observed in the prairies in the United States in the 1930s were due to wind erosion, and millions of hectares of productive land were lost. A single four-day storm in 1934 was reported to have blown away 300 million tonnes of soil.

Water erosion – Water erosion is damaging agricultural land in every developing country. It occurs wherever steep land is being farmed or sloping land is left exposed for any length of time. The Food and Agriculture Organization (FAO)⌂ estimates that about 25 billion tonnes of soil are lost worldwide each year. Just over 11 per cent of Africa north of the equator is subject to water erosion, as are 17 per cent of the land in the Near East and 90 million hectares of land in India. Two of the world's major rivers, the Yangtze in China and the Ganges⌂ in India, transport around 3 billion tonnes of soil annually.

Erosion occurs at a natural rate determined by a combination of climate, topography, soil type and vegetative cover. Human influence in exacerbating soil erosion is generally through the removal of stabilizing vegetative cover. Soil is washed into rivers, where it can silt up reservoirs and prevent hydroelectric power plants from operating. It is also washed into coastal waters, where it can destroy fishing⌂ grounds. Some estimates have placed the annual global loss of soil as high as 75 billion tonnes, although accurate soil-erosion data are scarce and erosion levels vary enormously within countries and geographical regions. In Africa, of those countries for which data are available, Madagascar, with a mean annual loss of 112 tonnes/ha, suffers most, followed by Kenya with 105 tonnes/ha. In North and Central America, El Salvador reports the greatest loss: 85 tonnes/ha. There is very little information on soil loss in Europe, but it has been estimated that Belgium loses around 17 tonnes/ha each year. In Asia, and in global terms, the worst-affected nation is Yemen, which is reported to be losing 275 tonnes/ha as the steep slopes throughout this mountainous country continue to be stripped of vegetation.

About half of the world's croplands are so badly managed that they are losing topsoil at the rate of 7 per cent per decade, with potentially catastrophic long-term consequences. Western farmers have abandoned fallow periods and crop rotation, and soil erosion has increased as a result. In the United States, which produces the bulk of the world's grain surpluses, the erosion of topsoil is seriously reducing the fertility of cropland. Despite corrective measures, on average over 12 tonnes/ha of topsoil are still being lost from cropland throughout the country, and 50 million tonnes of plant nutrients are also lost each year. Massive amounts of energy⌂, machine power, fertilizer⌂, pesticide⌂, and irrigation⌂ are needed to sustain US agriculture.

SOLAR CELLS Solar cells, also called photovoltaic cells, convert sunlight directly into electricity and are rapidly becoming a viable source of power production as their efficiency improves. Solar cells generate an electric current when light strikes a semiconductor: the solar energy⌂ excites electrons and promotes them to a higher-energy-state conducting level. Solar cells have traditionally been inefficient, but in 1989 a solar cell with an efficiency of 37 per cent was produced. This cell contains two energy-conversion layers, one gallium arsenide and one gallium antimonide.

Solar cells are clean, safe and silent, with no moving parts. They are manufactured from silicon, the second most abundant element in the earth's crust, and extremely thin cells can be made by depositing layers of amorphous silicon on to glass. At 1990 efficiency levels – which are still below what is required – it would take a 16-sq-km piece of desert to produce as much electricity as a large nuclear or coal-fired power plant. Solar cells are also relatively expensive, resulting in power-production costs of around $5 per peak watt hour, although their cost has been falling – 75 per cent over the 1980s – as their efficiency improves. Technical advances should bring down the cost to well under $1 per peak watt hour sometime during the 1990s, so that solar electricity will compete directly in price with grid electricity.

When the solar-cell industry began to flourish during the 1970s, the world's major oil⌂ conglomerates initially lobbied strongly against it. They then proceeded to buy out most of the independent producers and research and development organizations. By 1983, the four largest photovoltaic manufacturers, which commanded half the global market, were wholly owned by major oil companies.

SOLAR POWER The use of the sun's energy⌂ to provide heating or to generate electricity. A vast amount of solar energy falls on the earth each year. It can be used directly, the commonest and simplest method being the heating of water⌂ flowing through panels exposed to the sun's rays. The rise in temperature of the exposed water is fairly small, but it reduces the need for energy to produce hot water or space heating. Mirrors can also be used to concentrate the sun's rays to boil water, produce steam and drive a turbine, as at Odeillo in France where 20,000 mirrors are used to produce a temperature of 3,500°C.

Solar radiation can also be converted into electrical energy through the use of a solar (or photovoltaic) cell⌂. At present these are used mainly in small-scale operations, powering remote monitoring equipment in desert regions or spacecraft and satellites. The cells are expensive but research is continuing, and a drop in cost of a factor of between 5 and 10 would make them extremely economic. By the end of the 1980s costs of 25–35 cents (US) per kilowatt/hour (kWh) were already competitive and costs of 8–9 cents were considered to be achievable, one constructor of a 400 MW solar power plant in Southern California going so far as to guarantee power production for less than 9 cents per kWh.

SOUTH A crude demarcation, usually taken to refer to an area below a line drawn on the world map and emphasized by the Brandt Commission⌂ in 1980, identifying those countries with low *per capita* incomes and previously referred to as the 'less-developed' countries or the 'Third World'⌂. The terms 'North'⌂ and 'South' are widely held to be synonymous with 'rich' and 'poor', 'developed' and 'developing'⌂. In the South/North demarcation, the two 'rich' countries south of the equator, Australia and New Zealand, are considered part of the North. China was excluded from both North and South, but is now usually included in the South, or 'developing country' category.

SOUTHERN AFRICAN DEVELOPMENT CO-ORDINATION CONFERENCE (SADCC)
Aimed at promoting regional co-operation, lessening economic dependence on South Africa, fighting racial discrimination and improving stability in Southern Africa – the first SADCC Summit Conference was held in Lusaka in 1980. Membership includes all of the 'front-line' states, notably those that have borders with South Africa: Angola, Botswana, Lesotho, Malawi, Mozambique, Swaziland, Tanzania, Zambia, and Zimbabwe.

SOUTH ASIA REGIONAL CO-OPERATION COMMITTEE (SARC)
A regional organization established in 1983 to promote and develop matters of agriculture◇, telecommunications, health, population◇, sport, art and culture in participating states. Members are Bangladesh, Bhutan, the Maldives, Nepal, Pakistan and Sri Lanka.

SOUTH-EAST ASIAN FISHERIES DEVELOPMENT CENTRE (SEAFDEC)
A regional grouping of countries with the objective of improving the role and impact of fisheries◇ on the economy of member states. The inaugural meeting of the Council of SEAFDEC, the formation of which was originally agreed by six Asian countries – Japan, Malaysia, the Philippines, Singapore, Thailand and Vietnam – took place in 1968. SEAFDEC now has three major institutional departments of specialization, for training in fisheries (based in Thailand), for marine fisheries research (Singapore), and for fish◇ culture (Philippines).

SOUTH-EAST ASIA TREATY ORGANIZATION (SEATO)
A collective defence organization formed to protect South-East Asian countries from possible Communist aggression, analogous to the North Atlantic Treaty Organization (NATO)◇. The SEATO treaty was signed in Manila in 1954 by Australia, France, New Zealand, Pakistan, the Philippines, Thailand, the United Kingdom and the United States. The SEATO allies agreed to act together against any attack in the region upon any one of the signatories or on Cambodia, Laos or South Vietnam. Pakistan withdrew in 1973. No joint action was ever taken under SEATO, although American, Australian, New Zealand, Philippines and Thai forces all fought in the Vietnam War◇ during the 1960s. After the war the organization was phased out and formally ended in 1977, its non-military aspects being taken over by the Association of South-East Asian Nations (ASEAN)◇.

SOUTH PACIFIC BUREAU FOR ECONOMIC CO-OPERATION (SPEC)
SPEC is the executive arm of the South Pacific Forum◇. Through SPEC the Forum is actively associated with regional economic co-operation in such areas as shipping, exports and marketing. SPEC also services the Pacific signatories of the Lomé Convention◇ and has a responsibility for co-ordinating international regional aid◇, including aid provided from the European Development Fund.

SOUTH PACIFIC COMMISSION (SPC)
Established in 1947. Its purpose is to promote the economic and social welfare and advancement of the peoples of the South Pacific region. Participating governments include Australia, France, New Zealand, the United Kingdom and United States together with those of the independent developing countries◇ in the region.

SOUTH PACIFIC FORUM
Established in 1971 on the initiative of the Fijians to provide a forum for political consultation between heads of government of the independent and self-governing countries of the South Pacific, Australia and New Zealand. Virtually since its inception, the members of the forum – including Australia, Cook Islands, Fiji, Kiribati, Nauru, New Zealand, Niue, Papua New Guinea, Solomon Islands, Tonga, Tuvalu, Vanuatu and Western Samoa – have been particularly concerned with the problem of nuclear testing in the region. The Treaty of Rarotonga, agreed in 1985, established a South Pacific Nuclear Free Zone, but France refused to sign and the United Kingdom and the United States also withheld their signatures, although they agreed to bide by the terms of the Treaty.

SPECIAL DRAWING RIGHTS (SDR)
The accounts of the International Monetary Fund (IMF)◇ are maintained in SDR units. SDRs, which were originated in 1970, are the IMF's own currency equivalent. They are issued to members of the Fund from time to time in proportion to their quotas. SDRs are not money, they are potential overdraft facilities; their value derives from their convertibility to foreign exchange. They can be exchanged through the IMF for national currencies, or held by a

country as a reserve asset. They are not used commercially and are traded only between central banks and the Bank for International Settlements (BIS)⌬.

Member countries of the IMF have the right to draw on the Fund to finance balance-of-payments deficits. SDRs, unlike normal drawing rights, do not have to be repaid and therefore form a permanent addition to the drawing country's reserves, functioning as an international reserve currency and supplementing its holdings of gold⌬ and convertible currencies. SDRs are a composite currency unit computed as a weighted average of several international currencies, including the dollar (US), yen (Japan), deutschmark (Federal Republic of Germany), franc (France) and sterling (United Kingdom). Their advantage over gold and other reserve currencies is that their supply can be controlled and their worth is not so closely tied to gold deposits or dependent on the US balance of payments or the strength of the dollar.

SPECIAL PROGRAMME FOR RESEARCH AND TRAINING IN TROPICAL DISEASES
(TDR) A specialized programme run jointly by the World Health Organization (WHO)⌬, the United Nations Development Programme (UNDP)⌬ and the World Bank⌬. The TDR works with 5,000 researchers based in institutes and laboratories around the world to stimulate action, and develop vaccines, drugs, diagnostic tests and preventive methods to combat the six major tropical diseases: malaria⌬, schistosomiasis⌬, filariasis⌬, trypanosomiasis⌬, leishmaniasis⌬ and leprosy⌬.

Most people infected with – or at risk from – tropical disease live in countries with average incomes of less than $400 per person, where the government spends only $4 per person on health care. It costs in the region of $100 million to research and develop a drug or vaccine and make it readily available. Transnational⌬ pharmaceutical companies are reluctant to undertake this work where there is no likelihood of a financial return on their investment. Expenditure on biomedical research is mainly controlled by transnational companies or governments in industrialized countries outside the tropics. Consequently, less than 3 per cent of total global outlay goes towards research into tropical diseases, despite the fact that 10 per cent of the world's population suffer from one or more of them. During the 1980s, the TDR attempted to rectify this imbalance by focusing its efforts on funding and encouraging basic research into tropical diseases and strengthening the research

capabilities of national institutions in the affected countries. Over the decade, more than a hundred potential products for treating or controlling these diseases have been developed.

STABEX (SYSTEM FOR STABILIZATION OF EXPORTS) A mechanism designed to regulate export revenues for developing countries⌬ which are heavily dependent on agricultural commodities⌬. STABEX was established in 1975 under the original Lomé Convention⌬ and involves the European Community⌬ countries and those of the ACP⌬ group. It operates on a product-by-product basis and covers forty-eight commodities, although the list is altered during each round of Lomé Convention negotiations. Financing for the compensation given under STABEX is provided from the European Development Fund.

STABEX operates by first determining a reference level of export earning for each nation for individual products. Compensation is paid when annual export revenue for a commodity falls below the set level. Compensatory payments – usually in the form of loans repayable over the five-year life of the scheme – are targeted on the industrial sector of the product suffering the loss, or assigned to programmes aimed at product diversification. In certain cases the compensation is given on a repayment basis, but in general, for the twenty-six poorest and Least-Developed Countries⌬, it is provided in the form of non-reimbursable aid⌬. The recipient nation is free to choose how to use the resources transferred, informing the Community annually of its decision.

STABEX operated efficiently for the first five years of its existence but since an acute financial crisis during the first years of the second Lomé Convention, fewer than half of the ACP requests for compensation have been granted.

STAR WARS *See* Strategic Defence Initiative (SDI).

STOCKHOLM CONFERENCE Common name for the United Nations Conference on the Human Environment⌬, held in 1972 and attended by representatives from 113 nations. The Stockholm Conference is considered to be a major landmark in the development of global environmental concern. The Conference resulted in the formation of the United Nations Environment Programme (UNEP)⌬, although UNEP's achievements have been

somewhat limited by its inability to make pronouncements or take actions that might annoy the governments that provide its funds.

STOCKHOLM INTERNATIONAL PEACE RESEARCH INSTITUTE (SIPRI) An independent institute based in Sweden, with an international governing council and staff of experts, which conducts research into all aspects of peace and conflict, with particular emphasis on problems concerning disarmament◇ and arms regulation◇.

STRATEGIC ARMS LIMITATION TREATY (SALT) The forerunner of the United Nations◇, the League of Nations◇, first attempted to introduce a programme to reduce the weapons of war in 1930, albeit unsuccessfully. After the Second World War and the massive build-up of increasingly destructive armaments that followed, US President Lyndon Johnson put forward proposals for a Strategic Arms Limitation Treaty (SALT) in 1967, primarily concerning the two superpowers, the United States and the Soviet Union. Although agreement was delayed by Soviet intervention in Czechoslovakia, the first treaty, SALT I, became operational during 1972–7. An updated treaty, SALT II, concluded by Presidents Carter◇ and Brezhnev, was due to operate during 1979–85 but was never ratified, although both sides abided by it. Several further agreements have been made between the USA and the Soviet Union with regard to reducing the numbers of intermediate-range nuclear weapons, strategic arms and weapons in outer space, the Intermediate Range Nuclear Force Treaty (INF) of 1987 marking the first actually to require weapons to be removed and destroyed.

STRATEGIC DEFENSE INITIATIVE (SDI) More commonly known as Star Wars, the SDI is a research initiative taken by the United States which aims to produce weapons capable of destroying incoming nuclear missiles while they are still outside the earth's atmosphere. Announced by President Reagan◇ in 1984, the system is due to be operational by the year 2000. The programme involves the potential development of a variety of possible weapons systems, of which ground-based or satellite-mounted lasers are favoured. So-called 'directed-energy' weapons mounted on platforms in space could burn holes in incoming missiles to explode their fuel tanks or simply cause them to collapse. In addition to growing scepticism as to whether the technology for such weapons can ever

be developed, let alone deployed and operational, opponents complain that such systems represent the militarization of space, and so contravene international treaties banning the use of space for military purposes. Deployment – but not development – of such weapons is against the terms of the Anti-Ballistic Missile Treaty (1972). Proponents of the SDI claim that if a successful system can be developed, nuclear weapons would be rendered obsolete. The massive amount of funding and resources devoted to SDI research and development will be spread amongst the USA and its allies.

STRUCTURAL ADJUSTMENT LOANS Large loans given by the Multilateral Development Banks (MDBS)◇ with the objective of bringing about economy-wide reforms within recipient countries. These reforms usually include reductions in import restrictions and the introduction or promotion of 'free-market' policies, together with a relaxation of state controls on the economy. Together with Sector loans◇, Structural Adjustment loans make up 'programme lending', which includes all lending other than that devoted to specific projects.

In 1990 nearly one-third of the World Bank's◇ $22 billion budget went into Structural Adjustment Programme loans. To qualify for these loans, countries often had to cut social services, privatize public industries and, on occasion, devalue local currencies. As a result real wages fall, unemployment rises and the price of imported manufactured goods places them out of the reach of most of the local population.

SUGAR A sweet, crystalline substance, obtained from various plants and available in several forms, used in drinks, food and confectionery. Sugar production is believed to have begun in India around 5,000 years ago. It is produced essentially as a luxury in the human diet, but every country has developed a taste for it. The Western world consumes huge quantities, around 50 kg per person annually, although consumption is declining. In parts of Africa, such as Egypt, average consumption is 30 kg per person and rising, while India consumes around 80 per cent of its 9-million-tonne total production, the remainder being exported. World consumption has risen by 1 million tonnes a year over the last century.

Perceived by early colonialists as a cash crop◇, sugar is one of the most politically and socially sensitive of all crops. Commercially produced sugar was traditionally derived from sugar cane, which is grown extensively on large-scale plantations in tropical and subtropical regions, by-products from the processing of sugar cane, such as molasses and bagasse, being used for various purposes such as fuels, animal feedstuffs and paper-making. In the past, sugar was intimately associated with the slave trade and, later, indentured labour. In some parts of the developing world, cane plantation workers are still poorly paid and have very little employment protection. The general decline in global sugar trade and prices seen over the last twenty years, which led to 60,000 small-scale sugar-growers in the Philippines losing their livelihood, has had a similar effect elsewhere in the Third World⌂, and has arisen mainly as a result of policies adopted by industrialized countries.

When sugar from cane became too expensive for European importers, sugar beet began to be used, beet is produced mainly in Western Europe and the Soviet Union. Sugar beet is not economically competitive with sugar cane but tends to be grown as an import-substitution crop. By the mid-1970s the European Community⌂ was importing around 1 million tonnes of sugar annually. Now, as a result of the heavily subsidized production of sugar beet, it exports up to 5 million tonnes at reduced prices, depressing the price of sugar on the world market in the process.

There are at least a hundred major sugar producers in the world, leading to global overproduction and depressed markets – despite conventions which were supposed to impose export quotas and stabilize prices, but later collapsed. Declining terms of trade have adversely affected Third World sugarproducers for several decades and only about 20 per cent of the world's traded sugar is sold at the world price; many importing countries pay inflated prices for sugar from favoured nations.

Some countries, notably Brazil, faced with overproduction and depressed prices, put their sugar crop to other uses, converting it to alcohol for use as a fuel. Around 10 billion litres are fermented from sugar cane in Brazil and many vehicles, especially taxis, have been converted to run on the alcoholbased fuel known as gasohol.

Around 103 million tonnes of sugar are produced annually; both India and the Soviet Union produce over 9 million tonnes a year. Brazil produces 8.5 million tonnes, some of which is destined for gasohol production, while Cuba, which produces 7.5 million tonnes annually – most of which used to be sold to countries within the former Communist bloc

at prices up to ten times those seen on the world market – is facing major difficulties following the radical changes in Eastern Europe⌂ during 1989–90.

Artificial, man-made sweeteners, and those made by genetically engineered organisms, are beginning to have a major impact on the global sugar market, depressing it still further.

SULPHUR DIOXIDE One of the most prevalent atmospheric pollutants⌂ of industrialized countries, sulphur dioxide (SO_2) is emitted into the atmosphere chiefly through the combustion of sulphur-containing fossil fuels⌂. SO_2, a corrosive gas, is one of the major causal factors of acidic deposition⌂, both as rain and in dry, solid form. It is emitted by natural sources, such as volcanoes, decaying organic matter and sea spray, as well as through human agency. Global emissions of SO_2 have been rising for a hundred years: from less than 10 million tonnes in 1860 to around 150 million tonnes during the 1980s. The current total release from anthropogenic sources – approximately 90 million tonnes of sulphur per year – compares with the natural global release of 70 million tonnes.

The adverse environmental impact of sulphur dioxide emissions was recognized during the 1970s, and following international pollution⌂-control agreements, several countries managed to reduce their emissions between 1970 and 1985, mainly as a result of switching to cleaner fuels – especially low-sulphur coal⌂ – for power plants. Using lowsulphur coal and oil⌂ is the easiest way of reducing sulphur emissions, but supplies of both are limited.

Removing sulphur from fossil fuels⌂ is an expensive operation. Taking sulphur out of ordinary fuel oil costs around $30 per tonne of oil – actual costs vary according to the type of oil, amount treated, and other factors. The treatment effectively raises the price of electricity generated by 15 per cent. Washing coal, after first crushing and grinding it, removes one type of sulphur, pyrite (iron sulphide) at a cost of $1–6 per tonne. Chemical treatment, which removes organic sulphur as well as pyrite, is more effective but considerably more expensive. The costs of washing coal raise electricity prices by 1–6 per cent, whereas chemical desulphurization adds 15–25 per cent. Various advanced 'fluidized bed' technologies, in which coal is burnt more efficiently in a specialized bed of small particles that bubble like a boiling liquid, can allow sulphur emissions to be reduced through the simple addition of limestone or dolomite to the bed mixture.

An international protocol requiring nations to reduce sulphur dioxide emissions by 30 per cent of 1980 levels came into effect in September 1987. The protocol was signed by the members of the Convention on Long Range Transboundary Air Pollution⌀ – commonly referred to as the '30 per cent Club' – and the reduction was due to be completed by 1993. However, although both the United Kingdom and the United States are parties to the Convention, they refused to adopt the protocol. In 1988, the European Community (EC)⌀ passed the Large Combustions Plant Directive through which member nations were obliged to cut SO$_2$ emissions by 57 per cent by the year 2003.

Although the UK, with an output of 1.9 million tonnes of sulphur a year in 1989 remained the worst offender in the EC, its emissions were nowhere near as bad as those in Eastern Europe⌀, with the former German Democratic Republic heading the list of polluters, emitting 5.5 million tonnes annually.

SUPERFUND The name given to a fund which will be used to finance clean-up⌀ operations on sites in the United States that have been contaminated by the dumping of hazardous wastes⌀. Potentially dangerous chemical wastes are believed to have been dumped in at least 20,000 uncontrolled sites throughout the country. Properly known as the Comprehensive Environmental Response, Compensation and Liability Act, the legislation, passed in 1980, was the last bill signed by President Carter⌀. Following the Polluter Pays principle, the Superfund is to be financed by the industries that are deemed to have caused the problem. Industry is required to pay 88 per cent of the fund's total capital which, after much argument, was set at $1.6 billion. In 1986 this sum was raised to $8.5 billion, but this is nowhere near enough to pay for the clean-up of the 10,000 contaminated sites across the country which the US government's Office of Technology Assessment has identified as possibly being in need of clean-up operations to be financed from the Superfund. By 1986, the US Environmental Protection Agency (EPA) had designated 842 'priority' sites as being in need of remedial attention, estimating that $1.6 billion would be necessary for the clean-up programme on these sites.

SUPERPHENIX A fast-breeder⌀ nuclear power⌀ station in France. The reactor in the plant had to be shut down in 1987 when sodium coolant leaked from a fuel-transfer vessel. It was restarted in 1989, but only to allow power to be produced until

the fuel that was already within the reactor was exhausted. No new fuel was to be loaded until 1991, when a new system for transferring fuel, using argon as the coolant, was due to become operational.

SUSTAINABLE DEVELOPMENT Irrespective of social, political or economic structures, all human beings require a continuous and undiminished supply of untainted natural biological materials to satisfy their food, shelter, energy⌀ and medicinal needs. The World Conservation Strategy⌀ stated that for development to be sustainable it must not interfere with the natural functioning of life-support systems or natural ecological processes and equilibria. Ultimately, all economic development depends on the earth's natural-resource base. It is the net primary biological production generated by photosynthesis that sustains the human race and virtually every other life form on earth. Maintaining this biological productivity must therefore be the key to sustainability.

Sustainable development must meet human needs without depleting natural resources⌀ or irrevocably damaging the systems which produce those resources, whilst establishing equitable and viable patterns of living throughout the world. Any future social systems will therefore have to pay close heed to the limiting biological factors. Under present conditions, organic matter equivalent to 40 per cent of the primary production in terrestrial ecosystems⌀ is being consumed by humans as they follow varying lifestyles around the world. If marine primary production is also included, 25 per cent of the global total is being consumed. Consequently, if the world population⌀ doubles and current usage levels are maintained, humans will devouring as much biological material as is being produced each year. At this level of exploitation the quality of both environment and human population will quickly decline.

SWINE FEVER An infectious intestinal disease of pigs, usually overcome by slaughtering all animals in infected herds. In 1990 an outbreak occured in Belgium, which was believed to be free from the disease. Control measures introduced by the European Commission cost $42 million, mainly as a result of compensation paid for the 35,000 fattened pigs and 24,000 piglets slaughtered weekly over a seven-week period. The outbreak drew attention to the fact that under the Single European Market legislation, veterinary controls at borders between member nations of the European Community⌀ will be dismantled. All health checks on meat and meat products traded between member states will be abandoned, increasing the likelihood of the spread of disease, although live animals may well be excluded from the directive.

SYSTEM OF COMPENSATION FOR EXPORT RECEIPTS (COMPEX) A compensatory programme designed to benefit countries in the Third World⊃ by providing compensation for losses in their export of agricultural commodities⊃, established by the Finance Ministers of the twelve European Community⊃ nations in 1987. The system was due to cover the period up to the end of 1990, and involved a total of $59 million. The programme applied only to those countries classified by the United Nations⊃ as 'less-developed countries' but not signatories to the Lomé Convention⊃: Bangladesh, Bhutan, Haiti, Laos, the Maldives, Nepal, and Yemen. A total of fifty commodities are covered by COMPEX.

T

TAUNGYA A specialized form of agroforestry⊃, developed in Burma during the 1850s. Food or cash crops⊃ are grown among young trees until the trees reach a point where the canopy closes over. This usually takes between one to three years. Cultivation of crops helps to control weed growth and maximize production from the land⊃. Where landless farmers are allowed to practise this form of agroforestry, they are often obliged to move on after the trees have formed a closed canopy. Tenant farmers may also be allowed to produce crops only while the trees are still young, since they have been contracted to grow only the trees, the crops forming the payment for the contract. The needs of the landowner and the farmer are in direct opposition – one wanting to grow more food or cash crops, the other requiring more trees. As a result, trees have frequently been damaged to postpone the time at which they begin to shade out the crops, or simply neglected, as there is relatively little incentive for those farmers who never acquire any rights to the land⊃ being cultivated.

TEA The dried leaves and shoots of the evergreen shrub or tree *Camellia sinensis*, which yield a beverage when infused with water. The drink has a stimulating effect due to its caffeine content (about 3.5 per cent). Native to parts of India and China, the tea plant has three major varieties – Assam, Cambodia and China – and numerous hybrids. Tea is a valuable, labour-intensive export crop and, ideally, should be grown at 1,500–2,200 m altitude on well-drained soil⊃ in areas with an annual rainfall of 1,400 mm. Young leaves from the bushes are plucked by hand when the plant is about five years old. Black tea is made from dried leaves, broken up to release the essential oils and allowed to ferment before the moisture is removed in ovens; the leaves emerge in blackish form. Green tea is steamed and quickly dried before fermentation can occur and so remains partly green.

About half of the total annual production of tea – some 2.5 million tonnes in 1988 – is drunk in the countries of origin; the rest enters international trade. The world's leading exporting nations are India, which produced 690,000 tonnes in 1988, and Sri Lanka, where production was 225,000 tonnes. Most of the tea produced annually in China – around 540,000 tonnes – is destined for home consumption. Asian countries produce around 75 per cent of the world's tea, but eight European and American firms control both the growing and the sales of 90 per cent of the tea traded on the world market. Sri Lanka gains more than half its export earnings from tea and thus suffers most from fluctuations in the global market. In real terms, during the last decade the price of tea has fallen by 25 per cent.

The demand of consumers in the European and North American markets for cheap tea, and the stranglehold of the transnational⊃ companies that allows them to manipulate prices, ensure that workers on plantations in Asia, East Africa and other parts of the developing world where tea is grown are poorly paid.

TECHNICAL CENTRE FOR AGRICULTURAL AND RURAL CO-OPERATION (CTA)
The CTA was established in 1983 under the second Lomé Convention to facilitate the exchange of scientific and technical information between the European Community◇ and African, Caribbean and Pacific (ACP) countries◇. The main aim of the Centre, which is based in the Netherlands is to promote agricultural and rural development in an effort to assist these countries to move towards self-sufficiency in food production.

TECHNICAL CO-OPERATION AMONGST DEVELOPING COUNTRIES (TCDC)
A United Nations◇ Conference on Technical Co-operation amongst Developing Countries was held in Argentina in 1978. A plan of action for the promotion and implementation of TCDC was adopted, aimed at stimulating greater use of existing and newly emerging expertise and resources in inter-national development efforts. The recommendations contained in the plan call for action by both developed and developing countries◇ at national, regional and global levels.

TEHRI DAM
A controversial water-resource development project in the Himalayas in northern India. If completed, the 260-metre-high dam◇ will be the highest in India and the seventh highest in the world. Work on the rockfill dam system, which will generate an estimated 2,400 megawatts of electricity and irrigate◇ 660,000 hectare (ha) of land◇, began in 1968. The reservoir behind the dam will flood 45 sq km, causing the town of Tehri and twenty-three villages to be fully or partially submerged and forcing the resettlement◇ of over 70,000 people. In addition, 1,000 ha of cultivable land and 1,000 ha of forest◇ will disappear underwater.

The dam has been sited in an area with a history of intense seismic activity – earthquakes measuring up to eight points on the Richter scale have occurred since 1897. Although there have been very few tremors since 1951, critics of the scheme claim that the weight of the impounded water◇, around 3.2 billion tonnes, will trigger earthquakes. Six other Himalayan reservoirs with water levels of only 150–250 m have been found to cause earth tremors. Deforestation◇ further north in the Himalayas has already caused sedimentation◇ rates in the rivers Bhagirathi and Bhuilanga to rise, and if soil erosion◇ continues at the present rate, the build-up of silt◇ could render the dam useless in twenty-five years, although engineers claim that it will be able to operate for up to four times that long. In 1986 the Soviet Union agreed to provide finance and technical assistance for the dam at a cost of $15 million. The original costs were put at $135 million, but these had already escalated fivefold by 1980.

TERMITE
A soft-bodied, social insect which lives in large colonies, mainly in tropical regions. Termites construct extremely complex nests from soil particles◇ which they glue together. These nests, containing several galleries, are centrally heated and air-conditioned structures reaching as high as seven metres. Within these elaborate and almost self-contained structures, some species of termite cultivate fungus gardens using their own faeces; the fungus digest the faeces and thus recycle them. Although generally regarded as pests, termites play one of the most important roles in the biological cycle of living organisms, for they are one of the major means by which carbon◇ is recycled.

Termites eat all forms of vegetable matter, and numerous species of micro-organisms living symbiotically within their digestive system can break down cellulose. They may dispose of a quarter of the vegetation in any given area, and their fondness for wood◇ (especially that used to build houses and other buildings) brings them into conflict with humans. Digestion of the wood produces several gases, releasing both carbon dioxide◇ and methane◇. It is estimated that the world's population of termites is in the order of 2,400 million billion and they may be contributing heavily to the production of greenhouse gases◇, venting up to 150 million tonnes of methane into the atmosphere each year. In 1986 the US National Pest Control Association estimated that termites were responsible for $1 billion worth of damage in the country every year, and worldwide, especially in the developed nations, vast quantities of money, research and technology are being devoted to their destruction.

TERRORISM
Between 1975 and 1985 over 4,000 people died as a direct result of terrorist acts, and hundreds of thousands more were injured or made homeless. Terrorism, which has been steadily increasing since the 1950s, is a term originally used to describe any violent act perpetrated against civilians by the authorities in power in order to inspire fear for political reasons. The exact definition

has become somewhat blurred, and it can depend on which side the observer supports. Terrorism can best be described as 'the use of violence for political ends', and terrorist acts are perpetrated by a variety of groupings.

Ethnic or religious minorities in search of national self-determination, such as the PLO (Palestine), IRA (Ireland), ETA (Spain) and Tamil Tigers (Sri Lanka), have resorted to terrorism to further their cause. Anarchist groups, such as the Red Brigades (Italy), Red Army Faction (Federal Republic of Germany) and Sendero Luminoso (Peru) have also indulged in terrorist acts. State-sponsored terrorism also continues, with the goal of repressing political dissidents and enforcing social conformity. Over fifty Third World⟁ governments could be accused of state terrorism; Guatemala, Indonesia and Liberia are amongst the worst offenders.

THANT, U (1909–74)   Burmese diplomat. He served in an acting capacity as Secretary-General of the United Nations⟁ following the death of Dag Hammarskjöld⟁ in 1961 before holding the post in his own right from 1962 to 1971. During his tenure he played an important role in resolving the US–Soviet crisis over the installation of Soviet missiles in Cuba. He was also responsible for the controversial decision to withdraw UN peacekeeping forces from the Egypt–Israeli border in 1967 during the ongoing Arab–Israeli War.

THIRD WORLD   A collective term applied to developing countries⟁, usually taken as those nations of Africa, Asia, Latin America (including the Caribbean) and Oceania (excluding Australia and New Zealand) which are comparatively industrially underdeveloped and economically weak. Although Third World countries house about 80 per cent of the world's population⟁, they are responsible for less than 30 per cent of global industrial production.

The term 'Third World' may be attributable to the French *Tiers Monde* an academic quarterly launched in 1956 in Paris which suggested a parallel between the '*tiers monde*' – the world of the poor countries – and the '*tiers état*' – the third estate or common people of the French Revolutionary era. However, during the 1950s the world's nations focused themselves into three groups. Two of these – the Soviet-led 'Eastern bloc' and the North American–Europe (West) grouping – possessed between them the greater part of the world's economic and military power. The third group – the states of Africa, southern Asia, Latin America and the Pacific, most

of which had only just gained their independence – were concerned with their own plight and so devoted themselves to North⟁–South⟁ issues rather than the struggle between the other two worlds, East and West.

THREE GORGES PROJECT   The Three gorges hydroelectric power⟁ project in China, first proposed over thirty years ago, would have seen the construction of the world's biggest dam⟁ and created the potential for one of the world's greatest catastrophes.

The dam, to be built on the Yangtze River (now known as the Chiang Jiang River), was planned to be 2 km long and 175 m high. It was designed to produce 14,700 megawatts (MW) of electricity, with construction costs estimated to be between $11 and 20 billion.

Proponents argued that the dam would eventually produce 17,000 MW of much needed electricity. It would also prevent flooding⟁ and make the river navigable for ocean-going vessels up to Chongqing. The last major flood in 1954 reportedly killed 330,000 people.

Opponents continually pointed out that the Chiang Jiang is one of the most silt-laden rivers in the world, carrying 450 million tonnes of silt annually which would soon curtail or block hydroelectric production altogether. In addition, the dam's reservoir would extend for 500 to 600 km, flooding ten cities and partially flooding eight others. Over 440 sq km of fertile farmland would be inundated, 3.3 million people would have to be resettled, and up to 75 million people who live from fishing⟁ or farm areas normally flooded by the river each year may lose their livelihoods. Replacing lost factories and industry alone would cost in the region of $3.5 billion. In addition, the habitats of eighty fish⟁ species would be destroyed and downstream, rare river species such as the few hundred surviving Chinese river dolphins and alligators would be faced with extinction⟁. Not only did the dam threaten important wetland⟁ habitats and the Poyang Hu Nature Reserve, home to the world's largest surviving population of Siberian cranes, but the weight of the impounded water could trigger earthquakes in the region.

Successive Chinese governments have backed the scheme. Chairman Mao⟁ supported the project when it was first mooted in 1954, and Li Peng expressed favour when he revived it in 1984. A Canadian feasibility study completed in 1987 for the World Bank⟁, which was considering financing the

Project, concluded that the dam was 'technically, financially and economically achievable'. However, a book attacking the whole scheme was later published by over a hundred Chinese scientists, writers and academics – this, together with the intense international outcry against the Project, persuaded the Chinese government once again to shelve the plan in mid-1989.

However, in 1990 the Chinese authorities decided to reconsider the Three Gorges scheme, despite the fact that, due to inflation and interest on the huge loans that would be necessary to complete the Project should it get the go-ahead, the costs had spiralled to an estimated $135 billion.

THREE MILE ISLAND   Located in the Shenandoah River in the US state of Pennsylvania, the Island is the site of a nuclear power⌂ station containing two pressurized water reactors (PWRS). In 1979, several water-coolant feed pumps in the facility failed, but although the reactor closed itself down, emergency water-supply lines were blocked. The nuclear reaction in the fuel core continued and, compounded by human error, the heat continued to build up, to a point where the uranium fuel⌂ itself might even have begun to melt. The likelihood of a massive explosion and subsequent release of vast amounts of radioactivity⌂ forced the evacuation of people within an 8-km radius of the Island. The power plant itself was not safe to enter until two years after the accident and clean-up⌂ costs, originally put at $1 billion, are still rising. When the building is finally adequately decontaminated, the reactor will be entombed in concrete.

TIDAL POWER   The use of natural tides to generate electricity. Large tidal flows are needed to make the process economically viable, so schemes are usually restricted to bays or estuaries on ocean coasts. Tidal water enters a dam⌂ and gradually builds up behind it. When the tide ebbs, water⌂ flows out, turning turbines as it goes. Turbines can also be driven by the rising tide. Although capital costs for tidal power schemes are high, operating costs are low. The units generate large amounts of power and have long working lifetimes.

Throughout the world, six tidal power stations are operational: three in China, one in the Soviet Union – all of which produce less than 1 megawatt of power – and one in Canada. The biggest and best-known installation is across the Rance estuary in Brittany in France; which houses twenty-four large, reversible turbines in a barrage 750 metres

long. Operating on both ebb and flow tides, it produces in the order of 500 million kilowatt/hours per year. As it is a form of renewable energy, several more tidal power plants are planned, with the Severn estuary in Britain, one of the best-known tidal sites in Europe, holding significant potential as it has a tidal range in excess of 11 m, the second highest in the world. Although the use of tides would appear to be a non-polluting and sustainable source of energy⌂, virtually nothing is known of the long-term environmental effects of using tidal power.

TIN   Tin is a silvery-white, soft, pliable, generally non-toxic metal, mainly produced from cassiterite, which occurs in ores mineralized by molten rocks. It is used as a protective coating for mild-steel 'tin' cans and as tinplate for a variety of purposes. It is commonly used to manufacture useful alloys such as bronze, pewter and gunmetal. Although it poses little danger to humans, cases of food poisoning⌂ arising from food tins are frequently reported, and its use in anti-fouling paints on boats has led to cases of poisoning in estuarine wildlife. Airborne concentrations of tin are negligible, and in soil⌂ the metal is strongly absorbed by humus.

International trade in tin was governed by an International Commodity Agreement⌂, overseen by the International Tin Council⌂ until its controversial financial collapse in 1985. Around 210,000 tonnes of tin are mined each year, mostly in the developing world: Malaysia (48,000 tonnes), Brazil (43,000 tonnes) and Indonesia (30,000 tonnes) are the major producers.

TISSUE PLASMINOGEN ACTIVATOR (TPA)   Produced by Genentech, a US-based biotechnology concern, TPA is an anti-blood-clot drug used to treat victims of heart attack. It is the most commercially successful of the new range of products being evolved through the application of biotechnology⌂, commanding about two-thirds of the lucrative US market for thrombolytic or clot-dissolving therapies: sales in 1989 were well in excess of $200 million. TPA is also the first and most celebrated example of the problems that can arise through the patenting of biotechnological products. Genentech claimed that two other firms, Wellcome and Genetics Institute, had infringed three of their TPA patents, and sought to prevent these other firms from selling their versions of TPA. The case was unsuccessful in the British courts but proved successful in the United States. Despite this success, the makers of TPA are faced with an uncertain future as the drug is facing stiff competition from a similar compound marketed in the USA by SmithKline Beecham and research has shown that TPA is no more effective at prolonging survival after a heart attack than streptokinase, an older drug produced by the German-based Hoechst company, which costs 90 per cent less than TPA.

TITANIUM Titanium is a strong, lightweight metal, as strong as steel but half its weight. It is also very resistant to corrosion. The metal is extracted commercially from two ores, ilmenite and rutile, although the process is difficult and titanium is therefore very expensive. Its use is largely confined to jet engines, aircraft shells and other military and specialized industrial hardware. It is also used in surgery, to support broken bones inside the human body. Because of its strength, corrosion-resistance and superplasticity it is used to make *Synroc*, a synthetic rock used to store and contain high-level nuclear radioactive material. Annual global production has reached 6 million tonnes; Australia and Canada (1.5 million tonnes each) and Norway (780,000 tonnes) are the major producer nations.

TOBACCO The most commonly used and widely distributed drug in the world, yet one that damages both the natural environment and human health. Its use is legal in all countries even though it is potentially lethal.

The leaves of the tobacco plant (*Nicotiana*) are used for making cigars, cigarettes, snuff and chewing wads. Climate greatly affects the quality of the leaves, and optimum growth occurs where the temperature is between 18 and 27°C. Although tobacco originated in tropical America, the bulk of world production (6.6 million tonnes in 1988) now occurs outside the tropics. The world's major producers are China (2.3 million tonnes), the United States (604,000 tonnes) and Brazil (454,000 tonnes).

Tobacco is one of the world's most important crops, with considerable environmental, economic and social impact. It accounts for 1.5 per cent of total agricultural exports and is estimated to generate annual profits in the region of $3 billion. The total area of land⌂ under tobacco is declining, falling 10 per cent between 1976 and 1987, and now totalling 4.3 million hectares (ha), yet production is rising, increasing by 17 per cent over the same period. Today 0.3 per cent of the world's arable land is used to grow tobacco, with 72 per cent of the total

being in developing countries⌂. Tobacco has no nutritional value, yet it is grown in countries where hunger⌂ is rife. The crop ruins soil⌂, taking out 11 times more nitrogen, 36 times more phosphorus and 24 times more potassium than most food crops; therefore tobacco cultivation necessitates considerable nutrient input and regular applications of fertilizer⌂ and pesticides⌂. After harvesting, leaves are dried in the sun, hot air or smoke for up to two months and then fermented for four to six weeks. Around half the tobacco grown in the world is flue-cured over wood⌂ fires in an extremely wasteful process. Over 1 million ha of open forest⌂ are stripped for this purpose, and 55 cubic metres of wood are burned for every tonne of tobacco cured.

Commonly, dried tobacco leaves are made into cigarettes or cigars, burnt, and the smoke is inhaled – smoke which consists of droplets of tar, a mixture of gases including carbon monoxide, and nicotine. Nicotine is one of the most addictive drugs known. Each cigarette smoked is believed to shorten the life of a regular smoker by five minutes. As smoking increases, the chances of heart disease, lung and mouth cancer⌂, strokes, bronchitis and other life-threatening conditions also rise. Tobacco can also be chewed, and in regions where this form of consumption is commonplace, such as Asia, mouth cancers are also common. Tobacco-related diseases are estimated to kill 2.5 million people annually, making tobacco the largest single preventable cause of death. In comparison, some 10,000 people die worldwide each year from the effects of illicit drugs. Rising awareness of the dangers of smoking tobacco, coupled with a government-imposed suppression of promotional advertising, has led to a decrease in tobacco consumption in the developed countries of around 1 per cent per year. However, in the developing world governments have neither the power nor the wherewithal to suppress the promotional activities of the transnational⌂ tobacco companies; consequently, tobacco consumption in the Third World⌂ is rising by over 2 per cent annually.

TOKYO ROUND A set of multilateral trade negotiations, part of the General Agreement on Tariffs and Trade (GATT)⌂, launched in Tokyo in 1973. The Tokyo Declaration, agreed upon at the inaugural meeting, set out the major objectives: to reduce or eliminate both tariff and non-tariff barriers and secure additional benefits for the international trade of developing countries⌂. Negotiations really began in earnest only in 1975, when the

United States Congress passed a Trade Bill empowering the USA to participate. The fruits of the discussion, which provided only marginal benefits to Third World⌂ countries, were implemented in 1980.

TOURISM Tourism – international, interregional and domestic – has become one of the world's leading industries. It is set to grow more than 40 per cent during the 1990s, and to continue expanding well into the next century. International tourism leapt from 25.3 million people in 1950 to 286 million in 1980. By 1990, between 400 and 600 million tourists were criss-crossing the globe. Tourism and related activities are forecast to surpass all major economic sectors in business volume, and the money generated from the global interchange of people should soon exceed earnings from sales of oil⌂. Travel and tourism already constitute the largest industry in the world in terms of employment, with one in every sixteen workers owing their job to tourism. Receipts from international tourism alone rose from $18 billion in 1970 to $120 billion in 1985, and reached $166 billion by 1989. Europe still makes more money from tourism than any other region but developing countries⌂, particularly those in Asia, now receive a fifth of the total income. Cuba has decided to increase its tourism industry threefold, which would make it the nation's leading money-earner by the year 2000. India is also seeking to more than double hotel occupancy over the next decade, a move which will create at least 250,000 jobs.

Although the economic aspects of tourism have been realized, and a World Tourism Organization (WTO) has been set up with backing from the United Nations⌂ to help ensure that the benefits are felt by all, especially the developing nations, the cultural, social and environmental impacts remain to be evaluated. The social systems and physical or cultural resources of many developing countries⌂ have a limited carrying capacity⌂, able to cater for only a restricted number of tourists before deterioration sets in. Balancing levels of tourists with those that can be safely accommodated without inflicting environmental or cultural damage will be an important task for national and international bodies such as the WTO.

TOXIC CHEMICALS Toxic chemicals are properly defined as those substances which have proved poisonous to humans and wildlife, although the term is often used to describe compounds suspected of being potentially injurious. Several naturally occurring metals are known to be toxic to humans, including lead⌂, arsenic, mercury⌂ and cadmium⌂, but global anthropogenic emissions of these are 20–300 times higher than natural background levels. Many organic compounds, including most hydrocarbons, are carcinogenic, mutagenic, teratogenic or otherwise injurious to human health and vegetation.

In addition to naturally produced chemicals, thousands more are artificially made each year for industrial purposes. Ten million chemicals are now registered, and almost 80,000 are now in common usage around the world. In 1989 the Organization for Economic Co-operation and Development (OECD)⌂ announced plans to investigate almost 1,500 chemicals on whose toxicity there is little or no information. Virtually nothing is known about the impact of these substances on the environment, even though they account for 95 per cent by volume of all the chemicals used globally. Thousands of chemicals are being used for which no toxicological data exist, but the 1,500 named by the OECD represent those that are produced in bulk – in quantities exceeding 1,000 tonnes per year – and are therefore more likely to be damaging to both humans and the environment. All the chemicals in question had entered into common usage before regulations were introduced to force all new substances to be screened for safety purposes. Twenty-one member governments of the OECD will share the investigative workload. The first examination will be of 147 so-called 'mystery' compounds about which there is virtually no toxicological information whatsoever. This part of the investigation will cost $32 million and is due for completion by 1993. Seventy of these 147 compounds are produced in quantities exceeding 10,000 tonnes annually.

TRADE-RELATED INTELLECTUAL PROPERTY RIGHTS (TRIPS) The rights of companies and individuals to exploit intellectual property, a loose term covering patents, copyright, trademarks and trade secrets. During the 1970s and 1980s counterfeiting and piracy of goods and trademarks became widespread. Worldwide, companies were estimated to be losing up to $60 billion annually as a result of counterfeit goods or the unauthorized production of items using ideas, logos or technologies developed by others.

The questions of patents and intellectual property may prove to be one of the major issues of global economic concern during the 1990s – not just with regard to the illegal production of consumer goods, but also due to the advent of biotechnology⌂. By the end of the 1980s, several applications to patent genetically engineered plants and animals had been made, and the debate over ownership and use of germ plasm in the world's seed banks⌂ had reached a critical stage.

Industrialized countries generally give exclusive protection to patent-holders to exploit their idea or invention for between fifteen and twenty years. Developing nations frequently give far less protection, or none at all. In some countries, no product patents are given covering chemicals, drugs and food, and only very limited processing patents are granted, as the governments concerned put the nutritional and health requirements of their populace ahead of company profits.

Although granting exclusive rights encourages inventors and stimulates innovation, it also creates a monopoly. The developing countries⌂, which account for only 1 per cent of the world's patents, fear that if the industrialized-country system of granting patent monopolies is adopted globally, it will slow down the transfer of technology. Most of the patents will go to transnational⌂ companies, and the Third World⌂ will be exploited as a result.

TRIPS became an important factor during the Uruguay Round⌂ of the General Agreement on Tariffs and Trade (GATT)⌂ discussions, which ended acrimoniously in 1990. The new TRIPS regulations adopted by GATT will support and update those under the two conventions, both initially agreed in the late nineteenth century, which still govern dealings concerned with international patents and copyright. The Convention for the Protection of Industrial Property, covering patents and trademarks, was initiated in Paris in 1883 and has now been signed by ninety-one nations, representing an equal spread between the developing and industrialized worlds. Seventy-eight countries have also ratified a similar agreement, the Berne Convention for Literary and Artistic Works, originated in 1886. Both conventions call for an internationally standardized system for dealing with applications for the protection of intellectual property.

TRANSGENIC ANIMALS The name used to describe animals which have undergone genetic engineering so that they produce useful drugs or proteins in their milk. Small portions of DNA containing the genes which code for the production of the required substance are injected into embryos; the adult animals which then develop produce the required substance in the milk at differing concentrations. This process is gradually taking the place of bacterial cultures used for the same purpose. Researchers have transferred into mice, sheep and pigs genes which allow the new host to produce substances such as human insulin⌂, tissue plasminogen activator (a blood-clotting agent), factor IX (a substance missing in some haemophiliacs), and antitrysin (a chemical used to treat lung diseases).

Many of the companies working on transgenic animals that have been successful in producing genetically altered mice are attempting to produce the same effect in rabbits. Rabbits are commercially attractive because of their short gestation period and the relatively high concentration of protein in their milk. Human growth hormone is already being produced in rabbits at a third of the cost of material produced using bacterial cultures. Movement away from the relatively crude approach of microinjection of DNA, and development and refinement of techniques which use viruses to effect the introduction of genetic material, are expected to raise the yields of drugs and proteins in the animals by between ten- and a hundredfold. Arguments continue as to whether the genetically engineered animals and the chemicals they produce can be patented, and court cases have restricted the availability of several substances produced from transgenic animals.

TRANSNATIONALS/MULTINATIONALS
A term usually applied to major business corporations that have subsidiaries, investments or operations in more than one country. The term is often used to describe formally a company or enterprise, operating in several countries, with 25 per cent or more of its output capacity located outside its country of origin. Some transnational enterprises are viewed by some as a threat to individual national sovereignty and as capable of exerting undue influence to achieve the company's ends, or of sacrificing human and environmental well-being in the search to maximize profits. Annual sales of many of these organizations are in excess of $10 billion, far greater than the exports and imports of many Third World nations which are measured in millions. The term *transnational* is becoming more widely used as it is held to put greater emphasis on the cross-border difficulties in controlling their operations.

TRAP CROPPING A co-ordinated, low-cost crop-protection measure pioneered as part of an Integrated Pest Management (IPM)⌂ system. To prevent pests from carrying-over from one season to the next, fixed dates are given for harvesting and the ploughing-in of crop residues. Small sections of the crop are left standing in the corner of fields during and after harvest. Insect pests are attracted to these 'traps' and can be easily destroyed by pesticides⌂, which have to be used only in very small amounts. The system is particularly well developed on cotton⌂ and other crops in Nicaragua. Cotton is notoriously susceptible to a variety of pests but trap cropping reduced pesticide use by 30 per cent. Yields rose by 15 per cent compared with crops in other fields where no trap cropping was used but nine times as much pesticide was applied.

TREATY   In the context of international law⌂, treaties are transactions between states which govern a voluntary undertaking by the participating countries. The states may disagree on various parts of the treaty, but once a treaty is signed and ratified it becomes binding on the states which are said to be contracting parties. However, treaties are not legally binding, they are merely an obligation on the signatories.

TREATY OF BRUSSELS   A pact of economic, political, cultural and military alliance established in 1948, due to run for fifty years, between Britain, France, and the Benelux⌂ countries. The Federal Republic of Germany and Italy joined the grouping in 1954. The Brussels Treaty Organization, as it was originally called, provided for collective self-defence and economic, social and cultural collaboration amongst signatory nations. The treaty was the forerunner of the North Atlantic Treaty Organization (NATO)⌂ and the European Common Market.

TREATIES OF ROME   Two treaties, signed in Rome in 1957, by representatives of Belgium, France, the Federal Republic of Germany, Italy, Luxembourg and the Netherlands, which led to the establishment of the European Economic Community. Trade between the signatory nations was made duty-free by stages; the last tariffs were abolished in 1968. The treaties also led to the formation of the European Atomic Energy Community (EURATOM)⌂ and the European Coal and Steel Community (ECSC)⌂. The six nations eventually agreed to merge the three communities into one, and in 1967 the three executive bodies merged to become the European Commission. The six also formed a Common Agricultural Policy (CAP)⌂, removed barriers to trade and movement of labour,

and established the European Parliament. By 1990 several other nations had signed the Rome Treaties and the European Community⌂ had expanded to twelve nations: the original countries were joined by Denmark, Greece, Ireland, Portugal, Spain and the United Kingdom. Application for membership had also been received from a number of other European nations, such as Austria. Several of the Eastern European⌂ nations had also announced their intention to apply for membership following their move towards Western-style democracy and a free-market economy.

TRIBUTYL TIN (TBT)   A group of tin⌂-based organic compounds that are toxic to fungi, other plants and many invertebrates. They are common constituents of wood⌂ preservatives and antifouling paints, most notably for use on boats. The compounds are toxic at extremely low concentrations and have caused serious pollution⌂ problems in estuaries and bays. Even in concentrations of parts per trillion, TBT induces a phenomenon called imposex in many shellfish, causing females to change sex. Evidence from the United States Environmental Protection Agency (EPA) suggest that TBT may also attack the human immune system and cause birth defects. The use of TBT is currently under review in many industrialized nations and restrictions have already been introduced in several countries.

TRICKLE-DOWN DEVELOPMENT   Theory of development in which investment and resources poured into government coffers, industry and high-technology projects are supposed to create wealth. Under capitalist theory, the wealth is supposed to 'trickle down' to improve the life of the poor. Optimistic forecasts made during the 1960s that this practice would promote development in the Third World⌂ have proved false. In reality, small elites in the developing countries⌂, mainly those close to or related to those in power, have adopted a lifestyle similar to that prevailing in the donor countries, while the majority still live in conditions of poverty and suffering.

TRITICALE   Triticale is a fertile hybrid cereal, first developed in the 1950s, obtained by crossing wheat⌂ and rye. Triticale can be grown on land⌂ that is marginal for wheat. Where soil⌂ is poor, yields of triticale exceed those of wheat by about 30 per cent. Compared to wheat, triticale is comparatively disease-resistant, much hardier and well suited for growth in acid soils. In rice-growing areas, it performs well when grown out of season on rice⌀ paddies. In addition, triticale shows better nutrient efficiency than wheat. Triticale was much favoured during the 1960s but did not prove very productive. However, a discovery in the 1970s produced a variety with improved fertility, stronger stalks, earlier maturation and higher yields. By 1983 it was being grown on more than 1.5 million hectares in thirty-two nations including Europe, both North and South America and the Soviet Union.

TROPICAL FORESTRY ACTION PLAN

(TFAP) A collective Plan incorporating recommendations drawn up by the World Bank⌀, the United Nations Environment Programme (UNEP)⌀, the Food and Agriculture Organization (FAO)⌀, and the World Resources Institute (WRI)⌀. Launched in 1985, its principal objective is to secure a thorough review of the state of forests⌀ in every tropical country in terms of contribution to national economic development, the needs of the people, and the environment. These national reviews are meant to allow governments to decide their own priorities for action in forest conservation⌀ and development, and to secure the support and co-operation of the international donor community.

Following the basic premiss that the problems of deforestation⌀ cannot be solved in isolation from other aspects of development planning, the TFAP aims to co-ordinate efforts to save the world's tropical forests and improve the lot of the rural poor by identifying and promoting ways in which the forests can be exploited in a sustainable manner. Original costing estimates forecast that $8 billion would be needed to cover TFAP activities over the period 1987–91. During that time, $1.6 billion per year would be spent in forestry and related agricultural projects to control new deforestation in fifty-six developing countries⌀. Rescuing and restoring already damaged forests and watersheds could push the price up to $150 billion. Around half of the investment called for would be provided by the bilateral and multilateral development agencies and half from national goverments, foresters and the public sector.

Since TFAP was implemented, national sector reviews have been carried out and completed in twenty countries and are in various stages of completeion in thirty-four others, while a further eight have requested TFAP reviews. Requests from other governments were in the pipeline, and by 1990 over seventy-five Third World⌀ countries had become involved, although funding of only just over $1 billion a year was being received.

Ever since its inception, critics have complained that the TFAP is likely to lead to more deforestation rather than less. During the years of its existence they have claimed that the national reviews are not reaching their stated objectives and that the rights of forest people are being overlooked, mainly because national Action Plans are being dominated by the interests of the commercial forestry and timber industries. A major criticism of the TFAP is that the political factors that give rise to landlessness⌀ – itself a major cause of deforestation and the reason why more and more forest is being converted to agricultural land⌀ – are not being properly addressed.

In 1990 a critical review of the TFAP was put before the FAO. Claims that it was being a 'loggers charter' were reinforced. In addition, one of the major criticisms of the plan – that there had been lengthy delays in meeting requests for assistance from developing countries – was substantiated. Only six out of the seventy-five or so interested nations had had forestry plans inaugurated under the TFAP. Of those that had received assistance, both Cameroon and Colombia had problems implementing the plans through lack of funds. As a result of the review, the TFAP was reformed with the objective of 'conservation and sustainable development of forestry resources in the interests of the country concerned and the global community'. It is proposed that the plan, renamed the Tropical Forestry Action Programme, should be strengthened by an international forest convention. The convention, due to be launched at the 1992 World Conference on Environment and Development, would lay down regulations to govern all forests. It is favoured by the North⌀ but strongly opposed by developing nations in the South⌀.

TROPICAL TIMBER TRADE

Wood⌀ products are believed to be the third most valuable commodity⌀ in world trade after petroleum and natural gas⌀: global trade in wood and related products is projected to grow for the next fifty years. At present, 45,000 sq km of primary tropical forests⌀ are logged annually, and illegal fellings account for a further 5,000 sq km each year. Less than 1 per cent of all tropical timber traded comes from renewable sources. Consumption of the timber is balanced between the three major markets: Japan, the United States and Europe. Each takes about one-third of the tropical timber that enters international trade, worth $8 billion annually.

Exports of tropical hardwoods represent an extremely important source of foreign exchange for at least fifteen developing nations⌾. Yet over-exploitation and lack of proper management have left twenty-three tropical nations, many of which were once exporters of hardwoods, in a position where they now have to import manufactured forest products, paying out in excess of $50 million a year. By the end of the 1980s some thirty tropical countries were regularly exporting wood. If prevailing rates of depletion and reforestation⌾ were to continue, this number is forecast to fall to ten and the value of the export trade to slump to $2 billion within fifty years.

Malaysia heads the world's tropical hardwood exporting nations in all areas – sawn and planed timber and raw logs. In 1987 Malaysia exported 22.8 million cubic metres of logs, every other exporting country producing well under 1 million. Although Brazil is actually the largest producer of wood, most of it is used to satisfy local demand.

The world has woken up to the problem of tropical deforestation⌾ and campaigns are being organized at all levels to prevent the destruction of tropical rainforests⌾ – and of tropical dry forests. Tropical dry forests⌾, covering 8 million sq km, are also under severe threat: 38,000 sq km are cleared annually. The increasing success of the conservation⌾ lobby is reflected in the price of tropical hardwood timber. According to World Bank⌾ figures, the threat of scarcity is driving up the price of Malaysian logs: 1989 prices are 53 per cent higher than those of 1987. As campaigns intensify and gather increasing official backing, Papua New Guinea may well set a precedent to determine the future of tropical forests because its rainforest has been declared a World Heritage⌾ site. The World Bank is trying to arrange $75 million compensation for any economic losses incurred as a result of lost timber production.

TRUST TERRITORY A territory, previously known as a mandate, which is being prepared for self-government and for which the United Nations (UN)⌾ is responsible under its Trusteeship programme. Under the Treaty of Versailles, the administration of former German and Turkish possessions was entrusted to the Allied states by the League of Nations⌾, which later became the United Nations,

when the mandates became known as Trust Territories. There is now only one left, the Trust Territory of the Pacific Islands, administered by the United States. It originally comprised the Marshall Islands⌾, the Federated States of Micronesia and Belau⌾ (Palau), the latter being the last remnant.

The Marianna Islands were included until 1978, the territory was taken from Germany by Japan in 1914 and captured by the United States in 1944. The trusteeship was established in 1947, following the end of the Second World War. All other territories within the Trusteeship system have achieved independence, including Namibia (South-West Africa), which was always regarded as an exception within the Trust Territory system because the administering government, South Africa, did not recognize UN authority.

TRYPANOSOMIASIS A collection of debilitating, long-lasting diseases caused by infestation with microscopic single-celled *Trypanosoma* organisms. Included in this group are sleeping sickness amongst humans and ngana in cattle, both found in Africa and transmitted by tsetse⌾ flies. In the Americas, the incurable Chagas disease⌾ is a form of trypanosomiasis spread by triatomine bugs. At least 140 million people are affected by these diseases, mainly in the tropics.

African trypanosomiasis, or sleeping sickness, is a severe, often fatal disease. The infective parasites are transmitted to humans via the bite of the tsetse fly. Some 50 million people in tropical regions of Africa are at risk. Trypanosomiasis is a major constraint to human development in Africa. Ngana disease in cattle effectively renders commercial livestock⌾ production impossible over 10 million sq km of high-rainfall land⌾ on the continent, representing 45 per cent of the total land area in sub-Saharan Africa. Every year, 25,000 new cases of human trypanosomiasis are reported in thirty-six African countries: the disease kills at least 3 million animals and 10,000 people annually.

In humans, initial symptoms of sleeping sickness are a general ache and enlargement of the lymph nodes. After several months or even years, the parasites invade the blood vessels supplying the central nervous system, causing drowsiness and lethargy. Several different forms of the disease exist: Rhodesian sleeping sickness is the most virulent. Early diagnosis is difficult, yet once the disease is established it is almost invariably fatal. A few infected individuals do not develop the disease but remain carriers. Similarly, in many areas, indigenous cattle and wildlife have developed a degree of immunity to the parasite and therefore act as a reservoir of the disease. All drugs used to treat sleeping sickness have serious shortcomings and may themselves prove fatal; consequently, efforts to combat the disease have been directed at controlling the tsetse vectors. In 1990 a new drug, Ornidyl, was discovered that countered the disease. This drug, originally developed as a potential therapy for cancer◇, was found to act quickly and have very few side-effects, the common problem of all other drugs used to treat sleeping sickness. Ornidyl, also known as DFMO or eflornithine, is relatively cheap, costing $140 for a two-week course. However, its true cost is much higher, particularly as it needs to be injected several times daily for greatest effect, requiring both medical attention and materials.

TSETSE The tsetse is a relatively large bloodsucking fly of tropical Africa belonging to the genus *Glossina*. Several species exist, all restricted to Africa, which feed on humans and other warm-blooded animals, generally in areas close to river banks or on the edge of savannahs, their preferred habitat. The fly transmits sleeping sickness to humans and ngana to cattle, except to indigenous wild animals and game, which have evolved a degree of immunity. The fly has proved one of the greatest constraints to livestock◇ production in much of Africa and has effectively prevented the full utilization of vast areas of arable land◇.

The tsetse is a unique fly, and as such is difficult to control. As larvae develop to maturity in the female and are deposited directly into the soil◇ to pupate, later emerging as adults, control measures can be levelled only against the adult stage. The flies spend most of the day resting in the shade under branches and leaves, so aerial application of pesticides◇ is not particularly effective and must be supported by intensive ground-level spraying.

A multi-million-dollar scheme to clear the tsetse fly from thirty-eight African countries has been proposed. This, if successful, will open up 18 million sq km of land for agricultural production. The new scheme, like others in the past, has incurred criticism from a number of conservation◇ groups and ecologists who fear that the uncontrolled settlement, overgrazing◇ and soil degradation that have historically followed small-scale tsetse eradication schemes will be repeated, but on a much larger scale. The bulk of the funds for the new scheme will also be directed to the use of man-made pesticides, as opposed to safer control methods. Many dangerous chemicals have traditionally been used in the war against the tsetse, and over 300,000 sq km in Africa have been regularly sprayed with DDT◇, dieldrin◇ and lindane. Levels of DDT above the maximum safety limit established by the World Health Organization (WHO)◇ have been found in many people in areas which are regularly treated, and the chemical has also exacted a heavy toll on wildlife. Conservation groups argue that if there is to be an attempt at control, use should be made of safe odour-baited traps that have proved successful in eradicating tsetse in field trials. Furthermore, strict land-use◇ plans should be incorporated in any control programme.

TUBERCULOSIS An infectious disease caused by the bacillus *Mycobacterium tuberculosis*, characterized by the formation of nodular lesions, or tubercles, in body tissues. The bacillus is easily inhaled, and a primary nodule is quickly formed. The body's natural immune system may heal it at this stage; when this happens a lasting immunity develops. Other people may become infected but show no signs of illness, acting as carriers, transmitting the bacillus by coughing and sneezing. The highly infectious bacterium can remain dormant in the body for years, being triggered by a variety of factors: for example, it appears to become active when infection with the Human Immunodeficiency Virus (HIV) damages a person's immune system.

Symptoms of the disease include fever, night sweats, weight loss and spitting blood. Tuberculosis is curable with antibiotics, and inoculation with BCG vaccine affords protection to those who have not already developed immunity. Globally, there are 8 million new cases of the disease each year; the risk of infection is a hundred times greater in the developing countries◇ than in the developed world. Four million of these cases are infectious, and in total, there are 20 million active cases of tuberculosis. An estimated 3 million people die from it each year and another 1.7 billion are or have been infected with the tuberculosis bacillus.

Most of the fatalities occur in the developing world – only 42,000 deaths are reported from the industrialized nations, mostly among the elderly, ethnic minorities and migrants. However, progress against this disease has now come to a halt in several developed countries. The number of cases in the United States, which declined steadily for thirty-two years up until 1984, is now on the increase. One of the reasons behind the resurgence of the disease is the spread of infection with the Human Immuno-deficiency Virus (HIV) which, by weakening the immune system, increases the likelihood of infection with the TB bacillus. According to the World Health Organization (WHO)⌂, 95 per cent of tuberculosis patients can be cured within six months with anti-biotic treatment costing less than $30 per patient.

TUTU, DESMOND (1931–) South African clergyman, noted for his strong opposition to apartheid⌂. Ordained in 1960, in 1978 he became the first black general secretary of the South African Council of Churches, an organization with a membership of over 12 million. In 1984 he was appointed Anglican Bishop of Johannesburg, and in the same year he was awarded the Nobel Peace Prize for his tireless efforts to combat racial segregation.

TYRES Every year billions of rubber⌂ tyres are produced around the world for all forms of motor-ized vehicles and bicycles. Each year millions of used tyres are discarded, a situation that has been gather-ing momentum since the beginning of the twentieth century. Virtually all motorized vehicles have been designed to run on natural or synthetic rubber tyres. Most cars are produced with five tyres, and several more may be used during the vehicle's lifetime. These tyres do not degrade naturally, and at present there is virtually no viable way of large-scale recyc-ling⌂ of the rubber used to make them. Used-tyre 'mountains' already cover vast areas of land⌂. The situation is worst in the industrialized nations where car densities are highest; in the United States, where there is the greatest concentration of vehicles, 240 million tyres are discarded each year. Eventually every nation that has a sizeable fleet of automobiles or bicycles will be faced with a problem of what to do with their tyres.

Tyres can be burnt, but they produce clouds of acrid black smoke. In addition, their high sulphur content produces sulphur dioxide⌂ gas which, when converted to sulphurous acid in the atmos-phere, increases the likelihood of acid rain⌂. In 1990 a fire in a dump of 14 million tyres in Canada burned for two weeks, causing widespread air and water pollution⌂. A fire in a dump in Scotland, which contained 15,000 tonnes of tyres, produced 90,000 litres of black oil⌂ which seeped into a local river, killing all fish⌂ life up to 10 km downstream. A similar accident happened in England in 1975.

Several methods of reducing the used-tyre moun-tain have been tried, including using tyres as fenders on boats or recycling them to produce crude door-mats or aircraft landing mats. Yet more are ground down and mixed with asphalt to produce rubberized pavements. In Australia an inventor has proposed using old tyres to stop water⌂ evaporating from reservoirs. Suggestions have also been made to use tyres for lifebuoys and as lobster-pot indicators. An estimated 12 million tyres are discarded in Australia each year.

In the United States old rubber tyres are con-verted to a polymer composite that can be used to make hosepipes, gaskets, rollers, and new tyres. Using this method, new tyres can be produced at a lower price than tyres made from new rubber. Other US inventors have devised ways of turning old tyres into pipes, fenceposts and boards. The tyres are ground to a powder, which is then mixed with various plastics. The rubber-and-polymer mixture is then heated and extruded under high pressure. Piping produced by this method costs less than a quarter of plastic piping made using conventional methods, but the end product is not biodegradable.

U

UNITED NATIONS (UN)   An association of independent countries from all over the world with the aim of promoting international peace, security and co-operation. It replaced the League of Nations⌂ in 1945; the original membership comprised fifty-one states. With the independence of colonies and other territories, membership has risen to the present number – Namibia became the 167th member in 1990.

The Charter of the United Nations was drafted by the international community in San Francisco in 1945, and the headquarters are in New York. It has six principal organs: General Assembly, Security Council, Economic and Social Council⌂, Trusteeship Council, International Court of Justice⌂, and General Secretariat. Member states contribute funds according to their own resources, following a General Assembly assessment of what their contribution should be. The UN finances a variety of aid⌂ programmes through its own agencies and other international bodies. However, its activities are restricted by the flow of funds, with many nations failing to pay their dues. The total amount owed by member countries in 1990 stood at $660 million, the United States being the largest debtor to the UN regular budget, owing a total of $521 million. Of this, $288 million was for debts from earlier years with $233 million due for 1990.

The organization has six working languages: English, French, Russian, Spanish, Chinese and Arabic.

Over fifty states and territories are not members: some remain under foreign military occupation; others, such as Hong Kong and French Guiana, remain colonies of European states. A further four independent states with their own diplomatic missions – Switzerland, North Korea, South Korea, and the Cayman Islands – have chosen not to join. Taiwan was excluded from the UN when the People's Republic of China took over the seat in 1971. Byelorussia and Ukraine are both members, but they are part of the Soviet Union which thus, effectively, has three seats, and therefore three votes, in the General Assembly.

UN General Assembly – The General Assembly meets annually, and is composed of one representative from each of the member states. Each has a single vote and so the Assembly represents the one major forum where the South⌂ can outvote the North⌂. Major Assembly decisions require a two-thirds majority.

UN Security Council – When the UN Charter was adopted, the victorious Powers of the Second World War were in the ascendancy, and this was reflected in the composition and role of the Security Council. It was agreed that only the Security Council would have the power to recommend action to maintain international peace and security; this would involve either the use of armed force or other options such as economic sanctions or the severing of diplomatic relations. Five permanent members were elected to the Council – China, France, the Soviet Union, the United Kingdom and the United States. The other six members (increased to ten in 1965) serve a two-year term; retiring members are ineligible for re-election. Only the permanent members have the power of veto. Despite widespread conflict in the world since 1945, the Security Council has declared that a 'breach of the peace' has occurred only three times – the Korean War⌂ (1950), the Falklands Islands dispute (1982), and following Iraq's invasion of Kuwait in 1990. The power of the veto has meant that the Security Council could recommend no action over events such as the Soviet Union intervention in Hungary (1956), the Suez Canal incident (1956), Soviet intervention in Afghanistan (1979), US action in Grenada (1983), and the US bombing of Libya (1986).

UN Secretariat Based at the UN Headquarters in New York, the Secretariat is composed of the Secretary-General, the chief administrative officer of the UN, who is appointed by the General Assembly, and an international staff appointed by the holder of the post. There have been five Secretary-Generals:

Trygve Lie⌂ (Norway), 1946–53
Dag Hammarskjöld⌂ (Sweden), 1953–61
U Thant⌂ (Burma), 1961–71

Kurt Waldheim⊙ (Austria), 1972–81
Javier Pérez⌂ de Cuéllar (Peru), 1982–

UNITED NATIONS CAPITAL DEVELOPMENT FUND (UNCDF)

Established by the United Nations (UN) in 1966 as an autonomous organ of the General Assembly, but since 1967 operated as a special programme of the United Nations Development Programme (UNDP)⊙ and managed by the governing Council of the UNDP. Its purpose is to assist developing countries⌂ to promote their economies by supplementing existing sources of capital assistance by means of grants and loans. The UNCDF provides small-scale investment assistance to the Least Developed Countries⌂, both to help meet basic needs⌂ and to stimulate productive investments in development projects. By 1988 the Fund was supporting 245 operational projects totalling $241 million.

UNITED NATIONS CHILDREN'S FUND (UNICEF)

The mandate of UNICEF, originally established in 1946 by the General Assembly on a temporary basis to co-ordinate relief work in war-torn countries, was extended in 1953 for an indefinite period. Based in New York and Geneva, UNICEF devotes almost all its aid⌂ to long-term programmes, chiefly for mass health campaigns, directed at improving maternal and child welfare, child nutrition and education⌂. It provides and co-ordinates basic services for children, mothers and community development and promotes the concept of global interdependence and respect for other cultures. The Fund is financed entirely by voluntary contributions from governments, private groups and public and individual donations. There are at least thirty-five separate and autonomous UNICEF National Committees or Liaison Offices in individual countries around the world.

UNITED NATIONS CONFERENCE ON THE HUMAN ENVIRONMENT

Often referred to as the Stockholm Conference⌂, the first UN Conference on the Human Environment was held in the Swedish capital in 1972. In response to mounting global public concern about deteriorating environmental and living conditions, delegates from 113 nations met and produced an action plan of 109 separate recommendations. They also agreed a Declaration of 26 common principles on human rights⌂ and responsibilities in respect of the global environment which remain as a guide to influence human actions and policies:

1. The Human Race has the fundamental right to freedom, equality and adequate conditions of life and bears a solemn responsibility to protect and improve the environment for present and future generations. Policies promoting oppression and foreign domination must be eliminated.

2. The natural resources⌂ of the earth must be safeguarded for the benefit of present and future generations through careful planning and management.

3. The capacity of the earth to produce vital renewable resources must be retained and, wherever practicable, restored.

4. Man has special responsibility to safeguard and wisely manage the heritage of wildlife and its habitat which are now imperilled. Nature conservation⌂ must therefore receive importance in planning for economic development.

5. Non-renewable resources of the earth must be used in such a way as to guard against their future exhaustion and to ensure that any benefits are shared by all mankind.

6. The discharge of toxic⌂ substances, or of other forms of pollution⌂, in such quantities or concentrations that exceed the capacity of the environment to render them harmless, must be halted.

7. States shall take all possible steps to prevent pollution of the world's seas.

8. Economic and social development is essential for ensuring a favourable living and working environment.

9. Environmental deficiencies generated by conditions of underdevelopment and natural disasters can best be remedied by accelerated development through the transfer of substantial quantities of financial and technological assistance.

10. Stability of prices and adequate earnings for primary commodities⌂ and raw materials are essential to environmental management.

11. The environmental policies of all states should enhance and not adversely affect the present or future development potential of developing countries⌂.

12. Resources should be made available to preserve and improve the environment: additional international technical and financial assistance should be forthcoming in this respect.

13. States should adopt an integrated and co-ordinated approach to their development to ensure that development is environmentally sound.

14. Rational planning must reconcile any conflict between the differing needs of social development and the natural environment.

15. Human settlements and urbanization⌖ must be planned to provide maximum social and economic benefits for all with minimum adverse effects on the environment.

16. Demographic policies, without prejudice to human rights and which are deemed appropriate by governments concerned, should be applied in those regions where excessive population⌖ growth rates or concentrations may jeopardize the environment or development process.

17. Appropriate national institutions must undertake the planning, managing and controlling of a country's environmental resources.

18. Science and technology must be applied to the identification, avoidance and control of environmental risks and the solution of environmental problems, for the benefit of all.

19. Education⌖ is essential in order to promote enlightened opinion and responsible conduct by individuals, enterprises and communities with regard to protecting and improving the environment.

20. Scientific research and development on all aspects of the environment and development must be encouraged in all nations, and the free flow of data and information must be supported and assisted, with new technologies being made available to developing countries.

21. States have the sovereign right to exploit their own resources but must ensure that their activities do not damage the environment beyond the limits of their jurisdiction or control.

22. All states shall work toward developing viable and enforceable international law⌖ regarding questions of liability and compensation arising as a result of environmental damage or pollution.

23. The differing system of values prevailing in countries throughout the world must be taken into account in all respects, policies and judgements.

24. International matters concerning the protection of the environment should be handled in a co-operative spirit by all nations on an equal footing.

25. States shall ensure that international organizations play a co-ordinated, efficient and dynamic role in protecting the environment.

26. The human race and the environment must be spared the effects of nuclear weapons and other means of mass destruction.

UNITED NATIONS CONFERENCE ON TRADE AND DEVELOPMENT (UNCTAD)

The Conference arose following growing dissatisfaction amongst developing countries⌖ that were finding it impossible to close the gap between the economies of the developed countries and their own. It is an organ of the United Nations (UN)⌖ scheduled to meet approximately every four years, but its work continues between sessions through the Trade and Development Board, various standing committees and a Special Committee of Preferences, all based in Geneva. It covers a wide range of subjects, including monetary reform and debt⌖ problems, technology transfer, the 'brain drain', shipping and flags of convenience, commodity agreements⌖, tariff preferences and a comprehensive evaluation of world trade and economic performance.

The Conference or discussions are known as 'Rounds', attended in 1990 by 166 countries. UNCTAD conferences serve as the main international forum for dealing with North⌖/South⌖ economic issues and all UN member states may participate, along with Non-Governmental Organizations (NGOs)⌖ and intergovernmental organizations. The first meeting, held in Geneva in 1964, agreed in principle on a scheme for supplementary financial measures and a target of 1 per cent of national income for the flow of financial aid⌖ and resources from developed to developing countries.

The second Conference in New Delhi in 1968 increased the aid target to 1 per cent of Gross National Product (GNP)⌖ at market prices, and agreed a general system of preferences on exports of manufactured and semi-manufactured goods from developing countries.

The third UNCTAD meeting, in Santiago (1972), concerned itself mainly with commodities⌖, and developed a case-by-case timetable for negotiations and discussions of a Common Fund⌖ for financing buffer stocks, as well as addressing the debt⌖ problem.

The following conferences in 1976 (Nairobi), 1979 (Manila), 1983 (Belgrade), and 1987 (Geneva) dealt primarily with issues of protectionism and adopted well over twenty major resolutions by consensus in an attempt to promote freer trade.

The eighth series of negotiations, scheduled for 1991, was slanted to deal with strengthening international action and multilateral co-operation for a healthy, secure and equitable world economy, paying special attention to five areas: resources for development; international trade; technology; services; and commodities.

The outcome of many of UNCTAD's Rounds have been compromises which have not satisfied the majority of participating developing countries, and international harmony has not been particularly evident. Punta del Este in Uruguay, where the Uruguay Round⌾ of General Agreement on Tariffs and Trade (GATT)⌾ talks began in 1986, is the siting of the latest gathering, UNCTAD VIII. The location conforms with the principle of rotation, with developing regions taking turns in hosting the quadrennial meeting. However, the Uruguayan offer to host the talks came eight months after Cuba's decision, in March 1990, to waive the right to stage the Conference in favour of another Latin American nation. This followed prolonged opposition from the United States, which refused to endorse Cuba's candidacy to act as host for UNCTAD VIII and declared that it would not attend if the talks were held there. A similar situation had occurred before UNCTAD VI and UNCTAD VII, which were held in Belgrade (1983) and Geneva (1987) respectively.

UNITED NATIONS DEVELOPMENT FUND FOR WOMEN (UNIFEM)   One of the specialized programmes co-ordinated by the United Nations Development Programme (UNDP)⌾, UNIFEM promotes the involvement and participation of women in every aspect of development activities. It organizes workshops, provides funding for educational materials, and arranges meetings and visits.

UNITED NATIONS DEVELOPMENT PROGRAMME (UNDP)   Established in 1966 by the merger of the United Nations⌾ Special Fund and the Expanded Programme of Technical Assistance, the UNDP is the world's largest grant-awarding development co-operation organization. It serves as the central planning, funding and co-ordinating agency for all forms of technical and scientific co-operation carried out under the entire UN system. It orchestrates activities between the UN specialized agencies, governments and bilateral donors. UNDP and its associated special funds provide assistance in agriculture⌾, education⌾, health, employment⌾,

fisheries⌾, industry, science and technology, transport and communications. It is also concerned with promoting the New International Economic Order⌾, donor Round Tables for the Least-Developed Countries⌾, women in development, liberation movements and the activities of Non-Governmental Organizations (NGOs)⌾ in developing countries⌾. The Programme has built up field offices in 112 nations and serves 152 countries and territories.

The UNDP offers assistance in the form of experts, training and a small amount of equipment. It also organizes resource surveys, feasibility studies, and preparatory investigations intended to facilitate investment, and helps to establish technical, training and research institutions. UNDP does not itself provide capital for development projects. Resources are allocated to countries on the basis of need according to an internationally agreed formula using an Indicative Planning Figure (IPF), which is the projected amount that will be available for programme activities in a country over a five-year period. Calculations are made on the basis of population⌾, *per capita* Gross National Product (GNP)⌾ and other socioeconomic and environmental criteria. In 1989, $1.2 billion had been pledged to the UNDP and its associated specialized programme funds.

UNITED NATIONS DISASTER RELIEF OFFICE (UNDRO)   Founded in Switzerland in 1972, UNDRO is an overall co-ordinator of emergency assistance in disasters. It provides centralized information on relief needs and reports on what various donors are doing individually or collectively to meet those needs. UNDRO collaborates closely with such bodies as the Food and Agriculture Organization (FAO)⌾, World Health Organization (WHO)⌾, World Food Programme (WFP)⌾, and United Nations Childrens Fund (UNICEF)⌾, which themselves provide relief assistance. UNDRO receives its income from the UN regular budget, supplemented by a voluntary fund.

UNITED NATIONS EDUCATIONAL SCIENTIFIC AND CULTURAL ORGANIZATION (UNESCO)   A specialized United Nations⌾ agency established in 1946 with the aim of promoting international togetherness through education⌾, science and culture. Its work includes raising levels of literacy, improving facilities for teacher training and the promotion of international co-operation in scientific research programmes. It also collaborates in the application of natural sciences, technology and social sciences to development work of all kinds, promoting and preserving all aspects of culture and encouraging progress in the communication media.

Based in Paris, UNESCO has 158 members and two associate member states. It undertakes work on a broad range of topics including communications and informatics, human rights⌂, geological and environmental sciences, population⌂, women, youth, the preservation of cultural heritage, education and literacy, and the promotion of peace and international understanding.

UNITED NATIONS ENVIRONMENT PROGRAMME (UNEP)

UNEP was established in 1972 as a result of the United Nations Conference on the Human Environment⌂ held the same year. It was created essentially to provide a constant review of the global environmental situation, and to help safeguard the environment for present and future generations by ensuring that all environmental problems of international significance received appropriate consideration by the global community. It was the first UN agency with headquarters in a developing country⌂.

UNEP promotes environmentally sound economic and social development in both urban⌂ and rural areas and has identified environmental health, terrestrial ecosystems⌂, environment and development, oceans, energy⌂ and natural disasters as priority areas for its activities. It operates a worldwide surveillance system, Earthwatch⌂, which has three major components: the International Register of Potentially Toxic Chemicals (IRPTC)⌂, the International Referral System (INFOTERRA)⌂ and the Global Environmental Monitoring System (GEMS)⌂. Through information gathered via these conduits, and that derived from the Global Resource Information Database (GRID)⌂, UNEP is able to issue warnings against impending environmental crises. It also serves in a catalytic role, encouraging other UN agencies, governments and Non-Governmental Organizations (NGOS)⌂ to work towards environmental protection and conservation⌂.

UNEP also administers the United Nations Environment Fund, a fund supported by voluntary government contributions and used to finance specific environmental assessment and management projects.

UNITED NATIONS FUND FOR DRUG ABUSE CONTROL (UNFDAC)

Established in 1971 to develop short-term and long-term narcotics-control programmes and to provide funds for their execution. As resources have grown, UNFDAC has moved from specific projects designed to improve the research and information-gathering capabilities of drug-control agencies to the planning and implementation of technical co-operation programmes. These have involved pilot projects for crop substitution and rural development or community projects aimed at increasing individual governments' capacities to deal with drug-dependent persons and their rehabilitation and reintegration into society.

UNITED NATIONS FUND FOR POPULATION ACTIVITIES (UNFPA)

Originally established in 1967 as a Trust Fund of the Secretary-General, the UNFPA was placed under the administration of the United Nations Development Programme (UNDP)⌂ in 1969 and in 1972 transferred to the authority of the General Assembly. The UNDP Governing Council remains designated as its controlling body, subject to the policies and conditions established by the UN's Economic and Social Committee (ECOSOC)⌂. A condition of this arrangement was that the UNFPA should maintain a separate identity. It is supported by voluntary contributions from governments. Through the World Population Conference, UNFPA has been charged with the task of leading and co-ordinating efforts necessary for the implementation of the World Population Plan of Action. It helps governments obtain information on the growth, make-up and movement of their populations, promotes awareness of and research into the relationship between population⌂ and economic and social development, provides expert advice on population policy-formulation and supports family planning, education⌂ and training programmes. The Fund has attracted criticism from several governments for its position on the abortion issue.

UNITED NATIONS FUND FOR SCIENCE AND TECHNOLOGY DEVELOPMENT (UNFSTD)

One of the specialized funding programmes co-ordinated by the United Nations Development Programme (UNDP)⌂. The UNFSTD helps developing countries⌂ make use of up-to-date science and technology to further their economic development.

There are two specific programmes within the UNFSTD designed to improve the transfer of science and technology. Transfer of Knowledge Through

Expatriate Nationals (TOKTEN) is a programme which encourages expatriate skilled professionals to return to their native lands to undertake consultancy work. Short-Term Advisory Services (STAS) is a programme through which senior executive volunteers can be provided to solve commercial and public-service problems.

UNITED NATIONS GROUPS To ensure a balanced representation of regional and other interests in subsidiary United Nations⌂ committees, agencies and other organs with limited membership, the UN has evolved an arrangement, formalized in the constitution of the United Nations Conference on Trade and Development (UNCTAD)⌂, whereby UN member countries are divided into four groups. Each group has been allocated a set number of seats on the Trade and Development Board of UNCTAD and on the Industrial Development Board of the United Nations Industrial Development Organization (UNIDO)⌂. Group A consists of African and Asian countries, Group B comprises Western European, North American and other fully industrialized nations, Group C contains Central and South American Countries, with the Eastern European, former Communist bloc countries traditionally forming Group D. Members of groups A and C are sometimes referred to as the Group of 77⌂.

UNITED NATIONS HABITAT AND HUMAN SETTLEMENTS FOUNDATION (UNHHSF) Founded by the General Assembly in 1975, following recommendations from the Governing Council of the United Nations Environment Programme (UNEP)⌂ and the United Nations Educational, Scientific, and Cultural Organization (UNESCO)⌂. The Foundation's primary objective is to assist in strengthening national environmental programmes relating to all aspects of human settlements, particularly in developing countries⌂. This is carried out through the provision of seed capital for projects and essential technical co-operation.

UNITED NATIONS HIGH COMMISSIONER FOR REFUGEES (UNHCR) Established in 1951 to continue the work originally undertaken by the International Refugee Organization. UNHCR provides protection and assistance on a social and humanitarian basis to refugees⌂ who are not considered as nationals by the countries in which they seek asylum. It aims to facilitate their voluntary repatriation or their assimilation within

their new national communities. Its programmes are financed by voluntary contributions from governments, private agencies and institutions.

UNITED NATIONS INDUSTRIAL DEVELOPMENT ORGANIZATION (UNIDO) Originally set up in 1965 and based in Vienna, UNIDO now has 152 member states. Its purpose is to promote the industrialization of developing countries⌂ and to co-ordinate United Nations⌂ activities in this respect. It is an operational as well as advisory body, dealing with all aspects of factory establishment and management projects, regional and local industrial planning, institutional infrastructure⌂ development, technology transfer, and investment promotion. UNIDO derives the bulk of its technical co-operation funds from the United Nations Development Programme (UNDP)⌂, supplemented by voluntary contributions. Its Special Industrial Services (SIS) programme provides short-term assistance to deal with urgent industrial problems. Its governing body is the forty-five-member Industrial Development Board.

UNIDO administers the UN International Development Fund (UNIDF), a voluntary fund established by a General Assembly resolution in 1976 and aimed at enhancing UNIDO's ability to meet the needs of developing countries promptly and flexibly, where necessary supplementing the assistance in the industrial sector provided by the UNDP and other UN agencies.

UNITED NATIONS INSTITUTE FOR TRAINING AND RESEARCH (UNITAR) Established in 1965 by the General Assembly as an autonomous institution to train people, especially those from developing countries⌂, for assignment to the United Nations (UN)⌂ and all its specialized agencies or to their own government services, and to conduct research related to UN functions and objectives. It is financed by voluntary contributions from governments and non-governmental sources.

UNITED NATIONS NON-GOVERNMENTAL LIAISON SERVICE (NGLS) Established in 1975 and sponsored by several United Nations⌂ agencies, the NGLS co-operates with Non-Governmental Organizations (NGOs)⌂ to help advance work and projects on all aspects of development and promote a greater awareness of the need for a New International Economic Order (NIEO)⌂. Its field of work covers virtually all subjects of interest to the UN and its specialized agencies. It assists individual NGOs and networks in the design and implementation of their development education⌂ programmes, including supporting NGO regional and national meetings and conferences. It also services and links NGOs on the basis of their development information needs and interests, as well as facilitating links between NGO personnel and appropriate UN system officers. NGO involvement in thematic UN conferences is also promoted and co-ordinated, and informative background material and briefing documents from UN and other sources on NIEO-related issues are identified and disseminated. Specialized briefing programmes for NGOS on specific development issues are also organized.

UNITED NATIONS RELIEF AND WORKS AGENCY (UNRWA) Established by the General Assembly in 1949 to take over relief activities in the Middle East primarily concerning Palestinian refugees⌂. UNRWA promotes resettlement and helps channel resources and funds to facilitate technical co-operation in the countries in the region housing refugees. It caters to the needs of millions of registered refugees, helping to arrange the provision of basic shelter, food, improved health and welfare services, and educational facilities in Jordan, Lebanon, Syria and the occupied West Bank and Gaza Strip. It is financed from voluntary contributions.

UNITED NATIONS REVOLVING FUND FOR NATURAL RESOURCES EXPLORATION (UNRFNRE) One of the specialized funding programmes co-ordinated by the United Nations Development Programme (UNDP)⌂. The UNRFNRE was established in 1973 to meet the shortfall in high-risk exploration capital needed in the search for mineral resources and geothermal energy⌂. The Fund's mandate is to help developing countries⌂ to locate economically exploitable mineral⌂ deposits and geothermal reservoirs, assess the viability of developing these reserves, and promote investment towards future commercial exploitation. The UNRFNRE finances all exploration costs; governments receiving assistance commit themselves to make a small replenishment payment – but only if the project leads to commercial production within a specified period. The Fund derives its own resources from voluntary contributions by member governments of the UN, from co-financing contributions from donors for specific projects, and, eventually, from replenishment payments. To date, the bulk of the projects supported by the Fund have been devoted to the search for minerals and precious metals, with very little emphasis on the search for geothermal resources.

UNITED NATIONS SUDANO-SAHELIAN OFFICE (UNSO) One of the specialized funding programmes co-ordinated by the United Nations Development Programme (UNDP)⌂. The UNSO assists twenty-two countries in the Sudano-Sahelian region of Africa in their drought-⌂ and desertification⌂-control efforts. The majority of its operations, worth $100 million in 1988, centre on projects to reverse the effects of deforestation⌂, improve rangeland and water⌂ management, and encourage soil⌂ protection and sand-dune fixation. Support is also given to governments to achieve cohesion in planning and co-ordinating their anti-desertification and natural-resource⌂ conservation⌂ programmes.

UNITED NATIONS UNIVERSITY (UNU) Based in Tokyo, the University aims to promote research, advanced training and the dissemination of knowledge. It has no campus; the work is carried out through the University's support of individual researchers placed in a wide range of academic and research institutions around the world. The five major programme areas for research are peace and conflict resolution; the global economy; hunger⌂, poverty and environmental resources; human and social development; science and technology and the information society, within which the university operates nine programme areas. The UNU produces academic publications and journals, a regular series of reports and newsletters and a collection of audio-visual materials.

UNITED NATIONS VOLUNTEERS (UNV) One of the specialized programmes co-ordinated by the United Nations Development Programme (UNDP)⌂. The UNV programme provides a wide range of specialists in such areas as communications, engineering, community development, food production and processing, education⌂, environmental conservation⌂, management, and science and technology. Participants in the programme work with governments to train local personnel and help them to achieve self-reliance.

UNIVERSAL POSTAL UNION (UPU) Originally founded in 1874 as the General Postal Union, the organization became a specialized United Nations⌂ agency in 1948, changing its name in the process. Based in Switzerland, the UPU, with 169 members, aims to ensure the integrated organization and development of postal services around the world. It promotes the development of communication between peoples through attempting to ensure the efficient operation of postal services, contributions to international collaboration in the cultural, social and economic fields, and through participation in the provision of technical help as requested by member countries.

URANIUM The heaviest naturally occurring element. Uranium, a silvery white metal, is a radioactive⌂ element with several radioactive isotopes. It is now primarily used as a fuel in nuclear power⌂ plants: consumption doubled between 1980 and 1986. In nature it is often thinly dispersed and costly to extract, being fairly widely distributed throughout the world in vein deposits and in limestone. Pitchblende (uranite) is the major source. Natural uranium contains a mixture of three isotopes: U-238 (99 per cent), U-235 (0.7 per cent) and U-234 (0.006 per cent). Contemporary commercial nuclear power reactors use the isotope U-235, although 'fast-breeder'⌂ reactors are capable of using U-238, which is more abundant.

Nuclear fuel is frequently 'enriched' by increasing the concentration of U-235, which is much more radioactive, with a half-life of 713 million years, whereas the half-life of U-238 is well over 4 billion years. Like the plutonium⌂ produced in nuclear fuel reprocessing⌂ plants, enriched uranium tends to go missing. In the United States alone 4,500 kg of enriched uranium produced since 1950 cannot be accounted for.

Uranium deposits have been found in many countries and it is considered to be one of the three most strategically important minerals⌂, along with coal⌂ and oil⌂. There are estimated to be 1.5 million tonnes of economically recoverable reserves spread throughout the world. The biggest known reserves of rich uranium ores are in Australia (463,000 tonnes). Canada, Niger, South Africa, and the United States also house sizeable deposits. Uranium tends to occur in land⌂ that is unsuitable for agricultural use – land which is frequently set aside for minority groups. Consequently, native Indian tribes control more than half of the privately owned uranium in the United States, ranking fifth in the world as uranium owners. Similarly, most of the uranium in Australia is found on land which has been set aside as reserves for the Aborigines⌂. Over 70 per cent of the country's known deposits are in areas which are of importance to the Aboriginal

culture – which is under serious threat of disappearing as a result of mining and mineral exploration.

High-concentration deposits are rare; typically ores contain only 0.2 per cent uranium oxide, so extraction is an expensive process – often prohibitively so. Sweden has significant uranium reserves but has found it cheaper to import fuel for its nuclear reactors. Uranium in South Africa has been extracted and marketed at a competitive price, but only because it was a by-product of gold⌂-mining.

Global production is in the order of 37,500 tonnes annually, with the United States the largest producer (14,000 tonnes) followed by Canada (6,700) and South Africa (4,900), although figures for China and the Soviet Union are not available for comparison.

URBANIZATION Over the last decades, as rural migration has inceased and population⌂ levels have risen, urban centres have expanded dramatically. Between 1950 and 1980 the world's urban population almost tripled – from 701 million to 1,983 million. Globally, urban populations are increasing at an average of 2.4 per cent each year, over twice the level recorded for rural populations. However, increasing urbanization rates are much more common in the Third World⌂. Between 1970 and 1980 there was an influx of 320 million people into urban areas in developing countries⌂; it is estimated that this total will swell to 1 billion by the year 2000. In the developed world, urban populations grew by 2 per cent annually from the 1920s to the 1960s: it is predicted that they will grow by only 0.8 per cent during the 1990s. In the less-developed regions, urban numbers have been growing by over 3 per cent each year, reaching as high as 4.8 per cent in Africa, rates of growth two to three times faster than those experienced by industrialized countries in the past.

In 1920, only 360 million of the world's population lived in urban areas; by 2000 around 3 billion people will reside in cities or large towns. The situation is rapidly becoming critical in Third World countries where urban population amounted to 100 million in 1920, whereas it is now well in excess of 1 billion. Another 750 million people will inhabit Third World cities by the year 2000, by which time almost half the world's total population will live in large urbanized conurbations.

Urbanization in the industrialized world took many decades, allowing economic, social and political systems to evolve gradually to cope with the problems of transformation. In developing nations, the process is far more rapid and is occurring against a backdrop of high population growth and low incomes.

In 1950, only one city in the developing world, Buenos Aires, had a population of over 4 million. By 1960 there were eight cities that had reached or exceeded that size, compared with ten in the industrialized world. By 1980 there were twenty-two such cities in the developing world, six more than in the developed countries. It is forecast that by the year 2000 there will be sixty-one cities with a population of 4 million people or more spread throughout the developing world, compared to only twenty-five forecast for the industrialized countries.

Housing, water⇨, sanitation, power and other services need to be provided for all city dwellers. Food production has to be organized, together with transport and distribution networks. Employment and health requirements also need to be catered for. The governments of many developing countries do not have the resources to accomplish this, either now or in the future. In the past, illegal and dilapidated settlements sprang up within city limits, resulting in slums, shanty towns and ghettoes in which people have little or no access to drinking water or proper sanitation facilities and through which diseases such as typhoid and cholera⇨ can rapidly spread.

Over the coming decades, authorities will be faced with the problem of satisfying the ever-growing needs of urban dwellers and balancing this with the protection of natural resources⇨ and the local environment. Roads, factories, houses, shops will all be required in the new and expanding cities which, in turn, will require vast tracts of land⇨ and encourage urban sprawl, a process sometimes known as 'paving over'. In the United States 31 sq km of farmland are lost each year, covered over by concrete or asphalt. A further 11,000 sq km of prime agricultural land is lost each year to urban sprawl. This problem already exists in the Third World: in Egypt between 1955 and 1975, 4,000 sq km of fertile land in the Nile Valley were lost to urban expansion, more than was brought into new agricultural production. The situation will worsen as cities in the developing countries expand in number and size. Even if rural migration in developing countries is reversed, as is happening in several developed European countries, natural increases in the number of city dwellers, which have accounted for 60 per cent of urban population growth in recent decades, will ensure that urbanization remains a major problem well into the next century.

URUGUAY ROUND   The eighth round of the four-year-long trade negotiations under the General Agreement on Tariffs and Trade (GATT)⇨ which commenced in October 1986. This particular round of negotiations dealt with agricultural matters in far greater depth than any of the previous rounds. In an attempt to achieve greater freedom in the trading of agricultural products, the reduction of import barriers and schemes to discourage the use of government subsidies – specifically those in the European Community⇨ and the United States – to support agricultural products or production formed a major part of the talks. The developed countries spend $290 billion annually on measures to protect their farmers and guarantee artificially high prices for certain products.

Amongst the fourteen separate areas of negotiation, special attention was devoted to tropical agricultural products – in particular, how the plight of developing countries⇨ dependent on a single agricultural commodity⇨ for their export earnings could be improved. The talks also dealt with service-industry issues, trade-related investment issues and trade-related intellectual property rights.

A framework for liberalized trade in services, lower tariffs and access barriers for Third World⇨ exports, more efficient procedures for resolving trade disputes and closer monitoring of individual country trade policies were the avowed goals, but the United States and the European Community countries could not reach agreement over cuts in subsidies paid to European farmers and the talks ended in 1990 – the supposed deadline for completion – in deadlock.

In addition to the failure to reach accord over reductions in farming subsidies, the increasingly protectionist policies being practised by the USA and the EC looked set to threaten the future of GATT. Many Third World nations displayed annoyance that they were receiving no reward for implementing substantial-trade liberalization measures – in the face of often severe hardship – as part of major structural adjustment programmes. Mexico's maximum import tariffs, for example, had been reduced by 80 per cent during the time span of the Uruguay Round. Furthermore, developing countries' manufactured exports were facing 50 per cent more barriers in the markets of industrialized countries than in the mid-1980s. Anti-dumping legislation, agreed in previous rounds, were being used as a major tool of trade policy to the detriment of the Third World nations. Almost half of the 277 anti-dumping and anti-subsidy investigations undertaken (mostly by the USA and the EC) between mid-1987 and mid-1989 were targeted on developing countries, even though these nations accounted for little more than 13 per cent of world trade in manufactured goods.

V

VACCINE-PREVENTABLE DISEASES
Every year at least 46 million infants around the world are not fully immunized against the six major childhood killer diseases – polio⌂, tetanus, measles⌂, diphtheria, pertussis (whooping cough) and tuberculosis⌂. About 2.8 million children die as a result of contracting these diseases and another 3 million are disabled, mostly in the developing world. There are vaccines to immunize children against all these diseases but it is lack of money to purchase, store and administer them that is the problem. According to the World Health Organization (WHO)⌂, thanks to the Expanded Programme on Immunization (EPI)⌂, the level of immunization of the developing world's children has risen from 5 per cent to over 60 per cent during the last decade. It costs only $10 to vaccinate a child against all six diseases. The WHO reports that for less than $1 billion, or the cost of twenty modern military planes, the world could control all these illnesses. In total, 1.8 billion people are infected with the vaccine-preventable diseases, which have varying degrees of severity. For example, although well over 1 billion people are considered to be infected with tuberculosis, only 20 million are ill.

VIENNA CONVENTION FOR THE PROTECTION OF THE OZONE LAYER
A landmark conference, held in 1985, which produced an international agreement to combat an emerging problem, the continuing destruction of the ozone⌂ layer, and recognized the atmosphere as a natural resource⌂. The Convention, adopted by twenty-one states and the European Community⌂, called for the freezing of production of chlorofluorocarbon (CFC)⌂ gases for use in aerosols and further measures to reduce all emissions of CFC gases. Despite attempting to put legal instruments in place to protect the atmosphere, mostly concerning a cutback in CFC production and use, the Convention failed to do so. Third World⌂ producers such as China and India refused to sign the Convention until guarantees over compensation had been given. The Convention led to the Montreal Protocol of 1987⌂ at which the participating governments agreed to freeze CFC use at 1986 levels with a 20 per cent cut by 1994, the eventual aim being a 50 per cent reduction by the end of the century. Delegates agreed that sales of CFCs may increase in developing countries⌂, many of which received a ten-year exemption from the regulations. Nations with low rates of consumption were also allowed to increase CFC use by 0.3 kg *per capita*.

VIETNAM WAR
A war between communist North Vietnam and US-supported South Vietnam which lasted from 1954 to 1975. From 1954 onward, guerrilla warfare was waged by the northern-based Viet Cong, the North receiving direct support from China and taking full advantage of weak governments in Laos and Cambodia to move arms and soldiers into South Vietnam from the west. From 1961 onward the South was aided directly in the war by the United States, which had been providing South Vietnam's government with large-scale aid⌂ since 1954 in an effort to check the southward spread of Communism.

Following the Tonkin Gulf incident in 1964, in which two US ships were reputedly attacked by North Vietnamese forces, the United States intervened in the war and by 1965 US aircraft were regularly bombing North Vietnam. The South Vietnamese troops were also supported by contingents from Australia, New Zealand, the Philippines, South Korea and Thailand. However, by 1968, when there were an estimated 500,000 Americans engaged in the war, US President Johnson was offering to negotiate a peace settlement with the North. Further peace efforts in 1969 scaled down the conflict, which flared into life again following the US-South Vietnamese invasion of Cambodia the following year.

A massive Communist offensive in 1972, coupled with domestic opposition to continuing involvement, led to the USA making determined efforts to obtain a peaceful solution to the war. These initiatives eventually led to the Paris Agreement in 1973 and US troop withdrawal.

The North finally overran the South in 1975, all US help having been withdrawn. Following the Communist victory, North and South united in 1976. The war marked the first ever military defeat for the United States. Of the 2.5 million US troops dispatched to fight in the war, over 350,000 were either killed or injured and, in total, some 1.5 million people lost their lives during the conflict.

VILLACH CONFERENCE The second of a series of five-year reviews of climate change and the greenhouse effect⌂, held in Villach, Austria, in 1985. Experts from the United Nations Environment Progamme (UNEP)⌂, the World Meteorological Organization (WMO)⌂ and the International Council of Scientific Unions (ICSU)⌂, met to discuss all aspects related to climate change, including the build up of greenhouse gases⌂, sea-level rise⌂ and coastal impacts. The Conference made several recommendations for future action. Delegates concluded that the earth was likely to experience a warming of 1.5–4.5°C over the next fifty years, roughly equivalent to that which has occurred since the last Ice Age. Sea-level rises of up to 140 cm were forecast, together with disruptions in global patterns of rainfall and agricultural production⌂. The Conference's major recommendation was that more research be carried out to further greater understanding of the phenomena involved and so discover the best corrective methods.

VOLUNTARY SERVICE OVERSEAS (VSO) A British organization founded in 1958 by Alexander Dickson with the aim of sending skilled volunteers, all of whom must be British citizens, to work as teachers and doctors, on agricultural projects, industrial projects, or business schemes, as and when requested by overseas governments in the developing⌂ world. A VSO volunteer is employed for a two-year period by a government, which is responsible for providing accommodation and a living allowance, usually based on local rates of pay. The VSO administration provides relocation costs, a grant, full briefing and also ensures that health insurance and employment regulations in the UK are all in order. Similar organizations exist in most of the donor countries. Volunteers work in partnership with local communities, passing on skills, experience and knowledge, sometimes acquired through a working lifetime in health, agriculture⌂, technical services, education⌂ and commerce. Every post has a basic training element designed to allow the local community to achieve self-reliance.

W

WACKERSDORF The Federal Republic of Germany's first major plant for reprocessing⌂ nuclear fuel, sited near the Austrian border. The German nuclear industry became fully operational in 1969 and in order to comply with national legislation, plans were proposed to build a huge nuclear fuel processing plant, complete with vitrification and disposal facilities to deal with all the Republic's spent fuel. The construction of the original facility at Gorleben near the border with the German Democratic Republic began in 1969 with the aim of storing the vitrified waste in nearby salt mines. Work on the plant was halted in 1974 when the consortium formed to build it withdrew for economic reasons. In 1986, construction of the Wackersdorf complex began, accompanied by widespread opposition. The plant was designed to handle 350 tonnes of spent fuel per year, the waste products travelling 500 km north by road or rail to the salt caverns originally planned as disposal sites. In 1989, after a great deal of concerted pressure from environmental groups and the general public in Germany – and following the Three Mile Island⌂ accident in the United States and the Chernobyl⌂ disaster in the Soviet Union – the Wackersdorf project was finally cancelled. Germany has arranged to send around 4,000 tonnes (two nuclear flasks) of spent fuel weekly, for the next fifteen years, for reprocessing in Britain.

WALDHEIM, KURT (1918–) Austrian diplomat. He served in France and Canada before becoming Austria's representative to the United Nations⌂ (1964–68). Following a stint as Austrian Foreign Minister from 1968–70, he returned as the nation's representative at the UN in 1970 and succeeded U. Thant⌂ as Secretary-General in 1971, serving until 1981. He was elected President of Austria in 1986 but his tenure was troubled by repeated accusations of possible involvement in war crimes during his service as an army officer in the Second World War.

WAR The environmental impact of war, as opposed to the social and economic aspects, is beginning to be widely recognized and evaluated. In addition to destroying human life, most wars decimate agricultural productivity. The Second World War caused the loss of 38 per cent of the agricultural productivity in ten European nations. It took these nations at least five years to recover to previous base levels.

As weapons develop, so does the impact on the ecosystem⌂. The widespread use of defoliants, napalm and herbicides⌂ in the Vietnam War⌂ destroyed 1,500 sq km of mangrove⌂ forest⌂ and inflicted damage of varying degrees of severity on a further 15,000 sq km of land⌂.

Natural resources⌂, notably minerals⌂, are being depleted at increasing rates to fuel military development. Between 3 and 12 per cent of important minerals such as aluminium⌂, copper⌂, lead⌂, iron⌂ ore, nickel, platinum⌂, silver, tin⌂ and tungsten are used for military purposes, and the military account for 5 per cent of the global consumption of oil.

Armed conflicts also have a lasting, potentially disastrous impact on the health of both humans and local environments because of the so-called remnants of war. Since the end of the Second World War in 1945, 14.9 million landmines and 73 million bombs, shells and hand grenades have been recovered in Poland alone. During the wars in Indochina 2 million bombs and 23 million artillery shells were thought not to have exploded. In 1973, during the aftermath of the conflict between Egypt and Israel, Egypt discovered 8,500 non-exploded devices and cleared 700,000 landmines. In guerrilla conflicts in parts of Southern Africa, ordinary citizens are regularly being killed or maimed by landmines.

In the Kiskunsag National Park in Hungary, parts of which have been used by the Soviet army as a firing range, military activities caused fifty-two forest fires between 1975 and 1990. After forty-five years in the country the Soviets withdrew, but disposed of vast amounts of surplus military equipment and waste in areas which are subject to stringent environmental protection laws. A similar situation has been identified in Czechoslovakia, where up to 8,000 sq km of land have been polluted or despoiled by the Soviet army.

WAR ON WANT An independent international development agency, based in Britain, whose objectives include the relief of poverty, distress and suffering in any part of the world, whether they arise from natural disasters or other causes. War On Want funds long-term development projects in over thirty countries worldwide, and is particularly concerned with the provision of medical care, advice and treatment for the poor. It also promotes and conducts research into the causes and ways of relieving poverty, disease, sickness and disability, both mental and physical, as well as playing a role in informing the public at large of the nature and causes of poverty and the role of aid⌂ and development in the Third World⌂. During the late 1980s the organization was racked by scandal and the threat of bankruptcy, during which time the effectiveness of its work was severely reduced and its long-term future jeopardized.

WARD, BARBARA, BARONESS JACKSON (1914–81) British economist and conservationist. She wrote several books on ecology and political economy, including *Spaceship Earth* (1966) and, together with the US bacteriologist René Dubos, *Only One Earth* (1972). She became President of the International Institute for Environment and Development⌂ in 1973.

WARFARIN Produced by the Wisconsin Alumnus Research Foundation, from where it derives its name, warfarin was the first anticoagulant rodenticide. Initially introduced in 1950, it works by lowering the prothrombin level of the blood, disrupting the biochemical conversion of Vitamin K, an essential factor in the blood-clotting process. In addition, warfarin renders blood capillaries porous. Rats and mice eating the poison literally bleed to death internally. Unfortunately, within five years of its introduction many rat populations had developed immunity to its effects. So-called 'second-generation' rodenticides proved lethal to warfarin-resistant rats until 1990, when they were found to have developed resistance to these compounds as well.

Warfarin works cumulatively and is dangerous to all warm-blooded animals. The effects of accidental poisoning in humans can be counteracted by administering Vitamin K. Due to its effectiveness in preventing blood-clotting, warfarin is used in the treatment of coronary or venous thrombosis to reduce the risk of embolism and so prevent strokes or blocking of arteries. It is given orally or by injection; the toxic effects are manifest in local bleeding, usually from the gums and other mucous membranes.

WARSAW PACT   Established in 1955 between the Soviet Union and the Communist states in Eastern Europe⌂ after the Federal Republic of Germany had been admitted to the North Atlantic Treaty Organization (NATO)⌂. The Pact was basically a mutually protective military treaty between the Soviet Union, Bulgaria, Czechoslovakia, East Germany, Hungary, Poland and Romania. Albania was a member until 1968. The Pact was officially based in Moscow, headed by the Soviet army's highest ranking officer.

Following the fall of the Communist governments in many Central and Eastern European states in the late 1980s and the reunification of Germany, the Warsaw Pact became more or less defunct. At a Conference on Security and Co-operation in Europe (CSCE) meeting in Paris in 1990, the Warsaw Pact nations that were still in existence signed a declaration with the sixteen members of NATO to the effect that they no longer regarded each other as adversaries, thus signalling the virtual total collapse of the Pact.

WATER   Water, essential to life, is a renewable but limited resource. Some 97 per cent of the earth's water is salty and oceanic. Of the 3 per cent that is fresh, 77 per cent is locked up in icecaps and glaciers, 22 per cent is groundwater⌂, and the remainder is present in lakes, rivers and other watercourses. Much of the groundwater lies below 800 m depth and so is effectively beyond exploitation. Humans need fresh water to live and the only accessible source is therefore the runoff from land⌂, some 40,000 cubic kilometres (cu km) annually, that occurs as a result of the natural water cycle. Much of this is lost in floods⌂, or held in swamps, lakes or soil⌂, and only 2,000 cu km are readily available – but this is more than enough to satisfy the needs of the world's population⌂.

Every human being requires about 80 litres of water daily to support a reasonable standard of living; 5 litres are needed for basic survival. Average consumption varies from 5.4 litres per day in Madagascar to 500 litres per day in the United States, but total water use is continually increasing. By the year 2000, global consumption of water will be ten times greater than it was in 1900.

Global water use is generally broken down into three categories: agriculture⌂ (73 per cent), industrial uses (21 per cent) and public use (6 per cent). Water-use patterns differ from region to region and through time. In the developed world, industry accounts for 40 per cent of water use, while in the developing countries⌂ the bulk of the water is used for irrigation⌂, a total that is relentlessly increasing. In 1900, around 400 cu km of water were used. This total had risen to 2,200 cu km by the 1980s and was climbing at 6 per cent a year. It is expected to rise to 2,500 cu km by 2015. Overall, water for irrigation is projected to rise twofold and that for industrial use over fourfold by the end of the century.

The availability of fresh water around the world shows a very marked geographic variation, from the 110,000 cu m per person in Canada to absolutely no internal renewable water resources in Bahrain. As the demand for water increases, the supply of fresh water is likely to become of increasing global importance and contention over the coming decades. As supplies become overexploited and scarcer, conflicts will arise. Of the 200 rivers shared by two or more countries, the waters from several – including the Ganges⌂, Euphrates and Zambezi – have already been the subject of heated international conflict. Currently, around 10 cu km of fresh water are being withdrawn every day, but by the year 2000 it is expected that a total of around 4,500 cu km will be being withdrawn annually. As well as the 2,500 cu km for irrigation, 1,300 cu km will be for industry and the remainder for municipal and domestic use.

WATERBORNE DISEASE Name given to a group of diseases transmitted to humans in a variety of ways by bacteria, insects and other organisms that live or breed in water⟡. These diseases are usually classified as those caused through ingestion of contaminated water or food (waterborne): those caused by lack of or unhygienic washing (water-washed), and those transmitted by insects which breed or live near water (water-vector-borne). Most of the world's deadliest diseases are in these categories, including cholera⟡, amoebic dysentery, typhoid, hepatitis⟡, all diarrhoeal diseases and polio⟡ (waterborne); infections of the intestinal tract, trachoma, scabies, yaws and leprosy⟡ (water-washed); and malaria⟡, onchocerciasis, yellow fever⟡, schistosomiasis⟡, dengue⟡ and elephantiasis (water vector-borne). The World Health Organization (WHO)⟡ estimates that 80 per cent of all illnesses in the developing world stem from lack of safe water and adequate sanitation. These water-linked diseases kill at least 25,000 people each day, yet they can all be significantly reduced through the provision of safe water and adequate sanitation.

WATER CYCLE The natural circulation of water⟡ through the biosphere⟡. Water is continually lost from the earth's surface to the atmosphere either by evaporation from the surface of lakes, rivers and oceans or through the transpiration of plants and vegetation. Around 500,000 cubic kilometres (cu km) of sea water are drawn into the air each year by solar radiation. In the atmosphere, water forms clouds which, under the appropriate conditions, condense to deposit moisture and precipitation on land⟡ and sea in several forms, including rain, snow and fog.

Some 40,000 cu km of fresh water fall on the world's surface each year as part of the endless cycle of evaporation, precipitation and runoff. The water cycle is heavily dependent on the 97 per cent of the world's water found in the oceans, the salt content of which is lost during the evaporation process. The amount of water in the cycle is constant.

The cycle is dynamic, but water is replaced in the differing stages of the cycle at varying rates of replenishment. River water is replenished on average every nineteen days, whereas atmospheric moisture is replaced every twelve days. Deep groundwater⟡, which in many parts of the world is being unsustainably exploited to support agriculture⟡, requires hundreds of years to be renewed.

WAVE POWER The vertical motion and energy⟡ of sea waves can be harnessed to generate electricity. Many devices have been tested, but only one power-producing facility has been built and is operating commercially – in Norway. Several more devices are planned or are at the prototype stage, although arguments continue about the viability and cost effectiveness of wave power. The plant in Norway, an oscillating water-column (OWC) design, cost in excess of $2 million to build, has a peak rating of 500 kilowatts and initially began generating electricity at an estimated 5.4 cents (US) per unit. The OWC is basically a chimney: water⟡ rising in the chimney pushes air through a turbine and also drives the turbine when the air is sucked back down as the water flows out. The design is also said to be suitable for larger units. It is likely that wave power stations will first be used to supply the needs of coastal towns and villages, at least until much larger units are developed and have proven viability. The world's largest wave power project is being planned for India, designed by the United Kingdom's National Engineering Laboratory using technology developed in the early 1980s which the UK government decided was not commercially viable. The completed power plant will form part of a large harbour wall in a major port complex to be constructed at Encore, north of Madras, and when fully operational will generate about 5 megawatts of power.

The UK government decision in 1982 to abandon wave power on the grounds that it was not commercially viable was controversial, being based on sets of dubious data which, according to some, had been deliberately doctored to make the wave power option look unattractive. In 1987 Britain's first wave power station was announced, to be built on the island of Islay in the Inner Hebrides. Following a change in policy, the UK government committed itself to providing the bulk of the $450,000 construction costs. In 1990, after continuing controversy over the original decision to forgo wave power research, a full investigation and re-examination of the suspect figures was called for.

WEST AFRICAN ECONOMIC COMMUNITY (ECOWAS) An organization founded in 1975 to promote economic development amongst member nations. The sixteen-member community

includes members of the Francophone Commun-
auté Economique de l'Afrique de l'Ouest, founded
in 1972, the East and Southern Africa Preferential
Trade Area, and the Mano River Union, founded in
1973. ECOWAS comprises Benin, Burkina Faso,
Cape Verde, Ivory Coast, The Gambia, Ghana,
Guinea, Guinea-Bissau, Liberia, Mali, Mauritania,
Niger, Nigeria, Senegal, Sierra Leone and Togo.

WETLANDS   Wetlands, such as marshes,
swamps, bogs and fens, are amongst the most fertile
and productive ecosystems⌀ in the world. They can
produce up to eight times as much plant matter as an
average wheat⌀ field. Wetlands cover 6 per cent of
the earth's land⌀ surface and are found in all
countries and under all climates. They are important
breeding grounds for fisheries⌀, play a vital role in
maintaining the global water cycle⌀ and act as a
filtering system to clean up polluted water and
remove silt, encouraging plant growth and so
further improving water⌀ quality. They are particularly important in protecting coastlines⌀ from erosion and acting as barriers against storm surges,
protecting inland areas from flooding⌀, and providing people with a wide range of staple food
plants, fertile grazing lands and fuel.

Throughout the world, pressure for agricultural
land has led to increasing amounts of wetlands being
drained for cultivation. In the United States, for
example, 185,000 hectares of wetlands were lost
each year between 1950 and 1970. All dams⌀ and
barrages built for irrigation⌀ and hydroelectric
power⌀ generation reduce downstream wetlands,
destroying wetland fisheries and breeding-grounds
and removing coastal protection.

Wetlands are the only ecosystem to be protected
by a specific international convention. Under the
Ramsar Convention⌀ over 200,000 sq km of wetlands have been preserved but this represents only a
small proportion of the global total: some 8.5 million sq km.

WHALING   The hunting⌀ of whales began as
long ago as the fifteenth century and became widespread in both hemispheres during the eighteenth
and nineteenth centuries. Initially, the first species to
attract the attention of hunters was the right whale,
so-called because it was slow, easy to catch and a
good source of oil and whalebone. Victorian desires
for stiff whalebone corsets probably drove the right

whale close to extinction⌀, although population
estimates before to the 1900s were not reliable.

By the mid-1800s, sperm whales were being
hunted for their oil, which was used in lamps, and
the bowhead, grey and right whales for both oil and
whalebone, for use in corsets. In 1872, a cannon-
fired harpoon was developed which allowed the
faster species such as blue, fin and sei to be caught,
mainly for use as pet food⌀. Each species was
hunted until it became commercially unviable for a
nation to carry on, at which time attention was
switched to another species. Since 1900 whaling has
been concentrated in the Antarctic⌀ Ocean, where
the whales congregate in summer to feed.

Traditionally, whales were hunted offshore and
their carcasses – sources of meat, fats, oils and other
chemicals – taken to land for processing. The advent
of large 'factory' ships, introduced by the Soviet
Union and Japan following conscious government
decisions to fully exploit marine food resources,
enabled whales to be caught and processed at sea,
increasing whaling efficiency to a point where many
of the world's great whales became threatened with
extinction. Today several species – notably the blue,
humpback, bowhead and northern right whale, are
officially recognized as being 'endangered'⌀, as are
all species of sea cow and many species of seals and
dolphins.

Most common whaling is – or has been – concentrated on Atlantic whales. Figures for actual numbers of animals are a matter of conjecture and fierce
debate, mostly between whaling and non-whaling
nations. Local populations in various parts of the
oceans have become 'extinct', but individuals from
the same species surface elsewhere. In 1989 the
International Whaling Commission (IWC)⌀ estimated that there were 120,000 fin whales, 14,000
blue whales and 10,000 humpback whales still surviving. This reflected a drop in population numbers
over that estimated as existing before exploitation of
79 per cent, 94 per cent and 96 per cent respectively.
The IWC set a moratorium on whaling effective from
April 1988 to which all countries adhered, although
Iceland, Japan and Norway began to operate 'scientific' whaling programmes, ostensibly killing whales
to gather data on whale populations for the review
of the moratorium due at the end of 1990, the result
of which was an extension of the ban for a further
year. Opponents claimed that this 'scientific whaling' was merely a front for normal whaling activities.
Whale meat from the 1988 'scientific' catch was
selling in Japanese stores in 1989 for $51 per kg. In
1990, 600 tonnes of meat from the 300 minke

whales killed by Japan's whaling fleet for 'scientific' reasons sold for $10.6 million wholesale.

During the 1989–90 season Norway pledged to cut its whale kill to five. Pressure from the United States and consumers in Europe, led by the Greenpeace⌂ environmental group, resulted in Iceland suspending its whaling activities for two years. The international boycott reportedly cost the Icelandic fishing⌂ industry around $50 million in lost export contracts. The Japanese, originally stating that they would kill 825 minke whales and 50 sperm whales, later decided to restrict their 'scientific hunting' to a mere 300 minke whales.

With the decline in whale populations, whalers have only recently turned their attention to minke whales; hunting of this species commenced only in 1972. By 1986, the IWC reported that minke numbers in the North Atlantic had fallen by 80 per cent through overhunting, so they were included in the moratorium. In 1990 Norway sought permission to take between 1,700 and 2,000 minke per year in the North Atlantic, while Japan asked for a quota of 5,000 in the South Pacific. The Soviet Union also asked to be allowed to kill seventy Minke whales a year for two years for 'scientific' purposes.

It is believed that Japan will ignore the moratorium on whaling if it is extended for a lengthy period, and will commence commercial whaling again. There is no way to bring any real financial or political pressure to bear to stop the Japanese doing what they want with regard to whaling. US pressure on Japan to cease their whaling activities, for example, has been limited to excluding Japanese fishing vessels from US waters.

The IWC – set up to promote sustainable whaling, not prevent whaling altogether – has been the focal point and main forum for the international anti-whaling protests, but it is in dire financial straits, with several member nations years behind in paying their dues. In 1989 it was announced that the IWC's scientific budget was to be reduced by 36 per cent and member nation fees increased by 50 per cent.

WHEAT A cereal grass belonging to the genus *Triticum*, originally native to western Asia but now widely cultivated throughout subtropical and temperate regions. Wheat was amongst the first plants cultivated by humans and about 820 million people around the world currently live on wheat-based diets. Many different varieties have been developed and wheat is by far the most important cereal in the developed world. Grain is ground to produce baking flour and to make pasta. The quality and

quantity of gluten in wheat is good, the gluten helps bread dough to stick together. Wheat is also important in the brewing industry as a commercial source of alcohol, dextrose, and malt.

The world now grows more wheat than any other cereal and it is the most important cereal in international trade. In 1989, 537 million tonnes were produced, 29 million tonnes more than the previous year. There were estimated to be 130 million tonnes of wheat stockpiled around the world at the end of 1984. The glut was largely due to government policies in the industrialized nations – especially the United States and European Community⌂, where governments subsidize their wheat farmers heavily. EC farmers are paid ($2.5 billion in 1983) to produce grain which goes into storage: the increased supplies are dumped cheaply around the world. Having regularly encouraged the production of vast surpluses, the US government now pays its farmers not to produce (about $4 billion in 1983). This effectively helps to limit supplies and keep prices high. The drought⌂ in North America during 1988 had a similar impact, cutting US wheat production by 30 per cent and raising the price by more than 60 per cent. As a result, the bill for export subsidies to dump the annual 30 million tonnes of excess EC grain on the world market was reduced by 60 per cent in both 1988 and 1989.

By the end of 1989, reduction in industrialized government subsidies had begun to have an effect on world stockpiles, with global wheat carry over totalling only 116 million tonnes. This amount contributed to a world cereal stockpile of only 306 million tonnes, only 17 per cent of estimated global consumption, equal to the minimum level considered necessary by the United Nations Food and Agriculture Organization (FAO)⌂ to ensure world food security.

Wheat has become a major – but contentious – component of food aid⌂. Although it assists in warding off starvation, it also has a twofold damaging effect, disrupting the markets and peasant agriculture⌂ in recipient countries as well as creating a taste for a cereal that is not commonly grown in those countries. Wheat does not grow well in the climate prevailing in most developing countries⌂ without the addition of vast quantities of expensive agricultural chemicals. As an indirect result of food aid programmes, the Third World⌂ is consuming more and more imported wheat and less and less of traditional, locally produced crops – wheat consumption in developing countries⌂ is growing six times faster than that of native crops. On the global trading scene, subsidies maintain wheat prices un-naturally high, particularly as trading agreements to stabilize prices have broken down. So, as cereal prices rise, the food import bill for developing countries also escalates. Global annual production is already in excess of 500 million tonnes; the Soviet Union (90 million tonnes), the United States (65 million tonnes) and China (59 million tonnes) are the largest producers.

WHITLAM, EDWARD GOUGH (1916–)

Australian statesman, Labor Party leader (1967–77) and Prime Minister (1972–5). He ended conscription, relaxed rules on non-white immigrants and attempted to lessen United States influence in Australia. His administration also tried to cultivate closer relationships with Asian countries. Following the opposition blocking of finance bills in the Australian Parliament in 1975, Whitlam refused to call a general election and was eventually dismissed from the post of Prime Minister by the British Governor-General of Australia, Sir John Kerr. The dismissal represented an unprecedented step in the history of Australian politics, sparking controversy and a constitutional crisis. Heavily defeated in the subsequent election by opposition leader Malcolm Fraser, Whitlam retired from politics in 1978 and became a lecturer on political science. In 1983 he was appointed Australian ambassador to UNESCO⊙ and has maintained close links with the organization ever since.

WILDLIFE TRADE

International trade in wildlife, or in products derived from wild plants and animals, nets over $5 billion annually, more than double the total realized in 1984. The populations of hundreds of species of wildlife – including elephants⊙, rhinoceros⊙, tigers, crocodiles, turtles, whales, walruses and seals – have been severely diminished as a result of hunting⊙ and poaching to satisfy global demand. One of the greatest reductions in population numbers has been seen amongst rhinos: the demand for rhinoceros horn has caused their numbers to fall by 70 per cent since 1970.

Trade is officially banned for 675 species of wildlife, or products derived from them, listed on Appendix I of the Convention on International Trade in Endangered Species of Wild Fauna and Flora (CITES)⊙. Trade is also strictly regulated for at least 30,000 species listed on Appendix II. Monitoring of global wildlife trade is carried out by the Trade Records Analysis of Flora and Fauna in Commerce (TRAFFIC), an international network co-ordinated by the Wildlife Trade Monitoring Unit at the World Conservation Union⊙ (IUCN) Conservation Monitoring Centre. Figures indicate that the illegal trade in endangered⊙ or protected species of wildlife is worth $1.5 billion annually. This represents about 30 per cent of the total trade in wildlife and wildlife products. Most of the illegal trade occurs in South-East Asia, Japan, Africa and Latin America, although major steps are being taken in Africa and Asia to try and curb illegal trading activities.

WIND POWER

Wind has been used as a source of power to turn sails and work grain mills and water pumps for centuries. The use of wind energy⊙ to generate electricity, however, is a comparatively recent phenomenon that is becoming more attractive as an environmentally sound and cost-effective option for power generation. The main advantages of wind power are that it is pollution-free, uses no fuel, and times of peak output, on cold, windy days, frequently coincide with peak demand. Wind turbines harness a completely free resource. Furthermore, they do not add to the thermal pollution⊙ of the earth, as do the burning of fossil fuels⊙ and nuclear plants, both of which must dispose of vast amounts of heat as well as contributing various pollutants to the atmosphere⊙. Disadvantages include the fact that wind power sources must be supplemented by other systems as production is unpredictable. Wind farms also require large areas of land⊙ and generally need to be sited on open ground away from cities, where the power is needed most. Nevertheless, wind power is one of the most rapidly developing of all the so-called alternative sources of energy. The average size for a wind generator at the beginning of the 1980s was 25 kilowatts, it is now nearer 200 kW, and the production cost per kilowatt has fallen from $3,000 to less than $1,000 over the same period. Some manufacturers have gone so far as to guarantee production outputs, forecasting that over a twenty-year period, in favourable wind conditions, electricity can be produced costing less than 3 cents (US) per kilowatt/hour. At a cost of 3–5 cents/kWh, wind energy electricity would be one of the lowest-cost options available, competitive with oil⊙ and well below the cost of nuclear power⊙.

In the United States, wind farms are already sound commercial investments. Annual gross electricity sales per hectare from wind farms reach around $15,000 – a return per hectare some fifteen times greater than that obtained by a corn⊙ farmer and 100 times greater than that of a cattle rancher. A World Bank⌂ study has identified thirteen developing countries⌂ (Chile, China, India, Jamaica, Jordan, Mauritania, Morocco, North Yemen, Pakistan, Romania, Sri Lanka, Syria and Tanzania) as having good potential for economically viable wind energy developments. Of all these countries, India has the most ambitious wind energy programme, with plans to install 5,000 megawatts of capacity by the year 2000.

WINDSCALE   *See* Sellafield.

WOMEN'S WORLD BANKING (WWB) Established in 1980 and based in New York, WWB is a global network of more than forty-five financial institutions run by women. Each institution is committed to expanding the role and participation of women in economic development and the business sector worldwide. WWB was founded to advance and encourage the direct involvement of women and their families in the local economy, and help provide the means to achieve this goal, with particular emphasis on women in the developing world.

WWB intends to help create an environment in which women have equal access to the benefits of the modern economy; build local support bases in individual countries which can respond to the needs of entrepreneurs; establish a global network of influential women in the fields of banking, finance and commerce; and provide women everywhere with the confidence and encouragement needed to enable them to work towards economic self-sufficiency. WWB helps provide access to credit, assists in the purchasing of equipment, and promotes the marketing of goods and services. In addition, an international venture capital fund will invest in women's businesses in certain developing countries⌂.

WOOD Over 3 billion cubic metres of wood are harvested around the world each year. Of this, 55 per cent is hard wood and 45 per cent soft. Half of all wood harvested is used as fuel⌂, with 80 per cent of the total being burnt in the developing world. Of the remainder, the bulk (40 per cent) is used for construction, with three-quarters of this being used in the industrialized world. Only around 10 per cent of all wood harvested is used to produce paper⌂, with paper production and consumption in the industrialized countries far exceeding that in the developing world.

In many developing countries⌂, work connected with wood and timber extraction remains a major source of employment. In rural areas, felling, sawing, carpentry and other forms of woodwork sustain the local population, accounting for nearly 20 per cent of rural manufacturing and repair in Kenya and Sierra Leone, 16 per cent in India and over 30 per cent in Malaysia. Harvesting wood for use as fuel is also a major source of livelihood for thousands of people throughout the developing world. Supplying fuelwood⌂ to Burkina Faso's capital city, Ouagadougou, provides work for 16,000 people, with city residents paying out a total of $5 million for the service.

WORLD BANK Popular name for the International Bank for Reconstruction and Development (IBRD)⌂ and its affiliated bodies. It is a multinational development agency, with 151 member nations, formed to provide loans and technical assistance to the poorer countries of the world. Established in 1945 following the Bretton Woods Conference⌂ essentially to stimulate the flow of global capital and to facilitate the expansion of free markets⌂, its headquarters are in Washington. The Bank borrows on the commercial market and lends on commercial terms. It invests around $16 billion in the Third World⌂, mostly in long-term ventures, working in close association with its sister organization the International Monetary Fund (IMF)⌂, which concentrates on providing finance for shorter term budgetary problems.

The International Development Association (IDA) is a specialized arm of the Bank which collects donations and grants from wealthier Bank members, then lends the funds to poorer nations (considered poor credit risks by the IBRD) on concessional terms. Over 90 per cent of IDA loans are made to the poorest nations, with a nominal interest rate of 0.75 per cent to cover administrative costs. Loans are provided with a ten-year grace period, with repayment over 40 years.

The International Finance Corporation (IFC)⌂, another specialized unit of the Bank, provides financial support for private-sector investment in the Third World that is otherwise devoid of government repayment guarantees, and helps mobilize investment from the private sector. The bulk of its activities are in manufacturing, mining and heavy industry.

The Bank generally lends around $20–25 billion annually, but, as many critics quickly point out, mostly to projects with a negative impact on the environment and on human populations. Nearly 1.5 million people have been forcibly displaced by IBRD-supported hydroelectric power⌂ schemes, and a 1990 study suggests that another 440,000 people will be forced from their land⌂ under similar bank projects within the first half of the 1990s. Similarly, critics also point to the fact that although the Bank purports to lend on purely economic criteria, when Salvador Allende's⌂ left-wing government was elected in 1971 in Chile, the Bank effectively halted all loans to the country. Funding was resumed after the 1973 coup which toppled Allende. The direction of the Bank's lending is also open to question. In 1989, over a quarter of the total budget was spent in support of export agriculture⌂ and development finance companies, whereas only 11 per cent went on education⌂, health improvement, nutrition, water⌂ supply and sanitation projects.

Following the Third World⌂ debt crisis in the 1980s, the Bank began to follow a major programme of 'Structural Adjustment Programmes' (SAP)⌂. By 1990, around one third of bank loans were devoted to SAPs. To qualify for these loans, Third World nations are forced to implement a package of unpleasant economic reforms aimed at making their exports more competitive. This usually means that social services in the country in question are drastically curtailed, public industries privatized and currencies devalued.

A growing awareness of the social and economic importance of maintaining or improving the environment, together with a spirited and effective campaign by Non-Governmental Organizations (NGOs)⌂ to publicize the Bank's failings, has led to an attempt by the bank to improve its image. A three-year pilot programme was initiated in 1990 through which Middle-Income Countries (MICs)⌂ will be able to benefit from $400 million per year set aside for environmental improvement projects. This is in addition to a 1989 directive that called for environmental impact considerations to be applied to all funding applications and project evaluations. Furthermore, in 1990, the Bank approved loans of $161 million to support national environmental agencies in Brazil, Madagascar and Poland. In 1987 its President, Barber Conable⌂, announced that the most important facet of the new environmental lending focus would be a global programme to protect tropical forests⌂. If promised targets are met, the Bank will be directing $800 million a year to forestry projects.

Despite this progress, however, the Bank con-tinues to finance environmentally destructive projects, and by 1990 its newly formed Environment Department, which had never been allowed a great deal of input into project preparation or the build up of loan portfolios, appeared to be be in disarray. The new forestry projects appeared to be mainly designed to promote commercial logging. As an example, a $23 million forestry and fishery project in Guinea submitted for approval to the Bank's board in early 1990 was labelled by the Worldwide Fund for Nature (WWF)⌂ as a 'deforestation⌂ scheme in disguise'. The loan would finance the construction of 60 km of roads in areas of pristine rainforest⌂, two-thirds of which was to be opened up for logging. Despite widespread criticism from outside the Bank, the loan was approved. In addition to the Guinea project, as part of the Bank-supported and much maligned Tropical Forestry Action Plan (TFAP)⌂ a project in Cameroon will open up 14,000 sq km of virgin forest to logging, and another $80-million scheme planned for the Ivory Coast involved extensive felling of timber and the possible resettlement⌂ of 200,000 people – in flagrant violation of the Bank's own guidelines.

WORLD CHARTER FOR NATURE A charter adopted by the United Nations⌂ General Assembly in 1982, following an initative from President Mobutu of Zaïre. The Charter, drafted with the assistance of the World Conservation Union⌂ (then called the International Union for the Conservation of Nature [IUCN]⌂, incorporates a set of universal principles to govern the conservation⌂ of wildlife and outlines the functions and responsibilities of national governments, corporate organizations and individuals which will need to be fulfilled if these principles are to be implemented successfully.

WORLD COMMISSION ON ENVIRONMENT AND DEVELOPMENT (WCED) An independent commission established by the United Nations⌂ General Assembly in 1983 with the brief to give recommendations for a 'global agenda for change'. Critical issues of the interlinked crisis of the environment and human development were to be examined and analysed. Guidelines for concrete and realistic remedial action were to be proposed. Specifically, a strategy was to be formulated which would allow 'sustainable development'⌂ to be achieved by the year 2000.

Chaired by Gro Harlem Brundtland⌂, the Prime

Minister of Norway, the Commission was composed of twenty-one prominent political figures and leaders in environment and development from around the world, selected by Mrs Brundtland and the Vice-Chairman, Dr Mansour Khalid, former Foreign Minister of the Sudan. The Commission, with a budget of $6 million, travelled to ten countries on five continents, hearing evidence from government officials, scientists, industrialists, Non-Governmental Organizations (NGOS)⌂, and the general public. Seventy-five studies, written by experts, also formed part of the Commission's research.

The Commission's report, *Our Common Future*, published in 1987, put forward twenty-two new legal principles to help achieve sustainable development, recommending that these principles be incorporated into national laws or charters that specify the rights and duties of citizens and state, and into a world convention on the sovereign rights and responsibilities of all nations.

Policy-makers were asked to be guided on their objectives by eight major interdependent goals:

- The revival of economic growth.
- The improvement of the quality of growth, ensuring environmental and social soundness and meeting needs for employment, food, energy⌂, water⌂ and sanitation.
- The conservation and enhancement of the natural-resource⌂ base.
- The stabilization of population⌂ levels.
- The reorientation of technology and improved risk management.
- The integration of the environment and economics in the decision-making process.
- The reformation of global economic relations.
- The strengthening of international co-operation.

The report concluded that huge increases in expenditure would be needed to finance the repair of environmental damage, control pollution⌂ and invest in sustainable development. In one of its few really innovative recommendations, the Commission suggested that this finance should be raised by levying taxes on the use of the global commons, including 'parking' charges for geostationary communications satellites and taxes on revenues from seabed mining and ocean fishing⌂.

WORLD CONSERVATION STRATEGY In

1976, three conservation bodies, the Worldwide Fund for Nature (WWF), the then International Union for the Conservation of Nature and Natural Resources (IUCN) (now the World Conservation Union⌂) and the United Nations Environment Programme (UNEP)⌂, agreed the need for a globally unified conservation⌂ strategy. As a result, more than 700 scientists and 450 government agencies from over 100 countries contributed information and advice which resulted in the publication of the World Conservation Strategy in 1980.

The Strategy's main aim was to emphasize the need for the conservation of natural resources⌂ so that they can regenerate naturally and supply humans with their basic needs indefinitely.

It outlined the main aims of conservation, which were defined as:

- maintenance of essential ecological processes;
- preservation of biodiversity⌂;
- careful and sustainable use of the earth's natural resources.

As well as detailing the specific and vital importance of natural resources and identifying conservation priorities, it also proposed methods through which the recommended goals could be achieved. Although intended to convince everyone and encourage individual action, the Strategy was targeted on governments, industry and the world's decision-makers, identified as those most able to put its proposals into action. By the end of the decade, an updated, more practical version of the Strategy was being prepared. Entitled *Caring for the World: A Strategy for Sustainability*, the document is scheduled for publication in 1991.

WORLD CONSERVATION UNION (IUCN)

Formerly known as the International Union for the Protection of Nature and Natural Resources (IUPN) which was founded in 1948. In 1956, as a condition of United States conservationists joining, the objectives of the IUPN were broadened and it was renamed the International Union for the Conservation of Nature and Natural Resources (IUCN). In 1988, it was decided that the organization's name should be changed to something simpler, but the acronym should be kept. The IUCN is an independent, international organization representing a wide-ranging network of governments, Non-Governmental Organizations (NGOS)⌂, scientists and other conservation⌂ experts working together to promote the protection and sustainable use of the world's natural resource⌂ base. A total of over 500 national governments, state authorities, private nature conservation organizations and international conservation groups from 114 countries are members.

The IUCN promotes scientifically based action towards the sustainable use and conservation of natural resources⌀, ensuring that the potential of renewable⌀ natural resources is maintained for the present and future benefit of the world's population. It also helps to ensure that areas of land⌀ or sea which do not have special protection (the vast majority) are managed so that natural resources are conserved and the many species and varieties of wildlife can persist in adequate numbers, thereby maintaining the earth's genetic pool. Programmes aim to protect areas of land, and of fresh and marine waters, especially those which contain representative or exceptional communities of plants and animals. The IUCN also plays an important role at the forefront of devising special measures to prevent species of fauna and flora from becoming endangered⌀ or extinct⌀.

As part of its activities, the IUCN monitors the status of ecosystems⌀ and species of wildlife throughout the world. It plans conservation action at both strategic and programme level through schemes promoting conservation for sustainable development⌀. In addition to co-ordinating action by governments, intergovernmental bodies and NGOs, it provides the assistance and expert advice necessary for such action to be achieved. Its six specialized commissions contain over 3,000 experts working on endangered species⌀, protected areas⌀, ecology, environmental planning, environmental policy, law and administration, and environmental education⌀. It is funded through contributions from the Worldwide Fund for Nature (WWF)⌀, grants from governments, membership fees, donations from United Nations⌀ agencies and other independent foundations.

WORLD COUNCIL OF CHURCHES A grouping of over 200 Protestant and Orthodox Churches. A product of the ecumenical movement, it was founded at Amsterdam in 1948 and it is based in Geneva. Membership includes almost all the Christian Churches, Anglican and other Protestant denominations, Orthodox and Old Catholics, with increasing support from the Roman Catholic Church which, although never a full member, has always sent observers to Council meetings and co-operated more or less fully with it. The Council's

policy of providing support for liberation movements, some with terrorist associations, has been controversial.

WORLD COURT *See* International Court of Justice.

WORLD FERTILITY SURVEY (WFS) The WFS is an international population research programme aimed at helping countries, particularly those in the developing world, to conduct surveys to improve knowledge of human fertility behaviour, on both the national and international front. The WFS began in 1972, in response to the announcement by the United Nations⌀ that 1974 would be celebrated as World Population Year, and has since completed its work. Some 350,000 women from nineteen developed and forty-one developing countries⌀ were questioned about their childbearing practices, beliefs and desires. The Survey – undertaken by the International Statistical Institute, with the collaboration of several of the specialized United Nations agencies and in co-operation with the International Union for the Scientific Study of Population – concluded that in several countries, almost exclusively from the developing world, including Colombia, Indonesia, Peru, South Korea and Sri Lanka, the prevention of unwanted births would reduce fertility by one-quarter to one-third.

WORLD FOOD CONFERENCE The world food crisis prevailing in the early 1970s caused the world's governments to hold a World Food Conference in 1974, in order to discuss ways to avoid hunger⌀ and safeguard global food⌀ supplies. As a result of the Conference, the United Nations Food and Agriculture Organization (FAO)⌀ set up its Global Information and Early Warning System (GIEWS)⌀ in 1975. The Conference supported an FAO proposal that 18 per cent of the global annual consumption of cereals should be held in reserves to ensure food security; this resulted in the formation of an International Emergency Food Reserve (IEFR). The target of holding 500,000 tonnes of emergency food grain reserves was first met in 1981.

WORLD FOOD COUNCIL (WFC) The Council was established by the United Nations⌀ following a recommendation from the World Food Conference⌀ (1974) and is composed of thirty-six ministerial-level members from differing nations. Based in Rome, the Council works towards co-

ordinating the efforts of the global community on food matters, including those of governments and international agencies. It aims to keep under review major problems and policy issues affecting the world food⌀ situation, stimulate food production, improve food trade, and to put into action the recommendations of the World Food Conference.

WORLD FOOD PROGRAMME (WFP)

Established in 1963, the WFP attempts to stimulate and advance economic and social development through the provision of food aid⌀. Based in Rome and operating as a specialized programme of the Food and Agriculture Organization (FAO)⌀, to which governments can pledge food or money, the WFP concentrates its aid⌀ on those most in need. The majority of WFP projects are designed to increase food and agricultural production and promote rural development, particularly in low-income, food-deficit⌀ countries. A significant part of WFP resources and activities are devoted to improving nutrition in the most vulnerable groups – the poor, schoolchildren, and pregnant or nursing mothers.

The programme supplies food aid at the request of governments, one of the major governing criteria for all WFP projects being that the recipient country can continue them after the aid has ceased. The Programme is administered by the FAO Committee on Food Aid Policies and Programmes⌀, which consists of thirty representatives from donor and recipient governments. Since its inception the WFP has supplied well in excess of $7.5 billion worth of food to over 250 million people, assistance primarily being given to agricultural and rural development projects. Typically, labourers are given food rations for their work. Any money saved by using food aid goes to fund other development projects. The WFP also provides food aid to areas where the food supply has been terminated or destroyed by war⌀, other human activities or natural disasters. In large-scale emergencies, the WFP also plays a co-ordinating role to ensure a regular and efficient supply of food. Contributions of commodities⌀ and money from more than 131 countries are channelled through the WFP, which also buys, ships and monitors food aid on behalf of bilateral donors. The WFP now handles 25 per cent of global food aid – some 3 million tonnes in 1989 – and has become a major funder of agricultural and rural development projects in over 100 countries.

WORLD HEALTH ORGANIZATION (WHO)

The first specialized agency of the United Nations⌀ to be established; the WHO constitution came into being in 1948. The WHO works to help control disease and improve general standards of health⌀ and nutrition through international co-operation, with the overall goal of helping to facilitate 'the attainment by all peoples of the highest possible level of health'. Based in Geneva and comprising 166 member states, the WHO supports programmes to eradicate diseases, undertakes, co-ordinates and finances epidemiological research, trains health workers, strengthens national health services, and stipulates international health guidelines and regulations. It also provides aid⌀ in emergencies and disasters. It sponsors research through existing national organizations, universities and health services and promotes improved standards of teaching and training in the health and medical professions.

Most of its work focuses on health problems in the developing world, where tropical diseases are prevalent and diseases which have been eradicated in the richer world still prevail. Its supranational activities – which have led to the eradication of at least one major disease, smallpox⌀ – are based solely on national government co-operation. In addition to regular funding through the UN system, the WHO operates a Voluntary Fund for Health Promotion (VHFP). established in 1955, which has a funding level in the region of $55 million. More than a third of WHO's operating budget is now sustained through the contributions of donor goverments, international agencies, foundations and individuals to the VFHP, plus separate trust funds identified with specific donors.

WORLD HERITAGE CONVENTION

Common name for the International Convention for the Protection of the World Cultural and Natural Heritage, drawn up by the United Nations Educational, Scientific, and Cultural Organization (UNESCO)⌀ in 1972, which came into force in 1975. It has been ratified by at least 108 nations and provides an international framework for safeguarding humankind's cultural and natural heritage, including areas of outstanding natural beauty or scientific interest. It provides grant aid⌀ to projects in countries where resources are insufficient.

The World Heritage List contains at least 315 cultural properties and natural sites, in 67 countries. These entries are structures or natural areas of 'outstanding universal value' and can be nominated for inclusion by any countries who are party to the Convention. To be accepted for inclusion on the list, a site must be natural and must contain an example of a major stage of the earth's evolutionary history, a significant and continuing geological process, a unique or superlative natural phenomenon, formation or feature, or a habitat for an endangered species◌ of wildlife.

WORLD INTELLECTUAL PROPERTY ORGANIZATION (WIPO)

A specialized agency of the United Nations◌ set up in 1974, based in Geneva with 121 member states. The WIPO aims to promote co-operation in the recognition and enforcement of international agreements on industrial property, such as trade marks, patents, inventions and industrial designs. It also undertakes a similar role with regard to copyright and other rights, chiefly concerned with literary, musical and artistic works.

WORLD METEOROLOGICAL ORGANIZATION (WMO)

A specialized agency of the United Nations◌ established in 1951 with the aim of standardizing international meteorological observations and improving the exchange of weather information. Its chief activities are the World Weather Watch programme, which co-ordinates facilities and services provided by member states, and a research and development programme that aims to extend knowledge of the natural and human-induced variability of climate. The organization, with a membership of 154, is based in Geneva.

WORLD RESOURCES INSTITUTE (WRI)

The WRI is an independent policy research centre founded in 1982 and based in the United States. It aims to assist governments, international organizations, the private sector and others to determine how best to meet basic human needs and promote economic growth without undermining the natural-resource base upon which life depends. The Institute undertakes research and analysis of policy options for natural-resource◌ development and exploitation, drawing on an interdisciplinary staff of scientists and experts from around the world, with the objective of producing accurate information, identifying emerging issues and problems, and providing guidance and suggestions as to viable solutions. The WRI is funded by private foundations and receives contributions from various United Nations◌ agencies, the commercial sector and private individuals.

WORLD SUMMIT FOR CHILDREN

A 1990 meeting held in New York at which heads of state and representatives from a total of seventy-two nations signed the World Declaration on the Survival, Protection and Development of Children. The disappointing outcome of the Summit was a commitment to a ten-point programme designed to protect the rights of children and improve their lives. Initiated by Canada, Egypt, Mali, Mexico, Pakistan and Sweden, but attracting barely half the total number of United Nations members, the Summit called for concerted global action to achieve a set of twenty specified goals by the year 2000. These goals included a one-third reduction in infant deaths and a halving of maternal mortality◌. In a subsequent report, the United Nations Children's Fund (UNICEF)◌ estimated that it would cost $20 billion annually if the targets set by the World Summit were to be achieved by the end of the decade.

WORLD UNIVERSITY SERVICE (WUS)

WUS is an international development agency based on voluntary support in universities and colleges throughout the world. An assembly of the forty-five national committees meets every second year to review projects and developments, establish a programme of community-based projects in the Third World◌, and orchestrate scholarship programmes to combat all forms of discrimination.

WORLDWATCH INSTITUTE

An independent, not-for-profit research organization based in the United States. The Worldwatch Institute aims to identify and analyse global problems concerned with environment and development issues and bring these problems to global attention. The Institute undertakes in-depth research projects on a variety of topics and produces a range of informative papers and studies aimed at decision-makers, scholars and the general public. The Worldwatch Institute is funded by private foundations and through contributions from several United Nations◌ agencies.

WORLDWIDE FUND FOR NATURE (WWF)

Formerly known as the World Wildlife Fund (WWF), the organization changed its name to the Worldwide Fund for Nature in 1989 but retained the acronym. The WWF is an international organization dedicated to the conservation◌ of endangered species◌ of wildlife and their natural habitats. Founded in 1961, the WWF was seen by many as the fund-raising arm of the International Union for the Conservation of Nature (IUCN), (now the World

Conservation Union⌀) although WWF International was set up as a body completely independent of the IUCN. Since its inception the WWF, through its international office based in Switzerland and twenty-three national branches on five continents, has financed conservation⌀ projects around the world as well as remaining a major contributor to the IUCN budget.

Dedicated to the conservation of the natural environment and the ecological⌀ processes essential to life on earth, and to creating awareness of all threats to the natural environment, the WWF pays particular attention to endangered species of plants and animals and natural habitats which are of benefit to humans. It has channelled over $130 million to well over 5,000 projects in over 130 countries, served as a catalyst for conservation action, and provided a link between conservation needs, the scientific resources necessary to meet them, and the authorities whose participation or action is needed. The WWF works in conjunction with governments, Non-Governmental Organizations (NGOS)⌀, scientists, industry and the general public and maintains a strong educational component in all its programmes of activity. It has also adopted a crucial and pioneering role in bridging the gap between the conservation movement and the business community.

In 1990, following controversial revelations and an in-depth analysis of its operations, the WWF underwent major restructuring. Evidence emerged that WWF International had investments not only in companies that were producing conventional and nuclear weapons, oil⌀ and pesticides⌀, but also in transnational⌀ enterprises against whose activities it was campaigning. The review of past operations had found that of a sample of seventy-two campaigns run during the organization's first twenty-five years, 27 per cent had failed. Moreover, one of the most prestigious projects, the panda research institute in China was 'performing no useful role'. Although the WWF has long been identified with efforts to save particular species, the new thrust of its activities is to be on pursuing projects on sustainable development⌀, pollution⌀ control and reducing wasteful consumption.

In its report following on from the reappraisal entitled *Mission for the 1990s*, the WWF announced a decentralization of activities, the hiring of extra staff, and the creation of two new divisions to cover policy and institutional development.

WORLD WILDLIFE FUND *See* Worldwide Fund for Nature.

Y

YAOUNDE CONVENTION An agreement of association signed in the Cameroon between the European Community (EC)⌀ and the eighteen independent states in Africa which had previously been in association with the EC – as dependent territories – under Part IV of the Treaty of Rome⌀. It was superseded by the Lomé Convention⌀.

YE'EB BEAN A drought-tolerant plant, *Cordeauxia edulis*, native to the arid border region between Ethiopia and Somalia, which has the potential to become a profitable food and cash crop⌀ in tropical arid zones. The plant, which grows in areas where annual rainfall is as little as 150–200 mm, produces a nut which tastes like a sweet groundnut⌀ or macadamia⌀ nut. Its long roots enable it to tap deep-soil moisture and it remains green year round. It is a dwarf shrub: new plants will not bear pods for the first three years of growth, but from the fourth year on it will yield prolifically, if conditions are favourable.

Seeds from the pods can be cooked or eaten raw. Animals and small stock also readily browse the shrubby plant. The beans make an unusually nourishing and balanced diet, containing a relatively low level of protein and carbohydrate in comparison with other beans but boasting a significant fat and sugar⌀ content.

During the Sahelian droughts⌀ of the 1970s and 1980s, ye-eb beans were one of the few foods available to nomadic○ tribes and livestock○ in the Somali Desert. Consequently, ye-eb plants were devastated and the bean was threatened with extinction○.

YELLOW FEVER An acute viral infection transmitted to humans by the *Aedes aegypti* mosquito○. Once widespread, medical care, coupled with campaigns to reduce mosquito numbers, has significantly narrowed its range – it is now mainly confined to tropical regions in Africa and South America. The disease is commonly found in areas of tropical rainforest○ in Africa and northern regions of South America and, in common with all other mosquito-transmitted diseases, cases of yellow fever are on the increase. Most new cases and deaths are occurring in these regions. In human communities throughout endemic areas, discarded empty vessels such as cans, oil○ drums and tyres○ provide ideal conditions for species of the vector *Aedes*○ mosquitoes to breed.

After an incubation period of three to fourteen days the patient develops a chill, headache, aching muscles and a fever. In severe cases the virus affects the kidneys, heart and, most notably, the liver, when it causes jaundice. Death may result from heart or liver failure. Although yellow fever frequently proves fatal when no medical help is available, recovery from a first attack confers subsequent immunity. There is no specific treatment once the disease has developed, but two kinds of vaccine, the first produced in 1951, can successfully immunize people against the disease and have played a major role in limiting its range.

Despite the availability of viable vaccines, yellow fever is increasing in all areas in which it is endemic. Cases reported from South America have been steadily rising over the past few years: up from 50 in 1983 to 235 in 1987. In Africa, the picture is worse and the disease reached epidemic proportions in 1986–7, with over 5,300 cases being officially reported, half of which proved fatal. Of even greater significance is the fact that the number of cases 'officially' notified to the World Health Organization (WHO)○ represent only a fraction of actual cases in all the areas reporting.

In Bamako, the Malian capital, the 300 cases reported in 1986 were estimated to be about a fifth of the actual number. In Nigeria, experts calculated that in reality, over the 1986–7 period, 30,000 people were affected by yellow fever and 10,000 died, figures way in excess of those officially reported. The problem of diagnosis and patients not presenting themselves for examination, together with the relative lack of medical facilities and resources in the worst-affected areas, means that the incidence of the disease is likely to continue increasing. In South America during 1988, yellow fever increased in both incidence and severity, with over 280 cases reported from Bolivia, Brazil and Peru alone. Of these, 234 proved fatal. In Africa, Nigeria also reported 1,786 official cases during the year, with 1,497 deaths, all from a disease that can be prevented relatively easily. A campaign to vaccinate those at risk in Mali during 1987 reached 84 per cent of the 3 million target population○, and as a result the epidemic died out by early 1988.

Z

ZINC A bluish-white metal, occurring naturally in three ores: calamine, sphalerite and hemimorphite. Zinc is used to make galvanized iron○ and steel, and low-melting alloys. Zinc compounds also have a number of commercial uses. Zinc oxide is widely used in skin ointments and cosmetics, as well as in the manufacture of batteries, plastics, paint, glass and printing ink. Zinc sulphide is used in making television tubes and X-ray apparatus. Trace amounts of zinc are essential for growth in animals, including human beings. Zinc is also essential for healthy plant growth, helping plants to make full use of nitrogen○ and phosphorus. Zinc deficiency in soil○ depresses yields, and dietary deficiencies can lead to retarded growth or disrupted sexual development in humans and livestock○. Annual global production of zinc is around 6.9 million tonnes; the Soviet Union (1 million tonnes), Canada (700,000 tonnes) and Japan (675,000 tonnes) are the largest producers.